TELECOMMUNICATION SWITCHING AND NETWORKS

TELECOMMUNICATION SWITCHING AND NETWORKS

(SECOND EDITION)

P GNANASIVAM

Professor & Head
Department of Electronics and Communication Engineering
Agni College of Technology, Chennai
Tamil Nadu, India

Publishing Globally

NEW AGE INTERNATIONAL (P) LIMITED, PUBLISHERS

LONDON • NEW DELHI • NAIROBI

Bangalore • Chennai • Cochin • Guwahati • Hyderabad • Kolkata • Lucknow • Mumbai

Visit us at **www.newagepublishers.com**

Published by New Age International (P) Ltd., Publishers
First Edition: 2005
Second Edition: 2006
Reprint: 2016

GLOBAL OFFICES

- **New Delhi** **NEW AGE INTERNATIONAL (P) LIMITED, PUBLISHERS**
  7/30 A, Daryaganj, New Delhi-110002, (INDIA)
  Tel.: (011) 23253771, 23253472, **Telefax:** 23267437, 43551305
  E-mail: contactus@newagepublishers.com • Visit us at www.newagepublishers.com
- **London** **NEW AGE INTERNATIONAL (UK) LTD.**
  27 Old Gloucester Street, London, WC1N 3AX, UK
  E-mail: info@newacademicscience.co.uk • Visit us at www.newacademicscience.co.uk
- **Nairobi** **NEW AGE GOLDEN (EAST AFRICA) LTD.**
  Ground Floor, Westlands Arcade, Chiromo Road (Next to Naivas Supermarket), Westlands, Nairobi, KENYA
  Tel.: 00-254-713848772, 00-254-725700286
  E-mail: kenya@newagepublishers.com

BRANCHES

- **Bangalore** 37/10, 8th Cross (Near Hanuman Temple), Azad Nagar, Chamarajpet, Bangalore- 560 018
  Tel.: (080) 26756823, **Telefax:** 26756820, **E-mail: bangalore@newagepublishers.com**
- **Chennai** 26, Damodaran Street, T. Nagar, Chennai-600 017, **Tel.:** (044) 24353401, **Telefax:** 24351463
  E-mail: chennai@newagepublishers.com
- **Cochin** CC-39/1016, Carrier Station Road, Ernakulam South, Cochin-682 016, **Tel.:** (0484) 2377303, **Telefax:** 4051304
  E-mail: cochin@newagepublishers.com
- **Guwahati** Hemsen Complex, Mohd. Shah Road, Paltan Bazar, Near Starline Hotel, Guwahati-781 008,
  Tel.: (0361) 2513881, **Telefax:** 2543669, **E-mail: guwahati@newagepublishers.com**
- **Hyderabad** 105, 1st Floor, Madhiray Kaveri Tower, 3-2-19, Azam Jahi Road, Near Kumar Theater, Nimboliadda Kachiguda, Hyderabad-500 027, **Tel.:** (040) 24652456, **Telefax:** 24652457
  E-mail: hyderabad@newagepublishers.com
- **Kolkata** RDB Chambers (Formerly Lotus Cinema) 106A, 1st Floor, S N Banerjee Road, Kolkata-700 014
  Tel.: (033) 22273773, **Telefax:** 22275247, **E-mail: kolkata@newagepublishers.com**
- **Lucknow** 16-A, Jopling Road, Lucknow-226 001, **Tel.:** (0522) 2209578, 4045297, **Telefax:** 2204098
  E-mail: lucknow@newagepublishers.com
- **Mumbai** 142C, Victor House, Ground Floor, N.M. Joshi Marg, Lower Parel, Mumbai-400 013
  Tel.: (022) 24927869, **Telefax:** 24915415, **E-mail: mumbai@newagepublishers.com**
- **New Delhi** 22, Golden House, Daryaganj, New Delhi-110 002, **Tel.:** (011) 23262368, 23262370, **Telefax:** 43551305
  E-mail: sales@newagepublishers.com

ISBN: 978-81-224-1950-4

C-16-07-9526

Printed in India at Ajit Printing Press, Delhi.
Typeset at Goswami Associates, Delhi.

NEW AGE INTERNATIONAL (P) LIMITED, PUBLISHERS
7/30 A, Daryaganj, New Delhi-110002
Visit us at **www.newagepublishers.com**
(CIN: U74899DL1966PTC004618)

PREFACE

This text, 'Telecommunication Switching and Networks' is intended to serve as a one-semester text for undergraduate course of Information Technology, Electronics and Communication Engineering, and Telecommunication Engineering. This book provides in depth knowledge on telecommunication switching and good background for advanced studies in communication networks. The entire subject is dealt with conceptual treatment and the analytical or mathematical approach is made only to some extent. For best understanding, more diagrams (202) and tables (35) are introduced wherever necessary in each chapter.

The telecommunication switching is the fast growing field and enormous research and development are undertaken by various organizations and firms. The communication networks have unlimited research potentials. Both telecommunication switching and communication networks develop new techniques and technologies everyday. This book provides complete fundamentals of all the topics it has focused. However, a candidate pursuing postgraduate course, doing research in these areas and the employees of telecom organizations should be in constant touch with latest technologies. For this purpose, in each chapter end, the related websites, which provides sufficient informations's, are added.

Also, the subjects like signalling techniques, traffic engineering, billing and charging procedure, telecommunication system organizations are subject to changes according to the govt. policies and changing scenario in the economy of the country, political changes, pressures from people etc. Hence, the viewers of this subject should have constant touch with changes. For this purpose also, the related websites are added at all the chapters end.

Chapters 1, 2, and 3 are introductory chapters. A reader having previous exposure to these topics may start from the fourth chapter. Chapter 1 introduces the historical development of the subject. Chapter 2 deals with the telecommunication standards organizations and the standards, which are useful to provide worldwide telecom services with better interconnectivity. The telephone systems, transmission systems and the impairments related to the transmission of signals are described in the chapter 3. Also, the study of section 3.6 will be highly useful in understanding of digital switching and transmission.

Entire 4th chapter discusses the evaluation of PSTN, electromechanical switching system, SPC exchanges those are basics for digital switching systems, various switching procedures and the components used in switches.

Chapters 5 and 6 deal with the digital switching and computer controlled switching procedures and some currently available switching systems are discussed.

Chapter 7 concentrates on signalling techniques. The signalling system enables the quick path setup between calling and called subscribers. Chapter 8 is completely devoted to traffic engineering that covers various systems and blocking models. Chapter 9 helps the viewer to understand the numbering plan, charging and how to organize an exchange.

Chapters 10, 11 and 12 are devoted completely to communication networks. Chapter 10 deals with DSL technologies and SONET/SDH networks. The chapter 11 gives idea about OSI model, TCP/IP and ATM concepts. Chapter 12 is devoted to ISDN which is a network that supports various voice/data transfer related services.

For this book, I am greatly indebted to Dr. S.P.E. Xavier, who urged me and encouraged me to write it. I am pleased to acknowledge Mr. R. Vijayarajan, of my department who reviewed the manuscript and rendered valuable suggestions. I am indebted to Prof. A. Sivasubramanian, Head of ECE who has rendered his most gracious encouragement and very valued criticism. I express my thanks to the faculty members of my department and friends for their suggestions and encouragement. I thank Mr. A. Antony Rajasekar, Asst. Librarian for his help in identifying the books for reference

I express my sincere gratitude to our chairman Thiru. Jeppiaar, M.A., B.L., I extend my thanks to the directors Dr. B. Babu Manoharan, M.A., Ph.D., and Tmt. Sheila Babu Manoharan, M.A., B.L., and the Principal Prof. Jolly Abraham of St. Joseph's College of Engineering, Chennai for their constant encouragement. My wife Malathi and my son Tinu Rahul sacrificed in many ways to bring this book out. My wife took care of my personal needs and family needs and her support and encouragement made publishing of this book possible.

P. Gnanasivam M.E., M.B.A.

Assistant Professor

St. Joseph's College of Engineering

Chennai-600 119

CONTENTS

1

Introduction to Switching Systems

1

Introduction to Switching Systems

1.1. INTRODUCTION

Telecommunication networks carry information signals among entities, which are geographically far apart. An entity may be a computer or human being, a facsimile machine, a teleprinter, a data terminal and so on. The entities are involved in the process of information transfer which may be in the form of a telephone conversation (telephony) or a file transfer between two computers or message transfer between two terminals etc. Today it is almost truism to state that telecommunication systems are the symbol of our information age.

With the rapidly growing traffic and untegrated growth of cyberspace, telecommunication becomes a fabric of our life. The future challenges are enormous as we anticipate rapid growth of new services and number of users. What comes with the challenge is a genuine need for more advanced methodology supporting analysis and design of telecommunication architectures. Telecommunication has evaluated and grows at an explosive rate in recent years and will undoubtedly continue to do so.

The communication switching system enables the universal connectivity. The universal connectivity is realized when any entity in one part of the world can communicate with any other entity in another part of the world. In many ways telecommunication will acts as a substitute for the increasingly expensive physical transportation.

The telecommunication links and switching were mainly designed for voice communication. With the appropriate attachments/equipments, they can be used to transmit data. A modern society, therefore needs new facilities including very high bandwidth switched data networks, and large communication satellites with small, cheap earth antennas.

1.2. HISTORICAL DEVELOPMENT

By the early 1800's scientists had developed ways to generate and transmit electricity. In 1819, oersted discovered the relation between magnetism and electricity. Ampere, Faraday and others continued this work in 1820. In 1834, Gauss and Weber wired over the roofs of Gottingen to make a telegraph system.

Samuel F.B. Morse's developed the first significant work in telecommunication. F.B. Morse developed code telegraphy in 1837. In 1844, a 40 mile telegraph line was setup between Baltimore and Washington by F.B. Morse. In 1845, Morse formed a telegraph company based on his technology. In 1849, the first slow telegraph printer link was setup. In 1874, Ban dot invented a "Multiplexes" system which enables up to six signal from telegraph machines to be transmitted together over the same line.

Elisha Gray and Alexander Graham Bell contributed significant works and filed paper related to telephony. The early stages of the development of telecommunication were due to A.G. Bell, G. Marconi and C.E. Shannon. In 1876, Bell invented a telephone system. In 1897 Marconi patented a wireless telephone system. Teletypewriter service was initiated in 1931.

In early days, a very simple exchanges were used whose control was provided by a human operator and the elements of the switch assemblies are plugs and sacks.With increase in demand of service, human operator exchange was replaced by the invention of range of electromechanical switching devices. Of all the electromechanical switching devices that become available over the years, the step-by-step switching system invented by Almon B. Strowger in 1892 is still quite popular. The next automatic electromechanical switching system was crossber switching. First patent for crossbar device was granted in 1915 to J.N. Reynolds of wester Electric, USA. In 1919, two Swedish Engineers, Betulander and Palmgren got patent for crossbar switch. In 1938, AT & T laboratories in US introduced crossbar-switching system in the field.

The electromechanical switching systems have been replaced by computer controlled switching systems referred to as stored program control (SPC). In SPC, switching is controlled by software program. The first computer controlled switch was introduced in 1960. Till 1965, computer controlled switching used transistors and printed circuit technology. Since 1965 switching are based on microprocessors.

The use of computers to control the switching led to the designation "electronic" switching system (ESS) or Electronic automatic exchange (EAX). In 1970, first electronic switching system No. 1 ESS or No. 1 EAX was introduced. Digital electronic switching matrices were first introduced into the U.S. public network in 1976 with AT & T's No. 4 ESS digital toll switch. By the mid 1980's the interoffice transmission environment has changed to almost exclusively digital. Fig. 1.1 shows the various telephone networks.

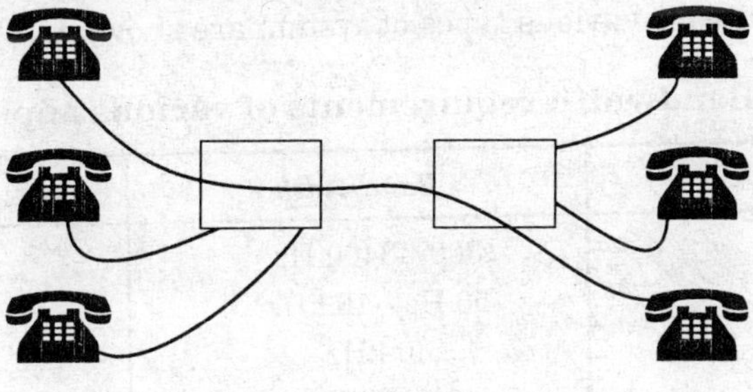

(a) Telephone network around 1890

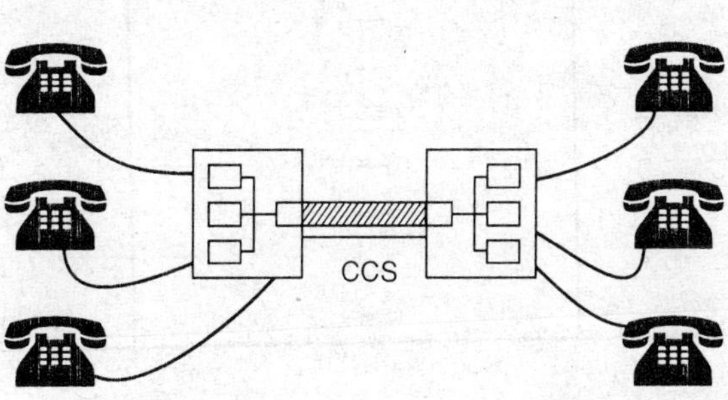

(b) Telephone network around 1988

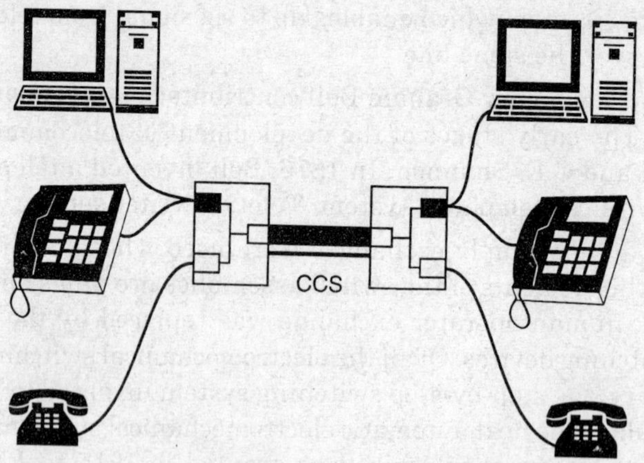

(c) Telephone network after 1990 with ISDN

Fig. 1.1. Various telephone networks.

1.3. SIGNAL CHARACTERISTICS

Telecommunication is mainly concerned with the transmission of messages between two distant points. The signal that contains the messages is usually converted into electrical waves before transmission. Our voice is an analog signal which has amplitude and frequency characteristic.

Voice frequencies. The range of frequencies used by a communication device determines the communication channel, communicating devices, bandwidth or information carrying capacity. The most commonly used parameter that characterizes an electrical signal is its bandwidth of analog signal or bit rate if it is a digital signal. In telephone system, the frequencies it passes are restricted to between 300 to 3400 Hz. Thus the network bandwidth is 3100 Hz. The bandwidth and bit rate for various types of system are shown in Table 1.1.

Table 1.1. Bandwidth requirements of various applications

Type	*Bandwidth*	*Bit Rate*
Telephone (speech)	300—3400 Hz	—
Music	50 Hz—16 kHz	—
Facsimile	40 kHz	—
Broadcast televison	0—55 MHz	—
Personal communication	—	300 to 9600 bits/sec
E—Mail transmission	—	2400 to 9600 bits/sec
Digitized voice phone call	—	6400 bits/sec
Digital audio	—	1 to 2 M bits/sec
Compressed video	—	2 to 10 M bits/sec
Document imaging	—	10 to 100 M bits/sec
Full motion video	—	1 to 2 G bits/sec

Speech spectrum. The telephone channel over which we wish to send data are designed to transmit electrical oscillations (microphone converts sound into equivalent number of electrical oscillation) of voice. Fig. 1.2 is described as a speech spectrum diagram. It illustrates human speech strength variations at various frequencies. Most of the energy is concentrated between 300 Hz to 3400 Hz.

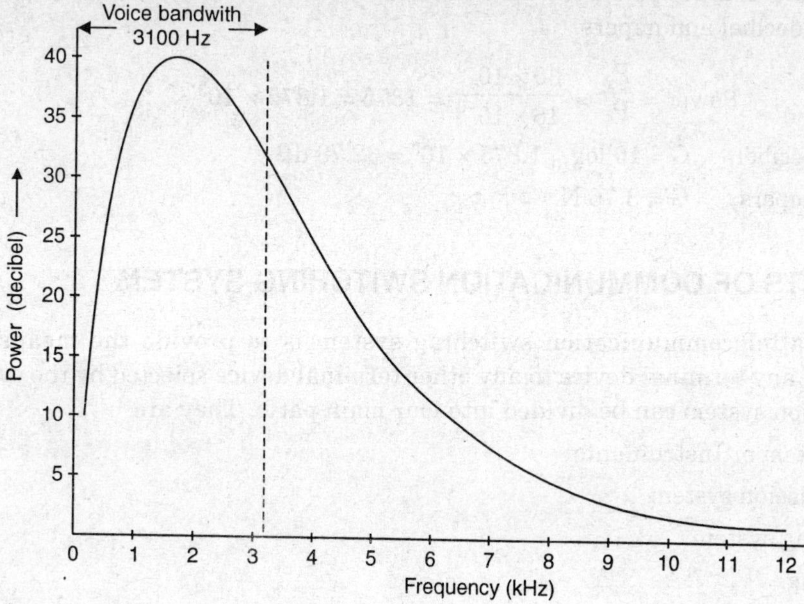

Fig. 1.2. Speed spectrum.

Decibels. The decibel is a valuable unit for telecommunication because losses or gains in signal strength may be added or subtracted if they are referred to in decibels. The signal strength at various frequencies is expressed by the unit of decibel (dB) in telecommunication. The decibel is a unit of power ratio. The power ratio is expressed as

$$G = 10 \log_{10} \frac{P_2}{P_1} \quad ...(1.1)$$

Where P_1 is input power (Normally) and P_2 is output power.

The decibel is also used to be defined as the unit of attenuation. One decibel attenuation means that a signal has dropped to 0.794 of its original power. One decibel gain means that a signal has increased to 1.259 of its original power. The decibel concept is further discussed in later chapter.

Voltage and current level can be quoted in decibel as follows

$$G = 10 \log_{10} \frac{P_2}{P_1} = \frac{V_2 I_2}{V_1 I_1} = \frac{I_2^2 R}{I_1^2 R} \quad ...(1.2)$$

$$G = 10 \log_{10} (I_2/I_1)^2 \quad ....(1.3)$$

$$G = 20 \log (I_2/I_1) \quad ...(1.4)$$

Similarly in voltage ratio,

$$G = 20 \log_{10} V_2/V_1 \quad ...(1.5)$$

$$\text{Gain in nepers} = \log_e \frac{I_2}{I_1} = \log_e \frac{V_2}{V_1} \qquad ...(1.6)$$

$$\text{Relation}: \quad 1\text{ N} = 8.69 \text{ dB} \qquad ...(1.7)$$

Example 1.1. If input power is 16 μW and output power is 30 mW, find the power ratio and express it in decibel and nepers.

$$\text{Power} = \frac{P_2}{P_1} = \frac{30 \times 10^{-3}}{16 \times 10^{-6}} = 1875 = 1.875 \times 10^3$$

Power in decibel, $G = 10 \log_{10} 1.875 \times 10^3 = 32.73$ dB

Power in nepers, $G = 3.76$ N.

1.4. ELEMENTS OF COMMUNICATION SWITCHING SYSTEM

The purpose of a telecommunication switching system is to provide the means to pass information from any terminal device to any other terminal device selected by the originator. Telecommunication system can be divided into four main parts. They are

1. End system or Instruments
2. Transmission system
3. Switching system
4. Signaling.

End Systems or Instruments. The end system or instruments are a transmitter or receiver that are responsible for sending information or decoding or inverting received information or message into an intelligible message. End systems in the telephone network have evolved from analog telephones to digital handsets and cellular phones. However, endless arrays of other devices are being attached to telephone lines, including computer terminals used for data transmission. Fig. 1.3 shows some of the end instruments.

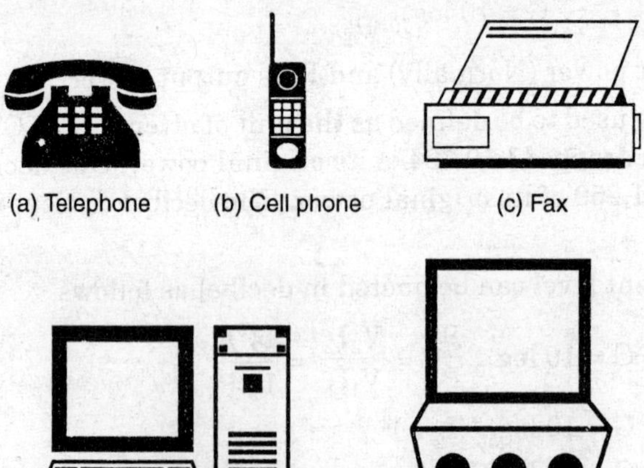

Fig. 1.3. Some of the end instruments.

Transmission System. Signals generated by the end system or the instruments should be transported to the destination by some means. The transmission on links conveys the information and control signals between the terminals and switching centers. A transmission link can be characterized by its bandwidth, link attenuation and the propagation delay. To maintain signal quality, the signal must be regenerated after a certain distance.

In general a communication path between two distinct points can be setup be connecting a number of transmission lines in tandem. The transmission iinks include two-wire lines, co-axial cables microwave radio, optical fibers and satellites. Functionally, the communication channels between switching system are referred to as trunks. Fig. 1.4 shows the various possible transmission media.

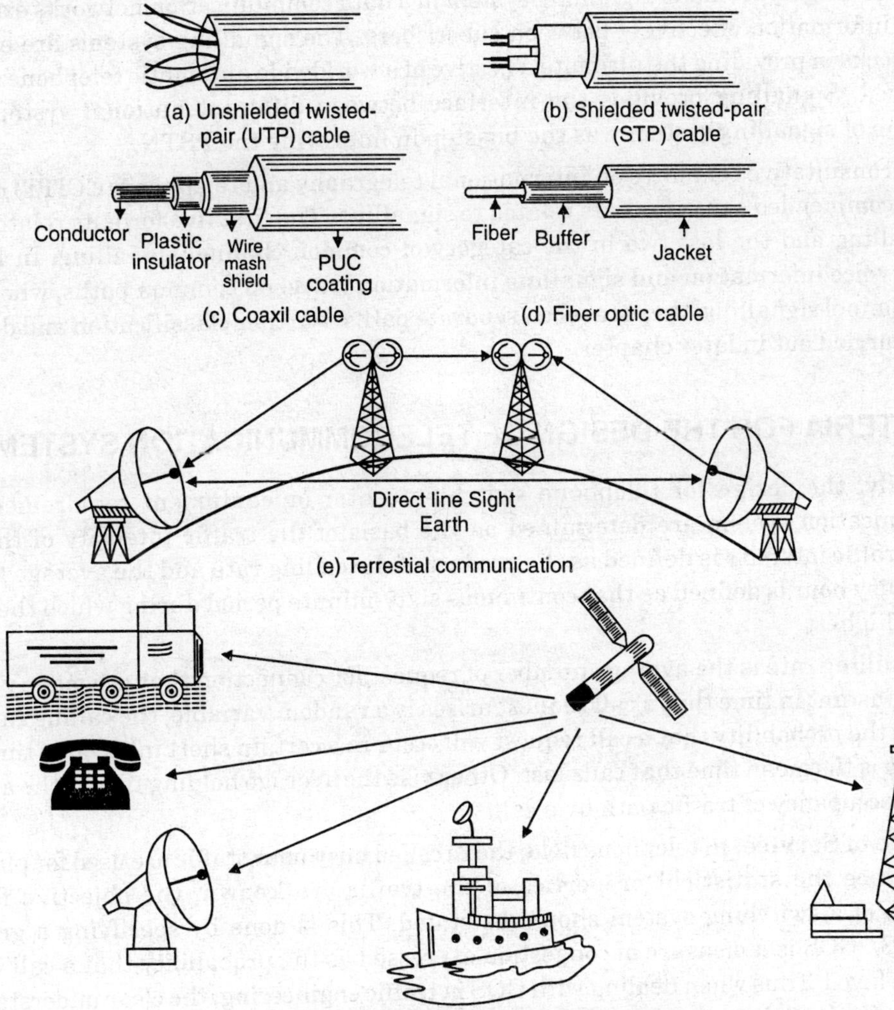

Fig. 1.4. Various transmission media.

Switching System. The switching centers receives the control signals, messages or conversations and forwards to the required destination, after necessary modification (link amplifications) if necessary. A switching system is a collection of switching elements arranged and controlled in such a way as to setup a communication path between any two distant points. A switching center of a telephone network comprising a switching network and its control and support equipment is called a central office.

In computer communication, the switching technique used is known as packet switching or message switch (store and forward switching). In telephone network the switching method used is called circuit switching. Some practical switching system are step-by-step, cross barred relay system, digital swtiching systems, electronic switching system etc.

Signalling Systems. A signalling system in a data communication networks exchanges signalling information effectively between subscribers. The signalling systems are essential building blocks in providing the ultimate objective of a worldwide automatic telephone services standardized. Signalling provides the interface between different national systems. The introduction of signalling system was the big step in improving the PSTN.

The consultative committe on international telegraphy and telephony (CCITT) based in Geneva, recommended seven formats related to signalling. The first five formats related to In-band signalling and the last two in the category of common channel signalling. In In-band signalling, voice information and signalling information travel on common paths, where as in common channel signalling, they travel on separate paths. Further classification and detained study are carried out in later chapter.

1.5. CRITERIA FOR THE DESIGN OF TELECOMMUNICATION SYSTEM

Traditionally, the design for telephone switching center or equipment requirement in a telecommunication system are determined on the basis of the traffic intensity of the busy hour. The traffic intersity is defined as the product of the calling rate and the average holding time. The busy hour is defined as that continuous sixty-minute period during which the traffic intersity is highest.

The calling rate is the average number of request for connection that are made per unit time. If the instant in time that a call request arises is a random variable, the calling rate may be stated as the probability that a call request will occur in a certain short interval of time. The holding time is the mean time that calls last. Otherwise the average holding time is the average duration of occupancy of traffic path by a call.

Grade of Service. In telephone field, the so called busy hour traffic are used for planning purposes. Once the statistical properties of the traffic are known, the objective for the performance of a switching system should be stated. This is done by specifying a grade of service (GOS). GOS is a measure of congestion expressed as the probability that a call will be blocked or delayed. Thus when dealing with GOS in traffic engineering, the clear understanding of blocking criteria, delay criteria and congestion are essential.

Blocking criteria. If the design of a system is based on the fraction of calls blocked (the blocking probaility), then the system is said to be engineered on a blocking basis or call loss basis. Blocking can occur if all devices are occupied when a demand of service is initiated.

Blocking criteria are often used for the dimensioning of switching networks and interoffice trunk groups. For a system designed on a loss basis, a suitable GOS is the percentage of calls which are lost because no equipment is available at the instant of call request.

Delay criteria. If the design of a system is based on the fraction of calls delayed longer than a specified length of time (the delay probability), the system is said to be a waiting system or engineered on a delay basis. Delay criteria are used in telephone systems for the dimensioning of registers. In waiting system, a GOS objective could be either the percentage of calls which are delayed or the percentage which are delayed more than a certain length of time.

Congestion. It is the condition in a switching center when a subscriber can not obtain a connection to the wanted subscriber immediately. In a circuit switching system, there will be a period of congestion during which no new calls can be accepted. There are two ways of specifying congestion.

1. **Time congestion.** It is the probability that all servers are busy. It is also called the probability of blocking.

2. **Call congestion.** It is the proportion of calls arising that do not find a free server. Call congestion is a loss system and also known as the probability of loss while in a delay system it is referred to as the probability of waiting.

If the number of sources is equal to the number of servers, the time congestion is finite, but the call congestion is zero. When the number of sources is large in comparison with servers, the probability of a new call arising is independent of the number already in progress and therefore the call congestion is equal to the time congestion. In general, time and call congestions are different but in most practial cases, the discrepancies are small.

Measure of GOS. GOS is expressed as a probability. The GOS of 2% (0.02) mean that 98% of the calls will reach a called instrument if it is free. Generally, GOS is quoted as P.02 or simply P02 to represent a network busy probability of 0.02. GOS is applied to a terminal-to-terminal connection. For the system connection many switching centers, the system is generally broken into following components.

(*i*) an internal call (calling subscriber to switching office)

(*ii*) an outgoing call to the trunk network (switching office to trunk)

(*iii*) The trunk network (trunk to trunk)

(*iv*) A terminating call (switching office to called subscriber)

The GOS of each component is called component GOS.

The GOS for internal calls is 3 to 5%, for trunk calls 1-3%, for outgoing calls 2% and for terminating calls 2%. The overall GOS of a system is approximately the sum of the component grade of service. In practice, in order to ensure that the GOS does not deteriorate disastrously if the actual busy hour traffic exceeds the mean, GOS are specified 10% or 20% more of the mean.

1.6. FUNDAMENTALS FOR THE DESIGN OF TELECOMMUNICATION NETWORK

A telephone network is composed of a variety of all processing equipments, interstate switching links and inters office trunks. Because of the random nature of the call request, the design of equipments switching links and trunks are quite difficult. Thus, the traffic analysis is the fundamental request for the design of cost effective, efficient and effective configuration of networks. The effectiveness of a network can be evaluated intermes of how much traffic it carries under normal or average loads and how often the traffic volume exceeds the capacity of the network.

Fundamental problem in the design of telecommunication networks concerns the dimensioning of a route. To dimension the route, volume of traffic required grade of service and capacity (in bits per sec) must be known.

Traffic. In telecommunication system, traffic is defined as the occupancy of the server in the network. There are two types of traffic viz. voice traffic and data traffic. For voice traffic, the calling rate is defined as the number of calls per traffic path during the busy hour. In a day, the 60 minutes interval in which the traffic is highest is called busy hour (BH).

Average occupancy. If the average number of calls to and from a terminal during a period T second is 'n' and the average holding time is 'h' seconds, the average occupancy of the terminal is given by

$$A = \frac{nh}{T} \quad ...(1.8)$$

The average occupancy is also referred as traffic flow of traffic intensity. The international unit of telephone traffic is the Erlang.

1.7. CENTRALIZED SWITCHING SYSTEM

Distributed Method

A simplest way of structuring the telecommunication switching is the terminal-to-terminal connection. This kind of switching is called distributed switching and applied only to small telephone system. Some examples of distributed switching are shown here. Fig. 1.5 shows the full interconnection of five terminals.

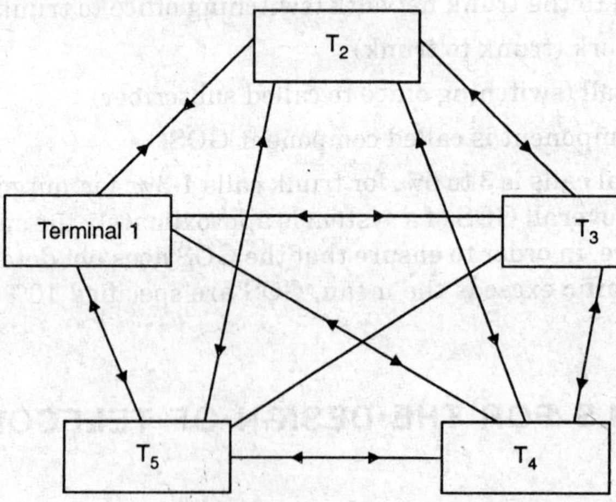

Fig. 1.5. Distributed model.

Each terminal have two kind of switches, one to make required link and other to connect a link to receive a call. By this method, for N terminals, the numbers of links required are $\frac{1}{2}$ N (N – 1) Fig. 1.6 shows the interconnection of four terminals but only with 4(N) links. Here each terminal is connected permanently to one channel and all other terminals may be accessed by operating a switch. Also it removes the need to connect a terminal to a link for an incoming call.

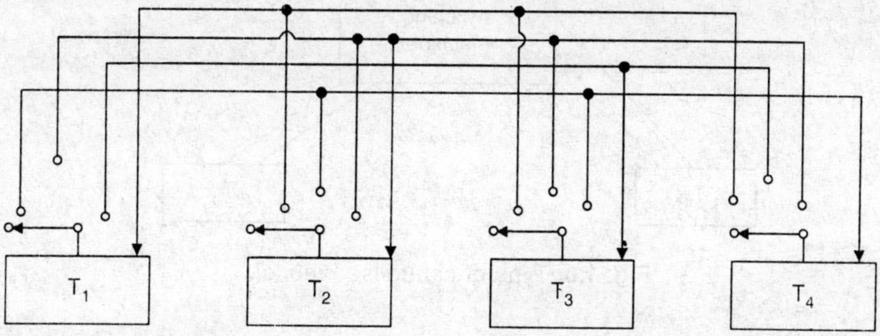

Fig. 1.6. Type of centralised model.

The circuit in Fig. 1.7 is similar arrangement of the figure, but with fewer than N links.

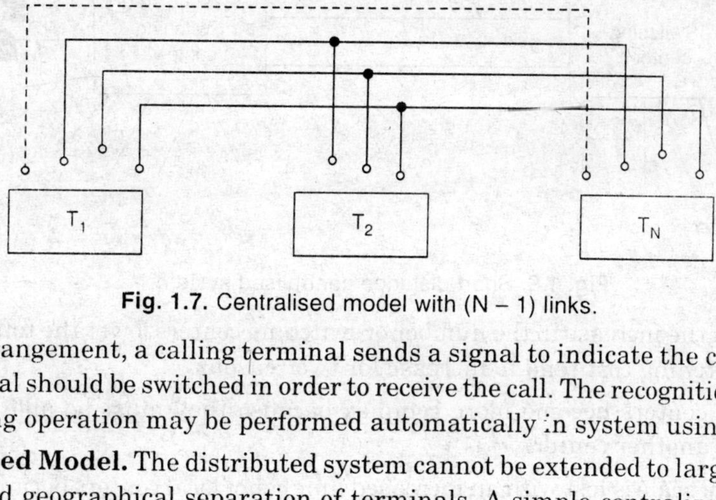

Fig. 1.7. Centralised model with (N – 1) links.

In this arrangement, a calling terminal sends a signal to indicate the called terminal to which the terminal should be switched in order to receive the call. The recognition of an incoming call and switching operation may be performed automatically in system using coded signals.

Centralised Model. The distributed system cannot be extended to large terminal cases and the increased geographical separation of terminals. A simple centralized system, which reduces the average length of transmission link, and hence the transmission cost is shown in Fig. 1.8. But this system increases the total switching costs.

Introducing more local centers instead of one national center switching machine can further reduce the transmission cost. Two local centers are connected by links called trunk. A trunk in telephone system is a communication path that contains shared circuits that are used to interconnect central offices. Fig. 1.9 shows the telecommunication system for short distances, with two exchanges (switching offices).

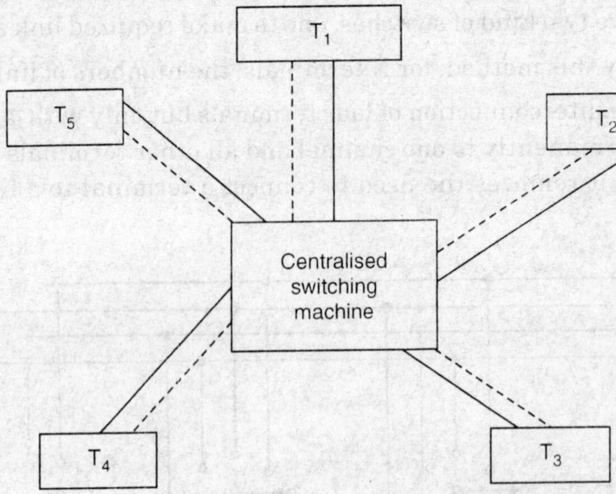

Fig. 1.8. Type of centralised model.

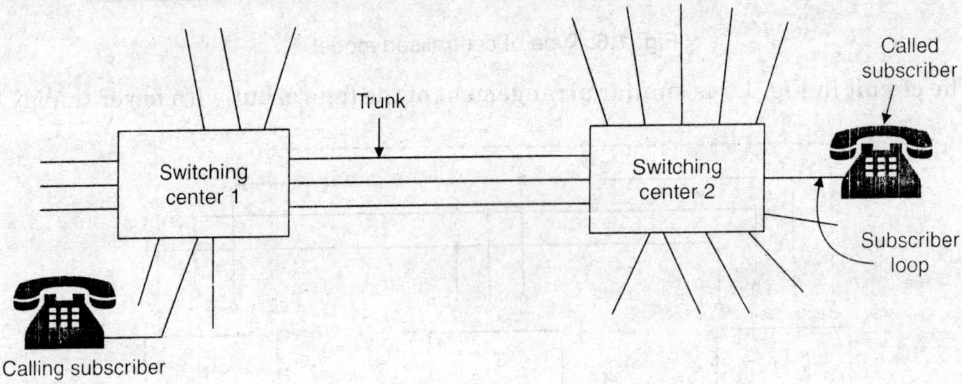

Fig. 1.9. Short distance centralised system.

Even though the increase in the number of switching centers lower the total transmission costs, the total switching cost tend to increase for two reasons.

1. The local centers become more complex because they must be able to decide on a suitable routing to another center.

2. Economy of scale is lost with an increased number of local centers because of additional numbers.

Hierarchical system. Central offices may be interconnected by direct trunk groups or through an intermediate offices known as a tandem, toll or gateway office. The process of centralizing switching centers can occur at several levels leading to the hierarchical network. Typical interconnection of central office is shown in Fig. 1.10.

Star connection may be used where traffic intensity is less. Mesh connection are used when there are relatively high traffic levels between offices such as in metropolitan network.

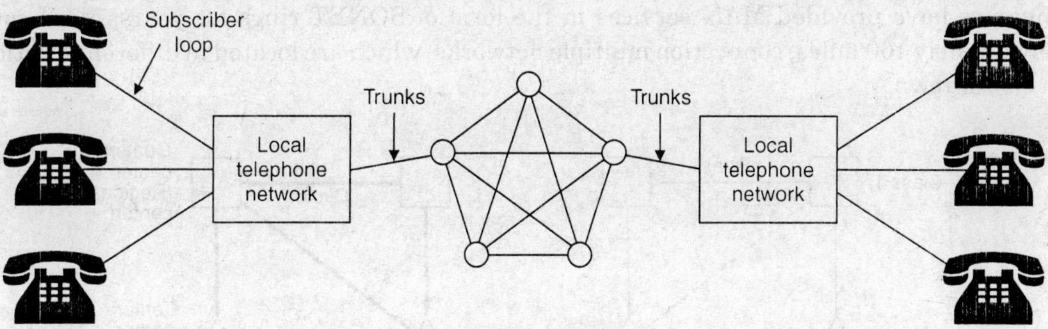

Fig. 1.10. Hierarchical network long level network (Star Connection).

1.8. AT & T AND ITU—(OR CCITT) HIERARCHICAL NETWORKS

The telephone network has developed dramatically in recent years so that it is now possible to make calls automatically between subscribes separated by thousands of miles. Long distance calls pass through several stages of switching and several possible transmission links before reaching their destination. In selecting the structure and layout, we must consider the trade-off between the cost of the network and the grade of service.

A popular concept in networking is to provide alternate routing when some parts of the network are congested. Also, the network must be adaptable to various traffic patterns, such as festival traffic, summer vacation traffic, weekend traffic etc. Alternative routing introduces switching complication. When alternate routing are sought, a call might be routed on many transmission links or routed in complex path. Thus the quality of the call may be affected severely.

In order to avoid unnecessary complications, to route traffic effectively and economically, AT & T and ITU-T developed two types of hierarchical networks. Each types serving about 50% of the wireless telephones. The hierarchical structures of the AT & T and the CCITT networks are shown in Fig. 1.11 on next page.

The AT & T (American Telephone and Telegraph in the united states) network is generally used in North America and ITU-T (international Telecommunication union—Telecommunication sector) network is used in rest of the countries.

1.9. TELECOMMUNICATION NETWORKS

Telecommunication netwoks have been evolving in the last 160 years and would continue to evolve to provide wider services. Telecommunication networks may be categorized according to networks emerged as a result of computer and communication technologies. The networks based on geographical area are characterized by the area (town, city or village), subscriber densities, traffic pattern, distance between the local exchanges and the subscribers character (residential area or business area or Govt. offices location)

Urban or metropolitan area networks (MAN). Metropolitan area networks, in general connect business to business and business to WAN and internet. The telephone companies have provided MAN services in the form of SONET rings for years. MAN cover approximately 100 miles, connection multiple networks, which are located in different locations of a city or town.

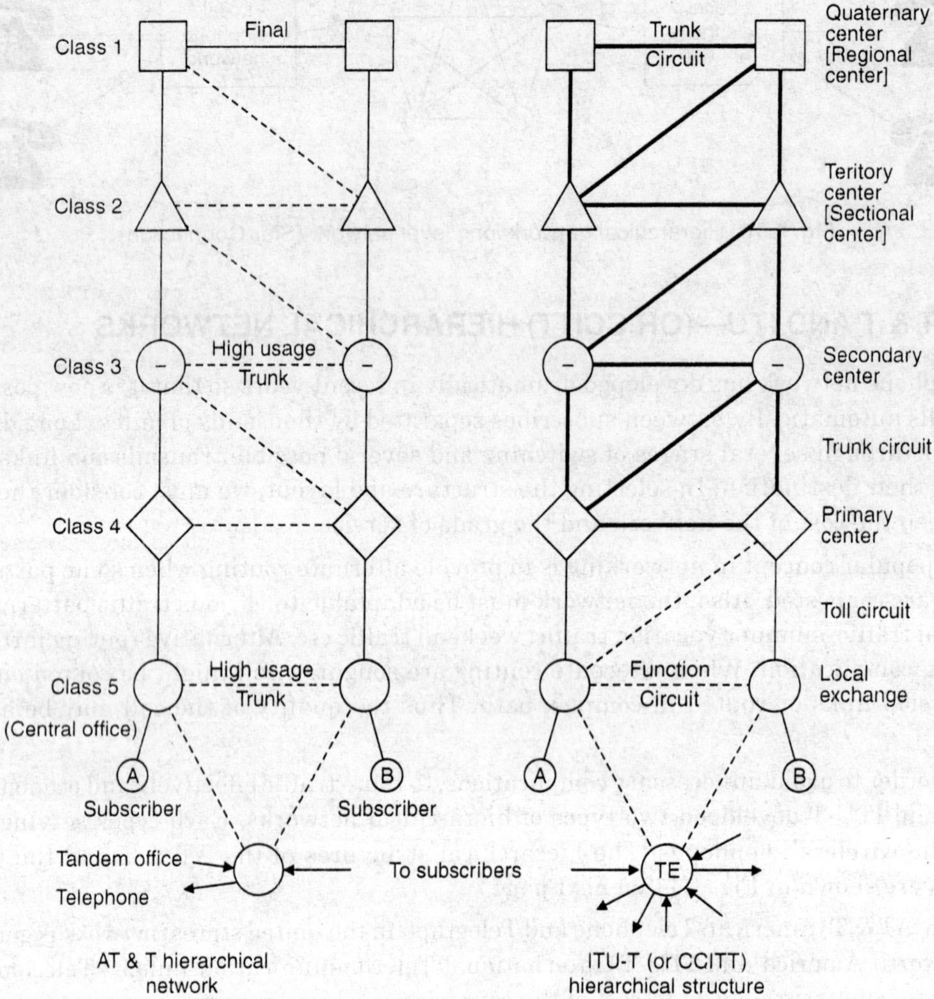

Fig. 1.11. Hierarchical structures of AT & T and ITU-T.

Rural networks, long distance or toll or wide area networks (WAN). WAN is used for long distance transmission of information. WANs cover a large geographical area such as an entire country or continent. WANs may use lines leased from telephone companies or public switched data networks (PSDN) or satellite for communication links. Rural areas are characterized by light traffic per subscriber, widely dispersed subscribers, local exchanges usually separated apart etc., these areas are served by rural networks or WAN.

The most common telecommunication network is Public Switched Telephone Network (PSTN) or sometimes known as plain old telephone system (POTS). The second major class of telecommunication is the data netwoks.

ARPANET. In 1968, the United States Department of Defence (DOD) created the Defence Advanced Research Project Agency (DARPA) for research on packet switching networks. In 1969, DARPA created the Advanced Research Project Agency (ARPA). ARPA built an experimental network (ARPANET). It is highly useful in interconnecting heterogeneous systems.

TYMNET. It is an another large scale, general purpose data network introduced in 1970, interconnecting geographically distributed computer systems, users and peripherals.

ISDN. Integrated Services Digital Network is now emerging as an major telecommunication network. It is capable of carrying multimedia services like voice, data, video and facsimile.

Intelligent Network. It is the public telephone network that contains the logic for routing calls, establishing connections and providing advanced features such as unique customer services and customer programming of the network. This consists of a signalling path that is separate from the central logic call circuit. Call setup information in handled by SS7 (signalling system 7), and the information transferred via packets across an overlay packet switching network.

New Public Network (NPN). It is the convergence of PSTN and the Internet. It allows Internet phone users to connect with PSTN telephone users and vice versa. If provides the reliability and 99.999% of availability of the PSTN. Internet engineers have developed their own set of protocols that provide telephony services over the Internet and interconnection with the PSTN.

ACRONYMS

ARPA	—	Advanced Research Project Agency
AT & T	—	American Telephone and Telegraph
CCITT	—	Consultative Committee on International Telegraphy and Telephony
DARPA	—	Defence Advanced Research Project Agency
DOD	—	Department of Defence
EAX	—	Electronic Automatic Exchange
ESS	—	Electronic Switching System
GOS	—	Grade of Service
IN	—	Intelligent Network
ISDN	—	Integrated Services Digital Network
ITU-T	—	International Telecommunication Union-Telecommunication sector
MAN	—	Metropolitan Area Network
NPN	—	New Public Network
POTS	—	Plain Old Telephone systems
PSTN	—	Public Switched Telephone Network
SPC	—	Stored Program Control
WAN	—	Wide Area Network
PSDN	—	Public Switched Data Networks

RELATED WEBSITE

Fundamentals of Telecommunication —*http://www.iec.org/tutorials/fund-telecom/*

Good overview of telecommunication technologies —*http://www.Paradyne.com/sourcebook_offer/Sb_Pdf_Hom/*

Good overview of telecommunication technologies —*http://www.us-epanorama.net*

Telecommunication magazine —*http://www-telecoms.jnag.com/*

Telephony world-com —*http://www.telephonyworld-com/*

REVIEW QUESTIONS

1. What is meant by telecommunication network ?
2. What is the range of voice frequencies ?
3. Define bandwidth.
4. Tabulate bandwidth and bit-rate of various applications.
5. Plot the speech spectrum.
6. Define decibel.
7. What are the elements of telecommunication systems ?
8. List the end systems of instruments of telecommunication systems.
9. List the various methods of transmission systems.
10. What are the various switching techniques in computer communication ?
11. What is meant by GOS ?
12. What is known as blocking criteria ?
13. What is meant by delay criteria ?
14. Define congestion.
15. Explain briefly the two types of congestion.
16. How the GOS can be expressed ?
17. What is meant by component GOS ?
18. Define traffic.
19. Define average occupancy.
20. Explain briefly with neat diagrams, the centralized switching and distributed switching.
21. Draw the typical hierarchical network structure and explain.
22. Compare with neat sketch, the AT T and CCITT hierarchical structure and explain.
23. Define MAN and WAN.
24. Write short notes on (*a*) ISDN (*b*) IN and (*c*) NPN.

2

Telecommunication Standards

2

Telecommunication Standards

2.1. INTRODUCTION

Traditionally, the telecommunication Industry around the world has been regulated by government, international organization and national organizations with the approval standards groups. In this chapter, the national and international organizations which develops the standards for telephony voice communications, data communication LAN hardware and software products developed for telecommunication and wireless communication are identified to understand the standards. This chapter enables one to understand the legislative activities of the standards organization, types of standards, requirements of standards, the advantages of standards to the user and suppliers and importance of standards.

More recently, telecom regulations are being liberalized and some regulations are even removed to provide good and speedy services to the subscribers and to promote healthy competitions. The deregulations approved by the standards organization are discussed in this chapter.

2.2. WHY STANDARDS

On a global scale, the primary standard organization are the ITU (International Telecommunication Union), ISO (International Standards Organisation) and the IEC (International Electro technical Commission).

In a large telecommunication network, such as the PSTN, the network structure is subject to continued changes with the increased rate of subscribers, change in subscriber behavior, the government policies, cost associated with switching office and the demand from the users. ITU-T recommendations are widely used because they generate the interconnectivity and interpretability of networks and enable telecommunication services to provide worldwide.

The rapid changes in technologies time to time causes the existing system to become obsolete in a short duration. The standardization guides and regulates the research bodies to undergo research into similar areas that results in better products and reduced costs. In the continuous expansion of telecommunication networks and facilities, the need of computers and its application are inevitable. The ISO's open system interconnection sever layer model (OSI model) is a set of communication standards and protocols that aims to establish an open environment for the movement of data between services.

Thus, standards are important for telecommunication for the better interconnectivity and interoperability. Standards are important for data communications. Standards are also plays an important role to assure product compatibility.

2.3. TYPES OF STANDARDS

The standards developed by various national and international organizations are classified into three ways based on the requirements of the public, telecommunication equipment manufacturers, independent network operators, national telecommunication companies, private exchanges, government and private administrators and so on.

De jure standard. These standards are developed by the national or international standards making organizations such as ANSI and CCITT. The standards developed are open to the public for scrutinizing, debate, modification, alteration or revision.

De facto standards. This type of standards refers to the set of procedures and protocols developed or defined by the individual organization or service providers. These procedures are protocols are developed for the companies welfare to perfect its products, improve the market and service facility and provide high competent to the related companies. Thus these protocols are not open to the publics and companies to revise or modify in any way. The organization, which developed the protocols, only have full right to make any amendment or total changes due to the expiry of technologies. For example the procedures and protocols developed by various computer organization for compatibility purpose during early 1980's to late 90's were in the stage to modify its procedures due to the obsolete technologies and newly implemented ones.

User or house standards. These types of standards are developed by the request of major suppliers or major users. Generally the major user will be the government organizations, and the major suppliers are referred to the equipment suppliers to military and other government agencies. House standards are produced only when de facto standards are not available. Some house standards developed for government or military are open to public (example TCP/IP – originally developed for the US defense for their internal data networks).

2.4. ADVANTAGES OF STANDARDS

The standards (any type discussed above) assures the interoperability of products and services among various telecommunication to the users, manufactures of the telecommunication equipment and the service providers.

The standards assure the quality of the products. It helps the user to have stability and confidence in a particular technology or application. There is a high possibility of reduced cost due to the mass manufacturing with the help of regulations. These are the advantages of standards to the users.

The standards pave the suppliers and service providers for designing, implementing and servicing their own telecommunication functions. The suppliers can work to a predefined specification. The new regulators and legislations by the standards organization and mastering the new technologies gives new business opportunities to the manufactures and thus the new innovations to the users.

2.5. STANDARDS-MAKING REQUIREMENTS

The telecommunications act standards regulations and policies should satisfy the following requirements to regulate networks as they evolve. The standards also should enable the various agencies, which are involved in setting up and operating data networks spread worldwide. The standards should be set at the appropriate time to avoid the failure of standards. The late introduction of standards is ineffective and the too early introduction of standards will be low quality, stagnated and with technological gap. Too many options, too many standards related to a particular activity, too complex standards, which is difficult to implement, and the standards which are not clear to understand are the general reasons for the failure of standards. Some of the standards setting requirements are.

Technical competence. The standards should be technically superior and easy to implement. It should be at the level of acceptance by any telecommunication serving agency to implement and maintain. The technology should pave way for the widest possible use.

A good occassion. The early occurrence of the standardization results in the restriction of research in the similar areas and results in waste of time and money. The late introduction in technological obsolation. Also the well emerged existing technology cannot be replaced by the late standards.

Consensus. The standards set by the standards organization should be agreed by the telecommunication service providing agencies, manufactures, government, policy makers, political forces and public. If all PTT's and network operator agree, then a standard is implemented.

Less options and less complex. The standards are to be easy to implement. This implementation should be cost effective and technologically feasible. The approach is easier from drafting to implementation.

Wide scope. Once the standards are accepted by various agencies and organizations, it can be implemented worldwide without any difficult, and any duplication.

2.6. HISTORY AND GROWTH OF TELEPHONE COMPANIES

Telecommunication industries are governed by private, public sector or government. In U.S. Telecommunication companies are largely privatized. When the patent of Bell on telephone ran out, there were around 4000 companies entered. With the merger of big companies and closing of small companies, at present there are about 1600 private owned companies in the U.S. that provide communication carries. The telephone companies governed by the government are usually known as the post, telegraph and telephone. In the following section, the history and growth of telephone companies are discussed.

Bell operating companies. In 1876, Elisha Gray and Alexander Graham Bell filed papers with patent office for an invention of telephone. In 1877. A.G. Bell formed the Bell telephone company. In 1893, the patent on A.G. Bell ran out. This led to the establishment of many telephone companies called independents. Between 1894 and 1900, in U.S. there were around 4000 companies. Bell operating companies (BOC's) served nearly 60% of the total subscribers.

AT & T. American telephone and Telegraph company (AT & T) is the successor to Alexander Graham Bell's phone company. In 1882, Bell brought the western electric company and in 1885 incorporated as AT & T. AT & T provides service in almost 100 major cities. AT & T acquired a number of companies, including TCG and TCI. AT & T is in a joint vendure with British telecom and acquired global network in December 1998. AT & T also involved in Inter net traffic and voice/data networking. AT & T also provides extensive support for VPN.

DOT. In India, posts and Telegraphs (P & T) department dealt with mail, telegraph and telephone communication till the end of 1985. From January 1985, the total responsibility of P & T was divided into two departments. Department of posts dealing with mail and the department of telecommunication (DOT) dealing with telephone, telegraph and data communication.

2.7. STANDARDS MAKING BODIES

In United States, regulation begins in 1866 with signing of the post Act, which gave the U.S. postmaster General to control over the telegraph industry. Since then standards for telephony, various standards groups develop data communication, wireless communication, software, hardware products etc. There are two levels of standards making bodies. National organizations establish standards for its own country. International organization formulates standards to interconnect data networks and organization worldwide.

International organizations. Standards are developed by the individulas and organizations over a period of time to fulfill various activities. Some of the international organization, which sets standards, are discussed below.

- **The international telecommunication union (ITU).** It was formed by the agreement of 20 countries to standardize telegraph networks. It involved with telephony regulation, wireless radio telecommunication and sound broadcasting. In 1927, the union was involved in allocating frequency bands for radio services. In 1934, the union was named as ITU. After world war II, the ITU became a special agency for the united nations and moved its head quarters to Geneva.

  ITU activities include co-ordination, development, regulation, and standardization of internal telecommunications, as well as the co-ordination of national policies. The ITU makes the recommendations about various technologies and publishers those recommendations for use by telecommunication industry. A complete list of ITU recommendations may be found at http:/www.itu.ch/itudoc/itu-t/red.html.

  The ITU also has specific study groups that gather information about technologies available. A study group related to telecommunications study the operations, tariffs, network management, telephone signalling and multimedia systems. For more information, refer to http://www.itu.ch/itudoc/itu-t

In 1993, the ITU went through reorganization into ITU-T (Telecommunication standard sector), ITU-R (Radio Communication sector) and ITU-D (Telecommunication development center).

- **ISO.** The international organization for standardization (ISO) is a worldwide federation of national standards bodies with representatives from over 100 countries. It is a non-governmental organization established in 1947. Its mission is to promote the development of worldwide the international exchange of goods and services and to develop co-operation in the spheres of intellectual, scientific, technological and economic activity. The open system interconnection (OSI) reference model is a seven-layer decomposition of network function published by the ISO. The structure of OSI is discussed later. OSI is a set of communication standards and protocols that aims to establish an open environment for the movement of data between devices. More information can be obtained from http://www.iso.ch/
- **ETSI.** European telecommunication standards institute (ETSI) formed in 1988. Its representatives are post, telephone and telegraph (PTT'S) ministries, computer and telecommunication vendors, manufactures, users and research bodies.

National Organizations. The following section gives the details of the American standards bodies and Indian standards bodies.

- **IEEE.** The IEEE is a technical association of industry professionals with a common interest in advancing all communication technologies. It is non-profit, technical professional association based in U.S. that develops, data communication and other things. The LAN/MAN standards committee (LMSC) developed the standards for the lowest two layers in OSI model is generally called IEEE 802 standards. IEEE 802 defines physical network interfaces such as interface cards, bridges, routers, connections, cables and all the signalling and access methods associated with physical network connections. The related information about IEEE and IEEE 802 can be obtained from

  http://www.ieee.org/ and http://grouper.ieee.org/groups/802/index.html
- **ANSI.** American National Standards Institute defines coding standards and signalling schemes in U.S. and a member of ISO and ITU. Some well known ANSI standards are EDI specifications (ANSI X 12), FDDI (ANSI X 3T9.5), ADSL (ANSI T1.4B-1995), security standards (ANSI X9) and SONET and programming and query languages (ANSI X3J16). The information about the above standard are available at http://www.ansi.org
- **EIA.** Electronic Industries Association (EIA) is a US organization of electronics manufactures. EIA published many standards related to telecommunication. The most popular serial interfaces by EIA standards are RS-232-c, RS-449, RS-422 and RS-423. The EIA RS-232 standards is also the CCITT standards V.24. For most cases EIA standards have CCITT Equivalent. For more information http://www.eia.org
- **TIA.** The telecommunication Industry association involved in International marketing opportunities, legistive efforts and standards development. EIA has recently joined with TIA to create commercial building telecommunication wiring standards (TIA/EIA 568 and 569). This standards facilitate the building designer to obtain data communication equipment wiring details. The TIA's website http://www.tiaonline.org provides educational material related to network design and cabling.

Other national standards including British standards institution (BSI), Association Francise de normalizations (AFNOR), Deutsches Institute Fur Normalische (DIN) and Bureau of Indian Standards (BIS) are members of ISO. COS is a group of major computer and telecommunication manufactures (more than 20 leading companies) that have joined together to adopt ISO standards.

2.8. DEREGULATION

Since 1968, telecommunication deregulations started due to pressures from various organizations. After 1996, some regulations are liberalized and even removed to provide good and speedy services to the subscibers. The BOCs and AT & T services were deregulated and prices were fixed for services. Deregulation of the telecommunications industry began with the deregulation of station equipment on customer's premises.

1996 Telecommunication Reform Act. The telecommunication act of 1996 was enacted by the US congress on February 1, 1996 and signed into law on February 8, 1996 by the President. The detail of the telecommunication reform act 1996 is available at the website http://www.fcc.gov/telecom.html

ACRONYMS

ADSI	—	Asymmetric Digital Subscriber Line
AFNOR	—	Association FranCoise de Normalization
ANSI	—	American National Standards Institute
AT & T	—	American Telephone and Telegraphy
BIS	—	Bureau of Indian Industries
BOC	—	Bell Operating Companies
BSI	—	British Standards Institution
CCITT	—	Consultive Committee for International Telephony and
DIN	—	Deutscches Institute Fur Normalische
DOT	—	Department of Telecommunication
EDI	—	Electronic Data Interchange
EIA	—	Electronic Industries Association
ETSI	—	European Telecommunication Standards Institute
FCC	—	Federal Communication Commission
FDDI	—	Fiber Distributed Data Interface
IEC	—	International Electro technical Commission
IEEE	—	Institute of Electrical and Electronics Engineers
ISO	—	International Standards Organization
ITU	—	International Telecommunication Union

ITU-D	—	ITU-Telecommunication development
ITU-T	—	ITU-Telecommunication sector
LAN	—	Local Area Network
LMSC	—	LAN/MAN Standards Committee
MAN	—	Metropolitan Area Network
OSI	—	Open System Interconnection
PSTN	—	Public Switched Telephone Network
PTT	—	Post, Telephone and Telegraph
SONET	—	Synchronous Optical Network
TCP/IP	—	Transmission Control Protocol/Internet Protocol Telegraphy
TIA	—	Telecommunication Industries Association

RELATED WEBSITES

American National Standards Institute	*—http: / / www.acm.org*
Electronic Industries Association	*—http: / / www.eia.org*
European Telecommunication Standards Institute	*—http: / / www.esti.org /*
Federal Communication Commission	*—http: / / www.fcc.gov /*
International Electrotechnical Commision	*—http: / / www.iec.ch /*
Institute of Electrical and Electronics Engineers	*—http: / / www.ieee.org /*
International Standards Organization	*—http: / / www.iso.ch*
International Telecommunication Union	*—http: / / www.itu.ch*
Telecommunication Industries Association	*—http: / / www.tiaonline.org /*

REVIEW QUESTIONS

1. What is the necessity of standards ?
2. What are the different types of standards ?
3. List the advantages of standards.
4. Name some of the standards making requirments.
5. Write a brief history on BOCs.
6. What are the major telephone companies in U.S. ?
7. Which department of India deals with Telecommunications ?
8. What are the International standards making organizations ?
9. List the national level standards bodies.
10. Explain briefly about various national and international standards organizations.
11. Explain briefly the concept of deregulation.

3

Telephone and Transmission Systems

3

Telephone and Transmission Systems

3.1. INTRODUCTION

The purpose of telecommunication switching system is to provide the means to pass information from one terminal (calling subscriber) to other terminal (called subscriber) somewhere. The telecommunication system is divided into four possible elements. They are end systems or instruments, transmission systems switching systems and signalling. In this chapter, the first two elements are explained in detail. Fig. 3.1 shows the telephone network and services.

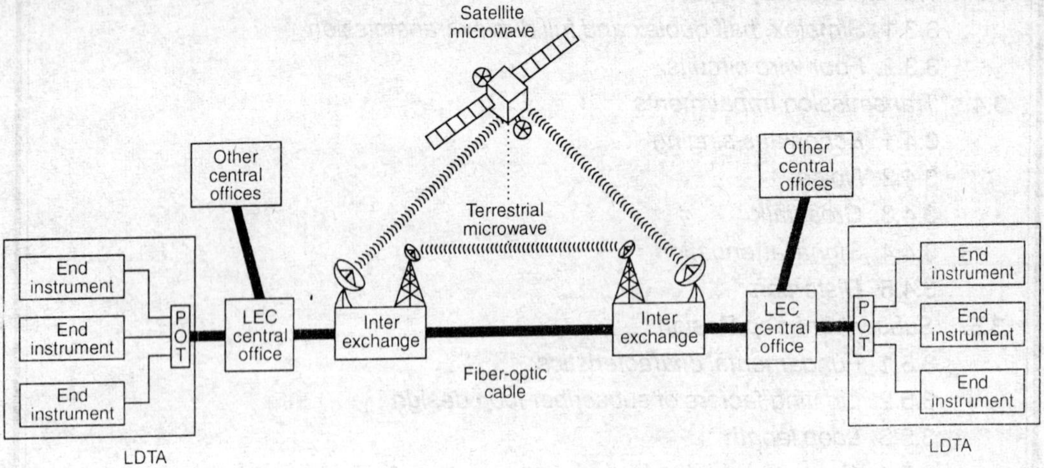

Fig. 3.1. Telephone network and services.

The end instruments are responsible for transmitting and receiving the sound, data, facsimiles, pictures, voice, video and other informations. The end instrument may be a telephone, fax, telex, computer or any other means. The instrument widely used for communication is telephone and other instruments are office related. Thus the general customer premise equipment (CPE) is assumed as telephone. Telephones come in variety of designs, colors, and styles. Many telephones are designed to work with the local telephone company. The telephone system section of this chapter explains the concept in detail.

The primary function of transmission system is to provide circuits having the capability of accepting electrical signals at one point and delivering them at destination point with good quality. The transmissions may take place over guided media (copper cables and fiber optic cables) and unguided media (wireless radio, microwave and infrared). Transmission components (links) define the cable or wireless infrastructure for transmitting signals. The wiring from the subscriber premises to the local exchange is called **local loop** or **subscriber loop.** The guided and unguided media used between the local exchanges and local exchanges and primary, secondary and teritory centers are called **trunks.**

Signal multiplexing is a technique of combining multiple channels of information over a single circuit of transmission path. To make cost effective use of the transmission system, multiplexing plays a very important role. There are several ways by which signals can be multiplexed. The most important ones are Space Division Multiplexing (SDM). Frequency Division Multiplexing (FDM) and Time Division Multiplexing (TDM) and they are discussed in detail.

Digital signals are transmitted over media by representing the binary digit as electrical pulses in which a pulse is a signal element. Bandwidth in digital transmission refers to the number of bits per second that can be transmitted over a link. In digital data communication a fundamental requirements is that the receiver knows the starting time and duration of each bit that it receives. The two schemes which meet these requirements are Asynchronous transmission and synchronous transmission. The circuits, techniques and applications of digital switching system are elaborated in the chapter 4.

3.2. TELEPHONE SYSTEM

The telephone is a familiar end instrument in telecommunication switching system. The development of telephone and circuits started around midst 1850. The table 3.1 shows the development of telephone.

Table 3.1. Development of telephone

Year	Description
Early 1854	Charles Bourseul discovered the vibration due to speech waves on the sufficiently flexible disc. Also he found this vibration can made and brake current from a battery.
1861	Philip Reis, a German physicist found that some membrane attached to the disc cab reproduce the sound wave.
1876	Sir Alexander Graham Bell discovered telephone. It contains both transmitter and receiver. It was made of diaphragm and armature for transmitter and receiver section.
1877	Berlinear and later Huges, Hummings found carbon granules as the most suitable material for transmitter.
1878	Thomas Watson, assistant of A.G. Bell invented the ringer device.
1886	Edison improved Hummings Telephone design.
1895	Pulse dialing originated.
1950	Touch tone dial telephone developed.
1964	Touch tone dial telephone are introduced into market after field trials by AT & T.

In this section the telophone handset basic telephone circuits, rotary dial telephone, touch tone dial telephone, simple telephone circuits and its related topics and finally standards for the telephone are discussed.

3.2.1. Telephone Handset

The telephone is a familiar end instrument in telecommunication system. The telephone is basically a transducer. Transducer is a device that converts one form of energy into a different form. The transmitter of telephone converts sound energy into electrical energy. The receiver converts electrical energy into sound waves. Fig. 3.2 shows the handset assembly of telephone. The handset consists of transmitter and receiver.

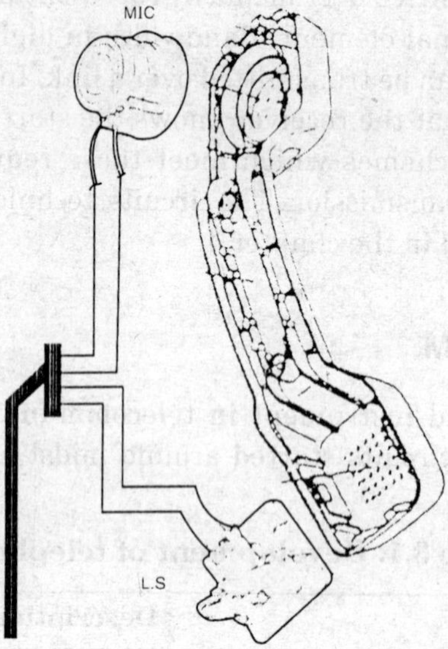

Fig. 3.2. Handset assembly of telephone.

Transmitter. The transmitter consists of a box containing a powder of small carbon granules. One side of the enclosure is flexible and is mechanically attached to a diaphragm on which sound wave impinges. The diaphragm causes the carbon granules to compress or allow them to expand. Consequently the resistance of the carbon granules decreases or increases in the box. The carbon granules conduct electricity and the resistance offered by them is dependent upon the density with which they are packed. If a voltage is applied to microphone, the current in the circuit varies according to the vibrations of the diaphragm. The varying electrical signal is similar to the varying sound signal. For this reason, it is called an analog signal. The theory of the carbon microphone indicates that the microphone functions like an amplitude modulator (Appendix A).

Receiver. The varying signal from the handset A (calling subscriber) is coupled by wires to a receiver of the handset B (called sibscriber). The receiver is an electromagnet with an accompanying magnetic diaphragm. The electromagnetic usually have two coils of about 100 turns with nominal resistance of 400 ohms. The receiver diaphragm must always be displaced in one direction from its unstressed position. It must be positioned with an air gap between it and the poles of the electromagnet. The diaphragm is made of cobalt iron and it is slightly conical shaped near the ear for uniform pressure distribution and hence the sound.

3.2.2. Central Battery System

The microphone requires to be energized in order to produce the electrical signals correspondig to the speech waveform. The minimum current required for proper operation of a modern carbon microphone is about 23 mA. In early system, local batteries at the subscriber premises were used. This local battery system uses a dry cells to power the microphone, a magneto exchange to generate required a.c to indicate the exchange for service and an autotransformer to provide impedance matching between transmitter, line and receiver. The local battery system is now practically obsolete due to the necessity of frequent replacement of dry cells and cumbersome procedure of magneto generator.

The local battery system was completely replaced by the central battery system where one battery of secondary cells is provided at the telephone exchange. Thus, a battery of 52 volts from the exchange powers the subscriber loop. The central battery system is shown in Fig. 3.3.

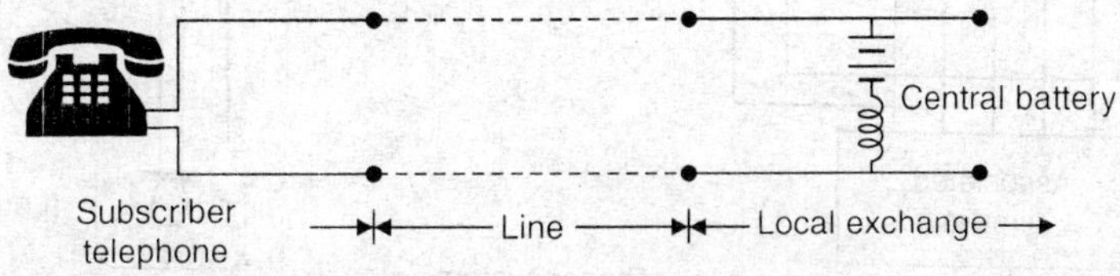

Fig. 3.3. Central battery systems.

In many situations, the loop length is extended while keeping the battery voltage constant. This causes reduced effectiveness in the signalling. Also, if the current is insufficient (more attenuation) due to increased length, the relay will not function properly and the line seizure cannot be effected. The amplifiers and / or loop extenders as well as the inductive loading are the method used to increase the effectivencss in this case. A loop extender is a device that increases the battery voltage on a loop and hence its signalling range. Amplifiers are used to counter balance the attenuation. Inductive loading consists of inserting series inductance in a loop at fixed intervals to reduce the transmission loss on subscriber loops.

The central battery system provides (*i*) an a.c ringing voltage for the bell on the telephone (*ii*) metallic path to provide power to the carbon granules (*iii*) a means to indicate the exchange for the need of service when the handset is removed from the cradle and (*iv*) generates the pulsating d.c. current when the dialing is made.

3.2.3. Telephone Base Unit

The modern telephone set circuits includes keypad, dialler IC, sound amplifier IC, associated circuits, ringer IC, buzzer, buzzer volume control and ringer switch which is operated during the on hook or off hook of the telephone handset. The above circuits are usefull in effective conversation between subscribers. Fig. 3.4 shows the typical structure of telephone circuits.

The basic single line telephone with many additional functions is referred as feature phone. The additional facilities includes redial (to redial the last number dialed), clearer channel (if noise interferes with the conversation), lighted keypad, ringer volume control, storing phone numbers in memory, storing a phone number and can be dialed with a one-touch operation, handset locator (cordless phone), call waiting service, automatic security code setting (used in cordless phone whenever the handset is placed on the base unit, the unit automatically selects one of many (in thousands) security codes. This codes prevent the unauthorized use of your telephone line by another cord less telephone user) etc.

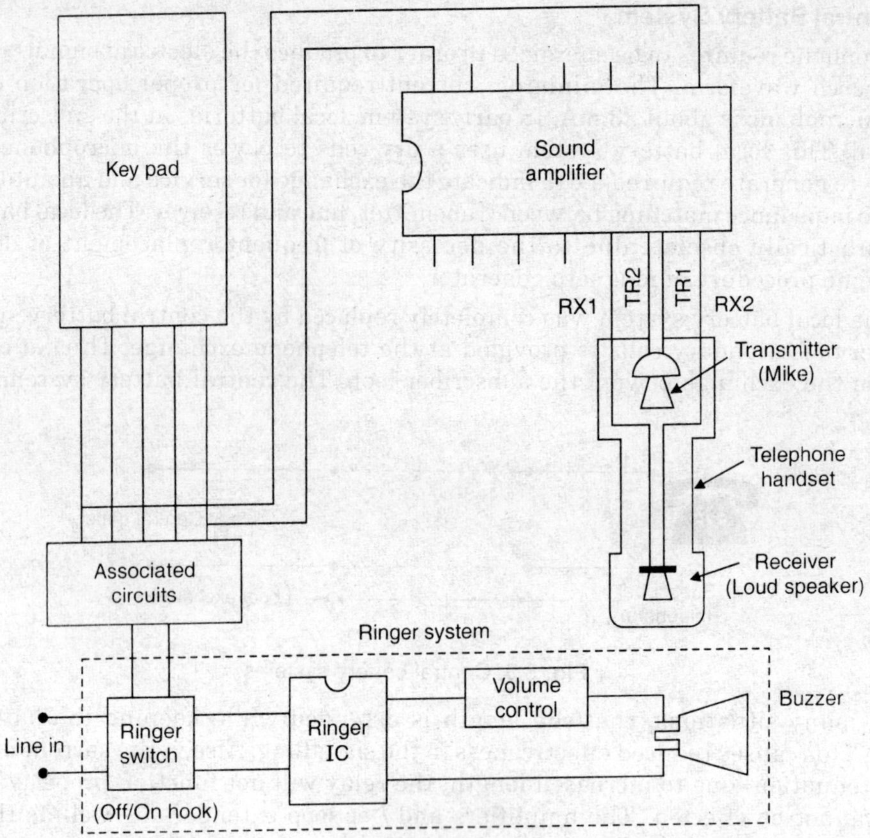

Fig. 3.4. Typical telephone base unit circuit arrangement.

The modern telephone has the answering machine facility, message recording facility which uses PC multimedia card attached to microphone. Present day electronic feature phones are proprietary telephone set, hand free phone, integrated telephone, ISDN telephone, key system, and hybrid key system.

Leading telephone manufacture/assembler companies in India are ANCO, Beetal, BPL, Crompton, Godrej, GCEL, ITI, Panasonic, Punwire, Siemens, Shyam, Tata Fone, Telemate Telerisccon, Tushaco, Webcom etc.

The local exchange and the line interface requirements found in Local Switching Systems Generic Requirements (LSSGR) by AT & T and Bell communications Research sets the standards for manufacturing telephone base units. This standards specify length of the local loop, type of ringer, minimum current for transmitter, DTMF pad etc.

Ringer. In 1878, Thomas Watson, assistant of A.G. Bell invented the device called ringer to alert the people with whom one want to talk. Original equipment includes a hard cracked electro magnet and a ringer to generate a.c signal. This signal caused the ring, but all phones connected to line. Today the ringing singnals sent by the exchange. When a subscriber calls another subscriber, the exchange relays are activated and the ringing frequencies are passed to the called subscriber.

Present modern telephones have ringer system, which includes ringer switch, ringer IC, ringer volume control and the buzzer. The ringing frequency varies for exchange to exchange. In India 16 2/3 Hz ringing frequencies are passed to the called terminal from the exchange.

Other possible ringing frequencies are 16 2/3, 25, 33 2/3, 50 and 66 2/3 referred as harmonic ringing. Telephones manufactured today are made with a ringer that will ring on any frequency of signal. This ringer is called a straight-line ringer. This eleminates the need to check ringer codes when buying a telephone. Popular ringer IC's used in modern telephones are 8-pin ringer IC KA 2418 B, 8 pin LS1240. 8 pin HA 31002P, 8 pin TCM 1536 P. Fig 3.5 shows the ringer IC of LS1240.

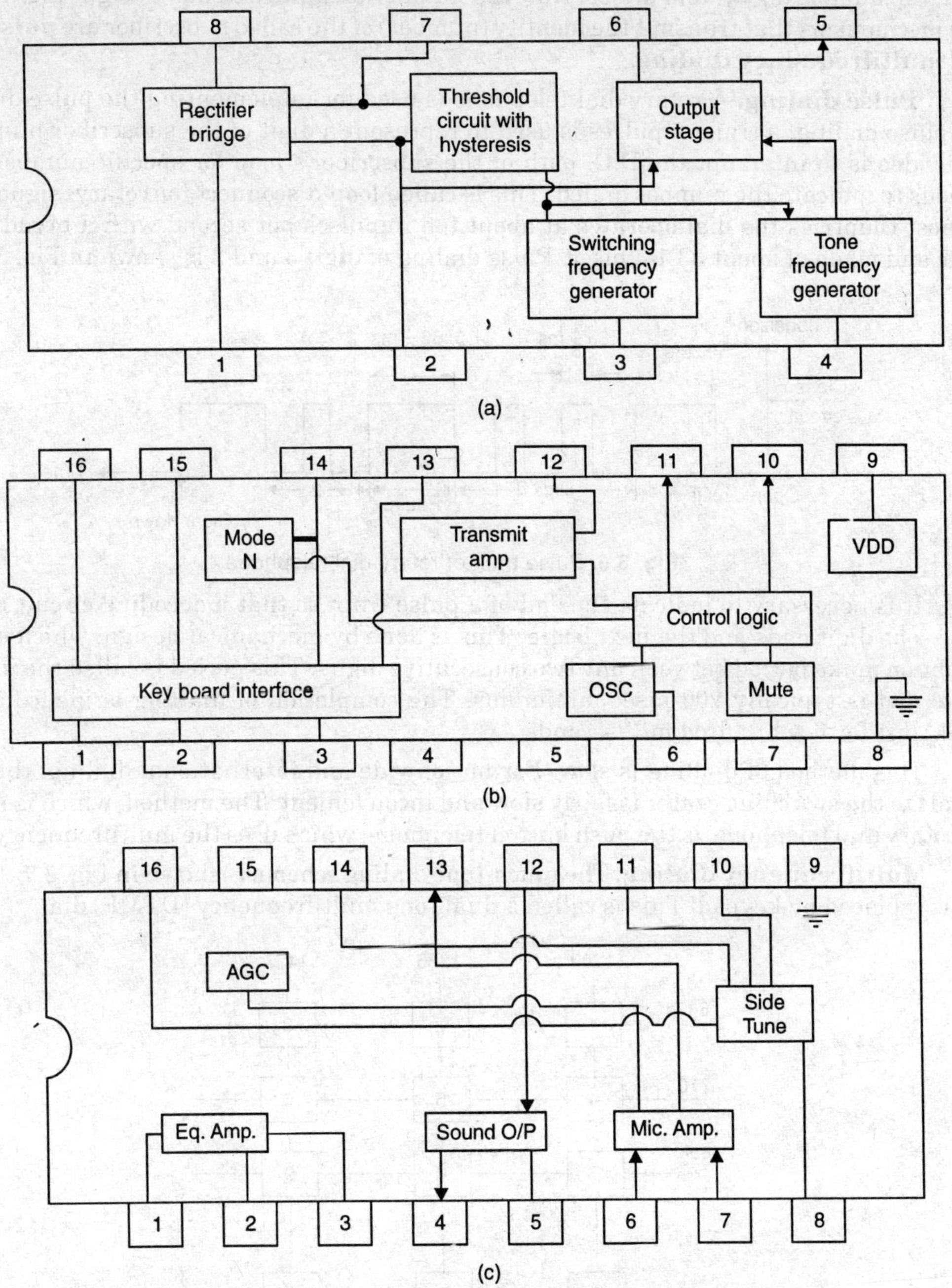

Fig. 3.5. (*a*) Ringer IC LS 1240 pin diagram (*b*) dialler IC HM 91C02 A (*c*) sound amplifier IC TEA 1062.

Dialling. At the early stage, there was no numbering systems. The calling person has to give all information about the called subscriber to the operator of manual exchange to make connection. During those times, a measles disease was spread over the Lovel town in United States. At that time, Moses Barker, a doctor raised a question that if the dicease affects the telephone operators of the exchange, who will give the connection to the subscribers. By his idea, the numbering system arises. Now the number of digits extends to eight digits in India. The mechanisms that transmit the identity (number) of the called subscriber are **pulse dialing** and **multifrequency dialing.**

Pulse dialing. A rotary dial telephone is used for implementing the pulse dialling. In the pulse dialing, a train of pulses is used to represent a digit of the subscriber number. The basic idea is to interrupt the D.C. path of the subscriber's loop for specific number of short periods to indicate the number dialed. This is called loop-disconnect (or rotary) signalling and in most countries the dial operates at about ten impulses per second with a bread of 66 2/3 msec and made of about 33 1/3 msec. Pulse dialing of digit 3 and 2 is shown in Fig. 3.6.

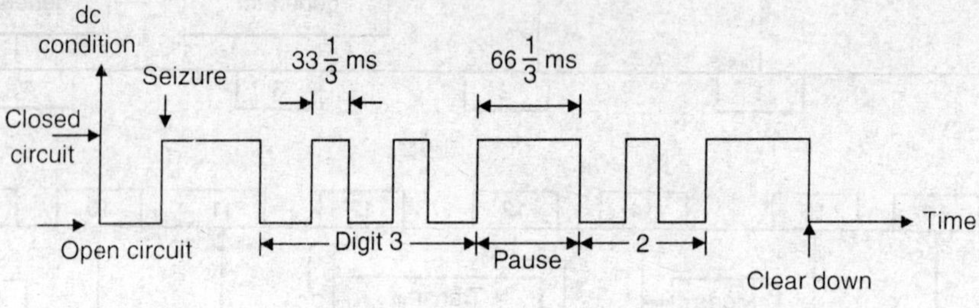

Fig. 3.6. Pulse train of rotary dial telephone.

It is necessary to indicate the end of a pulse train so that a decoding circuit can know where one digit ends and the next begin. This is done by mechanical design, which ensures a minimum make period between any two consecutive digits. This period is called the inter-digit pause and is typically 200 msec minimums. The completion of dialling is identified if any break, last for few hundred milliseconds.

This method of dialling is slow. For nationwide and international dialing, the routing signals to the switching center is fairly slow and inconvenient. The method, which is replacing the rotary dial telephone, is the push button telephone, which uses the multifrequency dialing.

Multifrequency dialing. The touch-tone dialing scheme is shown in Fig. 3.7 the rotary dial is replaced by keypad. This is called a dual tone multifrequency (DTMF) dial.

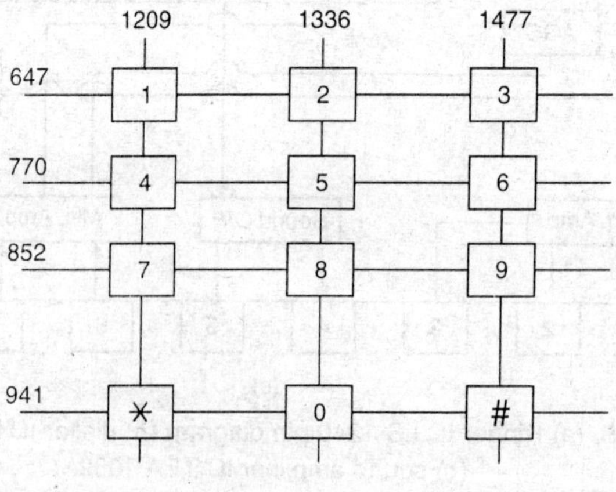

Fig. 3.7. Touch tone dialing scheme.

The principal method uses a pair of tones to signal each digit. These two tones are shown in Fig. 3.7. There is sixteen possible pair of tones. For the digits 0 to 9, ten combinations are used. Other six may be used for additional control signals. Allowing 40 msec interdigit pause, a maximum rate of about 12 digits/sec is feasible. When pressing a button (for example the digit 4), the corresponding tones are sent to the central exchange (for digit 4, 770 Hz and 1209 Hz). The central exchange decodes the different combinations of tones received into equivalent dialed digit.

The telephone set uses dialler IC and associated circuits. The dialer IC's of 16 pin HM 91C02A, 18 pin UM 91214B, 18 pin UM91214 E, 16 pin HT 9202G, 18 pin M 2560 G, 18 pin UM 1032 CP and 22 pin HM 91501, B are pupular. Fig 3.5 (*b*) shows the dialler IC HM 91C02A.

Sound amplifier. The two wires of transmitter and the receiver are connected to the sound and speech IC for the effective transmission and reception of speech signals. This IC performs various functions like Automatic Gain Control (AGC), dial tone, biasing etc. Popular speech or sound amplifier IC are 20 pins KMC 34214 P, 18 pin PBI 3726/11, 18 pin TEA 1060. The IC pin diagram of 16 pin TEA 1062 is shown in Fig. 3.5 (*c*).

3.3. TRANSMISSION SYSTEMS

The transmission system provides circuits and path between two subscribers. The transmission circuits are capable of receiving the electrical signals at one point and delivering them to a destination with good quality. If necessary, more transmission system can be setup for long distant transmission for good quality. The transmission path, which is also referred as telephone channel or transmission media is designed to provide voice-grade communication. It has evolved into a worldwide network that encompasses a variety of transmission media and switching systems. Thus the transmission system behaves as an excellent candidate for a data communication over long distances.

The transmission system involves analog transmission (voice communication) and digital transmission. The analog signals are characterized by frequency, amplitude and phase. In analog transmission system, signals propagate through the medium as continuously varying electromagnetic waves. The medium for an analog transmission may be twisted pair cable, coaxial cable, optical-fiber cable, microwave radio and satellites. The analog signals are subjected to deterioration due to attenuation and noise addition in the channel. Hence amplifiers, filters and necessary circuits are added in transmission system to upgrade the analog signal. Fig. 3.8 shows the voice signal and digital signals.

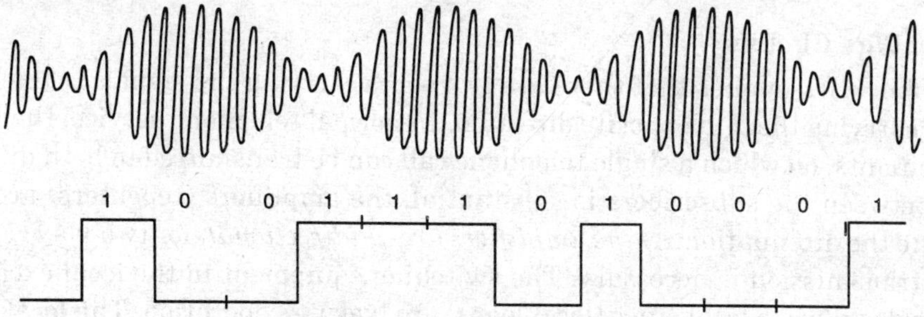

Fig. 3.8. Voice signal and digital signals.

In digital transmission system, signals propagate as a discrete voltage pulses (positive voltage represents binary 1, and negative voltage represents binary 0). The digital signals are measured in bits per second (bps). In data communications, analog signals are used to transmit information over the telephone system or over radio communication systems. A MODEM (modulator/demodulator) converts digital data to analog signals and analog signals can be converted to digital information. This process involves **sampling, quantizing**. This process is called **digitizing.** Now-a-days, the transmission between central exchanges and long distance sites are done by digitized voice cummunication. The analog transmission takes place only between the local exchange or end office and homes. In Fig. 3.9 digital to analog to digital conversion processing is shown as block diagram.

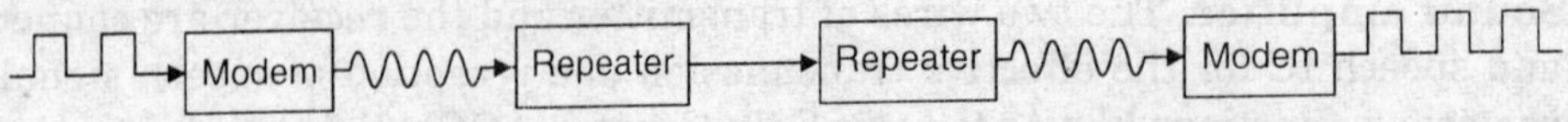

Fig. 3.9. Digital to analog to digital conversion.

Voice converted to digital requires a 64 k bits/sec channel, which happens to be worldwide standard called DSO (digital signal, level zero) for transmitting voice calls.

3.3.1. Simplex, Half Duplex, and Full Duplex Transmission

Transmission media may operate in simplex, half duplex and full duplex mode. Devices involved in transmission of signals (voice or data) may be transmitter or receiver or both. If one system only transmits and other only receives, the link is called **simplex.** Transmission of signals by Doordarson and All India Radio (AIR) are belongs to simplex transmission, as there is no possibility of reverse transmission. In certain cases, the control signals are returned to the transmitter to indicate the reception of signals or some portion of the signals or no reception. Considering this factor into account, the international telecommunication union (ITU) defines the simplex (circuit) as "a circuit permitting the transmission of signals in either direction but not simultaneously".

If both devices (transmitter and receivers) can send and receive, but only one device at a time, the link is called **half duplex.** The conversation is the example of half duplex mode transmission. The ITU defines half duplex (circuit) as "a circuit designed for duplex operation but which, because of the nature of the terminal equipment ; can be operated alternately only". A **full duplex** line allows both systems to transmit and receive simulatneously. The terms half duplex and full duplex are normally used when referring to computer and other devices connected to an analog telephone circuit using modern.

3.3.2. Four Wire Circuits

The term four wire implies that there are two wires carrying the signals in one direction and two wires carrying them in opposite direction. In normal telephone service, the local loops are two wire circuits, on which a single telephone call can be transmitted in both directions. If the distance between the subscribers is substantial, the amplifiers (repeaters) are necessary to compensate the attenuation. *As the amplifiers are unidirectional,* for two-way communication, four-wire transmission is necessary. The switching equipment in the local exchange and the line from subscriber to local office (local loops) are two wire operation. The local exchange will switch the subscriber loop to a toll conncecting trunk. This is also a two-wire transmission.

The toll offices are interconnected with inter tool trunks (which connects towns and cities). These trunks are of four-wire transmission. Fig. 3.10 shows the simple arrangement of the two wire and four wire transmission.

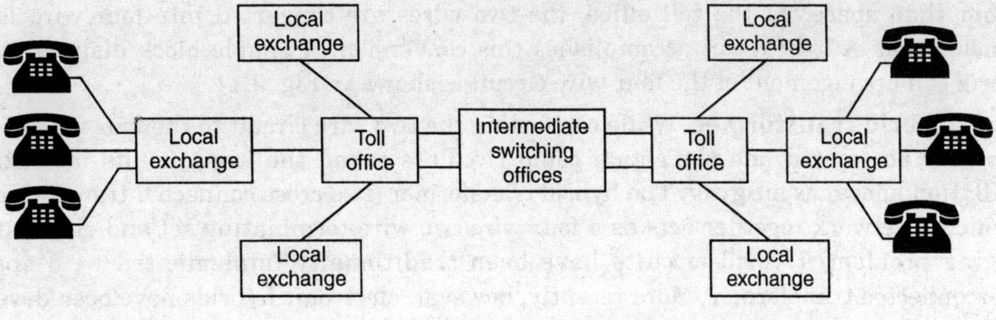

Fig. 3.10. The simple arrangement of the two wire and four wire transmission.

A four-wire circuit has amplifiers in its repeaters for each direction of transmission. The four wire circuits may be physical four wire or equivalent four wire. For short distances, actual

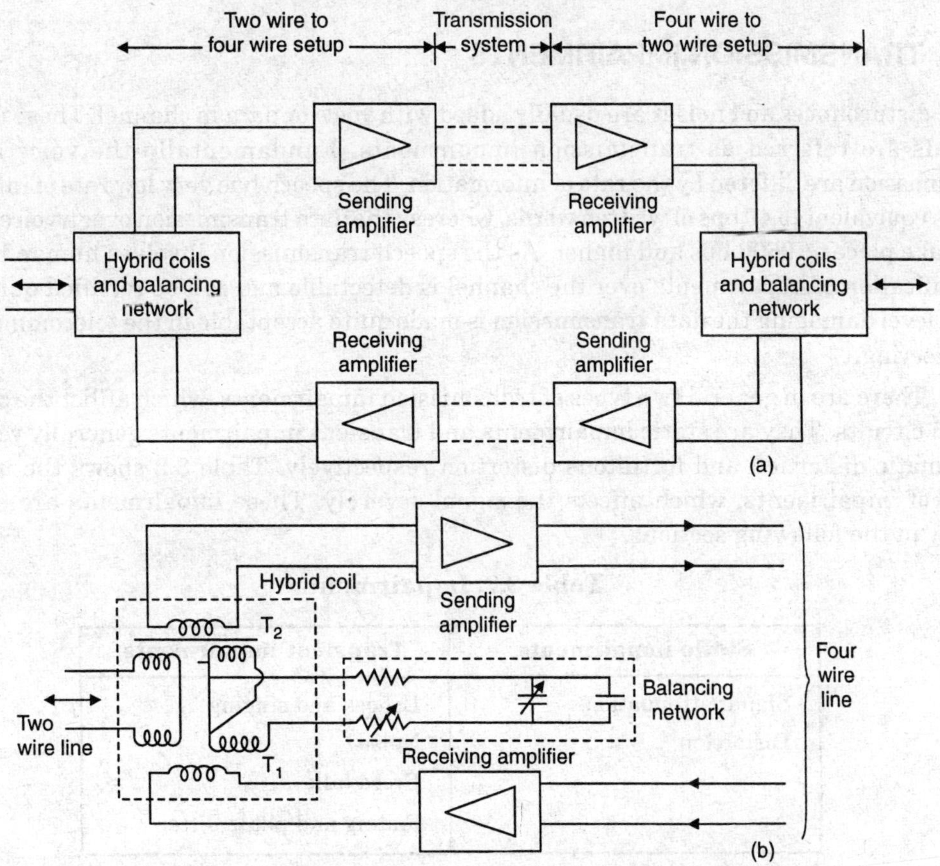

Fig. 3.11. (*a*) Block diagram of two wire to four wire conversion, (*b*) hybrid transformer circuit.

four wires used for transmission is referred as physical four wire circuits. But for long distance trunks physical four wire is undesirable and usually equivalent four wire transmission is used, needing one pair of wires only. The two directions of transmission use different frequency bands so that they do not interfere with each other. The two directions are separated in frequency rather than space. At the toll office, the two wires are converted into four wire for long transmission. A hybrid coil accomplishes this conversion. A simple block diagram and the hybrid coil arrangement of the four wire circuit is shown in Fig. 3.11.

Hybrid transformer. While connecting the two wire circuit to the four wire circuit, a loop may be created and the signal could circulate round the loop, results in continuous oscillation known as **singing.** The hybrid transformer (two cross connected transformer) and balancing network together acts as a four wire/two wire terminating set and eliminates the singing problem. Hybrid circuits have been traditionally implemented with specially interconnected transformer. More recently, however, electronic hybrids have been developed.

Cross-connected transformer windings results in zero current in the line balance impedance. The power thus divides equally between the input of the send amplifier and the output of the receive amplifier, where it has no effect. The price of avoiding singing is thus 3 dB losses in each direction of transmission together with any loss in transformers.

3.4. TRANSMISSION IMPAIRMENTS

Some disturbances and noises are usually added with voice or data in channel. These unwanted signals are referred as transmission impairments. Fundamentally the voice and data transmission are differed by the rate of information. The speech has very low rate of information and is equivalent to 40 bps of written words, whereas the data transmission over a voice channels can take place at 9600 bps and higher. As the speech transmission involves human being, the identification of impairments over the channel is detectable and can be rectified quickly. The noise level damaging the data transmission is made quite acceptable in the telecommunication engineering.

There are in general two types of transmission impairments, which afflict the communication circuits. They are static impairments and transient impairments generally referred as systematic distortion and fortuitous distortion respectively. Table 3.2 shows the important types of impairments, which affects the signal severely. These impairments are explained briefly in the following sections.

Table 3.2. Impairments

Static impairments	Transient impairments
Signal attenuation Distortion	Echoes and singing Noise Cross talk Fading and phase Jitter

3.4.1. Echos and Singing

Echoes and singing both occurs as a result of transmitted signals being coupled into a return path and fed back to the respective sources. Coupling will be zero only when perfect impedance matching occurs. Impedance matching between trunks and subscriber loop (two wire to four wire at hybrid) is difficult due to various subscriber loop lengths. A signal reflected to the speaker's end of the circuit is called **talker echo** and at the listener's end is called **listeners echo.** The talker echo is more troublesome. When the returning signal is repeatedly coupled back into the forward path to produce oscillations, **singing** occurs. Basically singing results if the loop gain at some frequency is greater than unity. An echo coming 0.5 msec after the speech is not much effect. The echoes with a round trip delay of more than 45 msec cannot be tolerated. Fig. 3.12 explains the path of the echo and the losses and gain of the signals at various parts of the system.

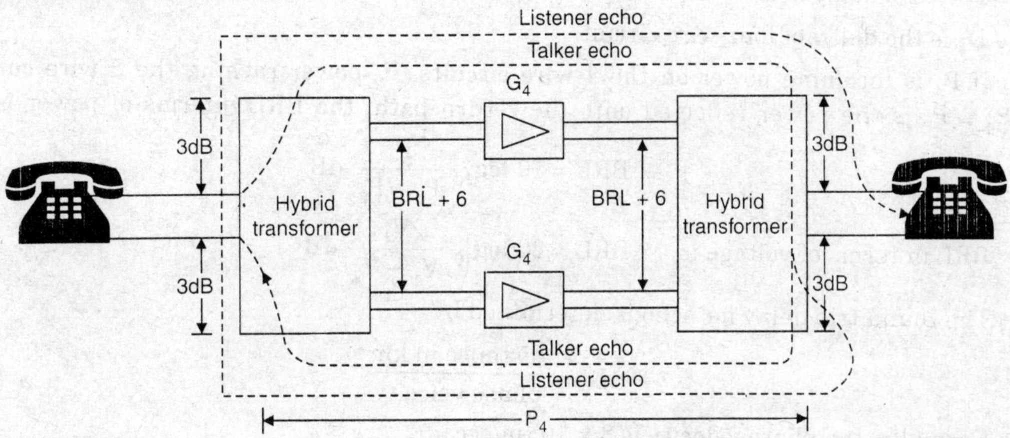

Fig. 3.12. Block diagram of echo path.

For the round trip delay of above 45 msec (representing approximately 2900 km of wire) attenuators are intorduced to limit the loudness of echo to a tolerable level. The attenuation required is related to the time delay. Hence long distance circuits require significant attenuation to minimize echo annoyance. The total attenuation from one two wire circuit to the other (α_2) is

$$\alpha_2 = \alpha_{24} + \alpha_{42} - G_4 \qquad ...(3.1)$$

where α_{24} = attenuation between 2 to 4 wire line

α_{42} = attenuation on between 4 to 2 wire line

G_4 = total gain of one side of four wire circuit (in dB)

The α_{24} and α_{42} are normally 3dB. Thus

$$\alpha_2 = 6 - G_4 \qquad ...(3.2)$$

The **transhybrid loss (TL)** is the attenuation through the hybrid coil from one side of the four wire circuit to the other and given as

$$TL = 6 + BRL \qquad ...(3.3)$$

The loss of reflected signal due to the mismatch of the network is given by **balance-return loss (BRL)**. If impedance of the four wire circuit is Z_B and the impedance of two wire circuit is Z_2, the BRL is given as

$$BRL = 20 \log_{10} \left| \frac{Z_B + Z_2}{Z_B - Z_2} \right| \text{ dB} \quad ...(3.4)$$

If Z_B and Z_2 matches BRL = ∞. It indicates the attenuation on echo is infinite and no necessity of any attenuators.

The total attenuation on echo is given by net attenuation offered by the four-line circuit minus the gain of the amplifiers. Hence by clockwise movement from talker to talker circuit,

From eqn 3.2, $\alpha_t = [3 - G_4 + (BRL + 6) - G_4 + 3]$ dB

$$\alpha_t = 2\alpha_2 + BRL \quad ...(3.5)$$

The echo delayed is $D_t = 2D_4$ ...(3.6)

where D_4 is the delay of four-wire circuit.

If P_4 is incoming power on the 4 wire circuits, P_2 power raching the 2 wire curcuit and $P_4 - P_2$ is the power reflected onto the return path, the BRL interms of power is

$$BRL = 10 \log_{10} \frac{P_4}{P_4 - P_2} \text{ dB} \quad ...(3.7)$$

BRL in terms of voltage is

$$BRL = 20 \log_{10} \frac{V_4}{V_4 - V_2} \text{ dB} \quad ...(3.8)$$

The round trip delay for echo is determined by

$$D_e = \frac{\text{distance in km}}{\text{phase velocity}} \quad ...(3.9)$$

Normally, the phase velocity is 3×10^7 m/sec.

Echo canceller/Echo suppressor. These devices are used to control the echo. If round trip delay exceeds 45 msec. A block diagram of echo suppressor is shown in Fig. 3.13. An echo suppressor operates in four wire circuits by measuring the speech power in each leg and inserts a large amount of loss (35 dB typically) in the opposite leg when the power level exceeds a threshold. Thus a returning a echo is essentially blocked by the high level of attenuation.

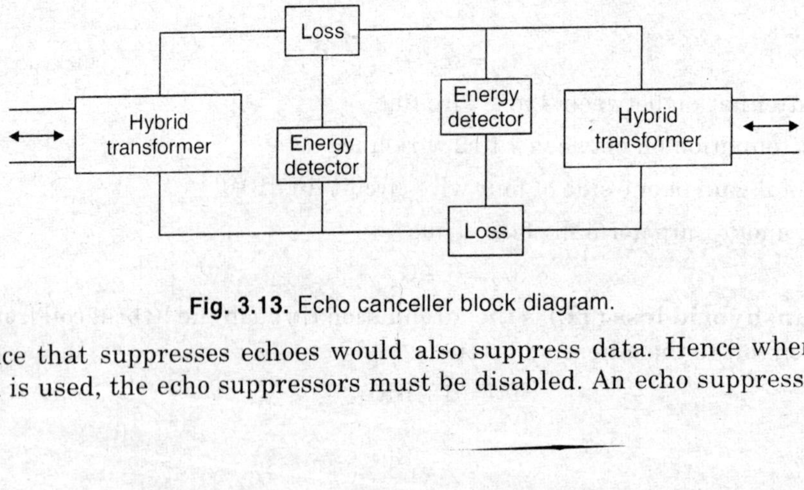

Fig. 3.13. Echo canceller block diagram.

A device that suppresses echoes would also suppress data. Hence when full duplex transmission is used, the echo suppressors must be disabled. An echo suppressor converts a full duplex circuit into a half duplex circuit with energy sensing being the means of turning the line around. The echo suppressors must be disabled when full duplex transmission is used. The user's modem can disable the echo suppressors by transmitting a single frequency tone in the band 2010–2040 Hz for at least 400 m sec. During this time, no other signal or tone should be transmitted. The echo suppressor will remain disabled if a signal transmitted within 100 msec of disabling tone being removed. Thus there may be a clip at the beginning portions of speech segments. This is another drawback of echo suppressors for voice circuits.

Echo control or echo cancellation is the recent electronics technology. An echo canceller operates by simulating the echo path to subtract a properly delayed and attenuated copy of a transmitted signal. Echo cancellers do not physically insert attenuators. A typical echo canceller block diagram is shown in Fig. 3.14.

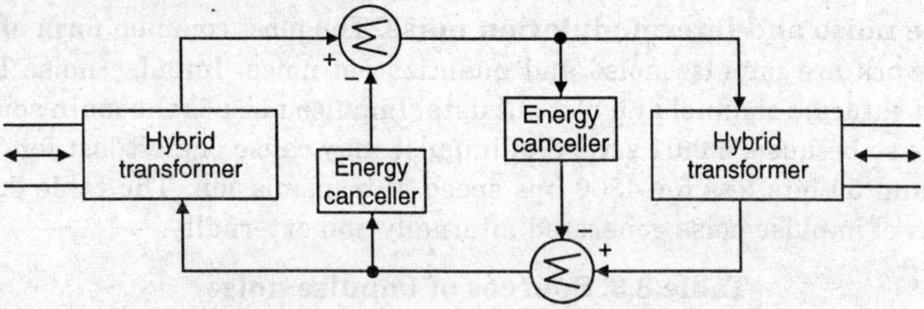

Fig. 3.14. Block diagram of Echo canceller.

The transmittted speech is stored for a period of time equal to the round trip delay of the circuit. The stored signal is attenuated and then subtracted from the incoming signal. This kind of circuits are available in satellite circuits.

3.4.2. Noise

Noise is an unwanted electrical energy. In any real physical systems, the signal arrived at reccive may be accompanied by a unknown waveform which varies with time in an entirely unpredictable manner. This unpredictable waveform in a random process is called noise. At receiver whether the signal arrives over a communication channel or is received by an antenna, is accompanied by noise. At each stage of the switching system (local exchange, toll office, repeaters) additional noise added on the signal. Transmission lines have some amount of background noise generated by external sources. This noise combines with and distorts a transmitted signal. This noise generator may be fluorescent lights, motors, ovens, phones, copiers etc. The common noises in telecommunication system are white noise, impulse noise and intermodulation noise. Fig. 3.15 shows the signal plus noise at receiver.

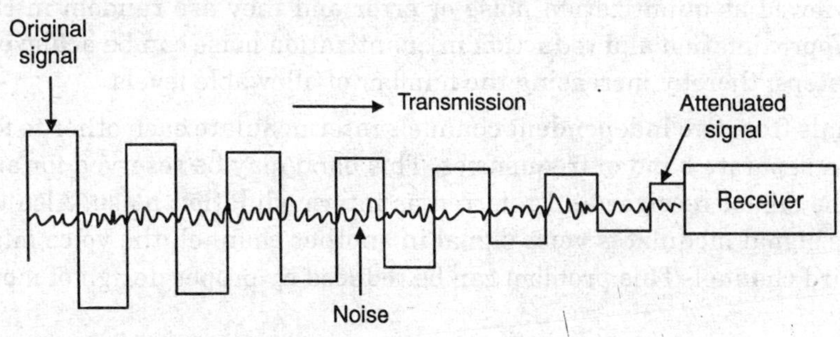

Fig. 3.15. Signal plus noise at receiver.

White noise. This is the most common noise in communication. This noise is easy to analyze and easy to find since it arises as thermal noise in all electrical components. The term white is used in analogy with white light, which is a superposition of all visible specturm components. As the noise voltage and hence the noise is caused by the random movement of electrons in any of the active or passive device found in receivers, it is referred to variously as thermal, agitation, white or Johnson noise. White noise is truly random in the sense that a sample at any instant in time is completely uncorrelated to a sample taken at any other instant in time. Power spectral density of white noise is uniform over the entire frequency range of interset. Battery systems used to power subscriber loops are also a source of this type of noise. The amplitude of the signal is kept sufficiently above the white noise to prevent an excess of hiss on radio or telephone circuits.

Impulse noise and intermodulation noise. The most common form of noise in the telephone network are impulse noise and quantization noise. Impulse noise has peaks of amplitude that saturate channel and blot out data. Impulse noise is the main source of errors in data. This is so because, a 0.01 second of impulse may cause one bit lost for 75 bps speed transmission and 50 bits loss for 4800 bps speed of transmission. The table 3.3 shows the various sources of impulse noise generated internally and externally.

Table 3.3. Sources of impulse noise

Internal sources	External sources
1. Poor quality soldering 2. Relay contacts and jacks 3. Non soldered twisted joints	1. Inductance and capacitance effects 2. Sharp voltage changes in adjacent wires 3. Open wire pairs hanging between telegraph poles can pick up atmospheric static 4. Radar interference, electric trains and electric machinery

Effects of impulse noise. As the duration of impulse is too long in comparison with the speed of transmission, a sharp click like sound will be heard to the human listener. This noise removes two or more adjacent bits and thus the parity checking devices may not detect the error.

By quantization, we create a new signal, which is in approximation to the original signal. If the actual signal is compared with quantized signal, some difference can be noticed. This error may be viewed as quantization noise or error and they are random in character. The quality of the approximation and reduction in quantization noise can be achieved by reducing the size of the steps, thereby increasing the number of allowable levels.

The signals from two independent channels intermodulate each other to form a product that falls in to a separate band of frequencies. This band may be reserved for another signal. This distorts the signal receiver and referred as intermodulation noise. Also when a single frequency voice signal modulates voice signal in another channel, the voice might be clearly audible in a third channel. This problem can be reduced by proper design of modem.

3.4.3. Cross Talk

The current from the battery in the subscriber loop (when telephone handset is off hook) is limited to the range of minimum 20 mA to maximum of 60 mA. The current variation depends on the length of the subscriber loop. ((In long loops the current is less and in short loops the current may exceed 60 mA (an electronic component varistor in telephone set is used to limit the current with in 60 mA). The large current flow causes electromagnetic fields and thus creates signal distortions in adjoining wires. This distortion is called cross talk.) Some of the major sources of cross talk are coupling between wire pairs in cable, inadequate filtering or carrier offsets in older frequency division multiplexing (FDM) equipments and the effects of non-linear components on FDM signals.

Cross talk is one of the most disturbing and undesirable imperfections that can occur in a telephone network. In analog system, as the power levels of voice signal considerably varies (40 dB range) the cross talk in this system is difficult to control. In fact cross talk is more noticeable during speech pauses, where the power level of the desired signal is zero. In digital system, by the pulse amplitude modulation, pulse length modulation and pulse position modulation, which are used in TDM system results in attenuation and delay distortion. This causes dispersion of the transmitted pulses. They spread in time and interfere with the pulses of adjacent channels and causes interchannel cross talk. Pulse code modulation is used to overcome the interchannel cross talk problem.

NEXT AND FEXT. The basic forms of cross talk concern to telecommunication engineers are near end cross talk (NEXT) and far end cross talk (FEXT). NEXT occurs near the transmitter and creates distortions that affect the signal on adjacent receive pairs. This type of noise can be generated when a transmission line carrying a strong signal is coupled with a transmission line carrying a weak signal. NEXT is measured in dB, with higher values being better. NEXT is measured for all frequencies between 0 and 100 MGz. NEXT is measured by injecting a signal on a wire pair and measuring its cross talk on another wire pair. NEXT should be measured at both ends. NEXT is more trouble some because of a large difference in power levels between the transmitted and receive signals. Twisted wire pairs reduce this type of cross talk.

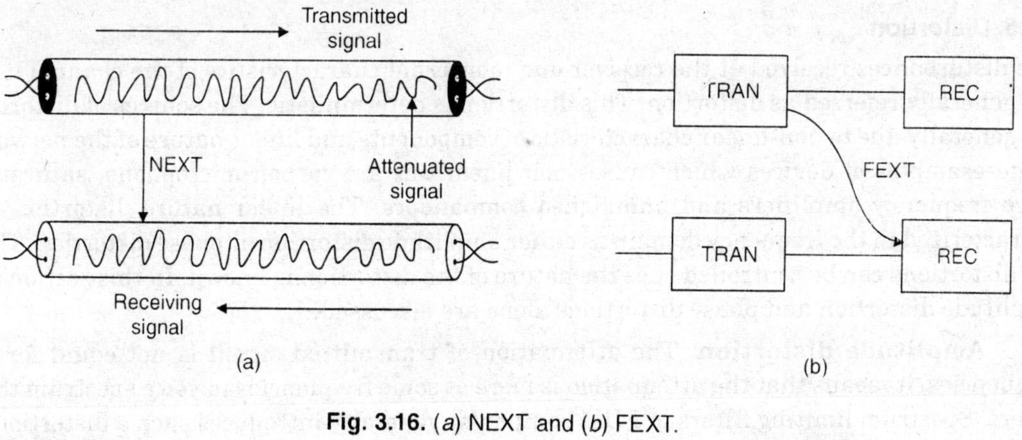

Fig. 3.16. (*a*) NEXT and (*b*) FEXT.

FEXT is a measure of the cross talk that exists at the receiver end of the cable. FEXT refers to unwanted coupling into a received signal from a transmitter at a distant location.

This noise is relevant on networks that transmit on multiple pairs in the same direction in the same time. For example Gigabit Ethernet, which enable the organizations to upgrade to 1000 Mbps while using the same operation system and software. One example of transmission media of Gigabit Ethernet is 1000 Base T and it uses four pairs of CAT–5 GB cable with RJ–45 connector. The ELFEXT (Equal Level Far End Cross Talk) test provides standard way to measure far end cross talk (irrespective of cable length) and is required for Gigabit Ethernet cable certification.

3.4.4. Signal Attenuation

The attenuation of signal varies with frequencies over the transmission line. The attenuation of a typical cable pair is approximately proportional to the square root of the frequency. Attenuation is also increases with temperature. Metal conduit also increases the attenuation. The preferred loss in a telephone connection should be in the neighborhood of 8 dB. A local phone connection have only 0–6 dB more than ideal. An average toll connection had an addition 6–7 dB of loss. The standard deviation in toll connection was 4 dB. The figure (*a*) shows the weakening of the signal due to attenuation.

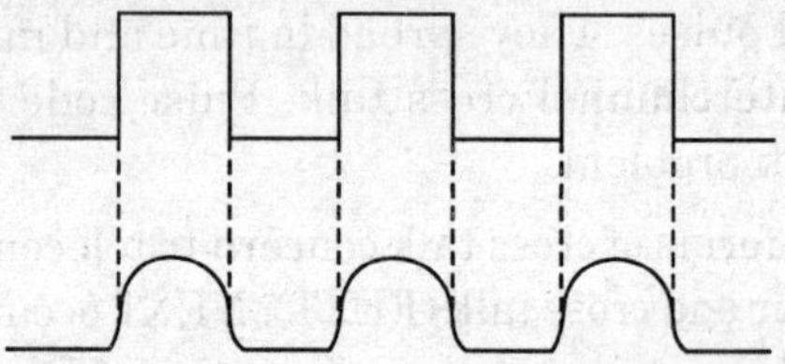

Fig. 3.17. Attenuation of the signal.

Attenuation is measured in decibels (dB) of signal loss. For every 3dB signal loss, a signal loses 50 per cent of its remaining strength. The frequency, temperature, and metal conduit should be considered when planning cable length. To compensate the attenuation, the cable may be loaded by adding inductance at intervals. Cable testers that inject signals with a known power level at one end of the line and measure the power level at other end of the line can measure the attenuation. Fluke tester is used for attenuation measurement.

3.4.5. Distortion

The disturbances received at the receiver due to internal characteristics of the channel itself are generally referred as distortion. This distortion is deterministic. The sources of distortion are generally due to non-linear characteristic of components and linear nature of the network. Some examples of devices which causes non linearities are carbon microphone, saturation voice frequency amplifiers and unmatched compandors. The linear nature distortion are characterized in the frequency domain as either amplitude distortion or phase distortion. Thus the distortions can be controlled once the nature of the distortion is known. In this section the amplitude distortion and phase distortions alone are discussed.

Amplitude distortion. The attenuation of transmitted signal is not equal for all frequencies, it means that the attenuation is more at some frequencies in voice spectrum than others. Spectrum limiting filters in FDM equipment generally introduces such a distortions.

Phase distortions. It is a serious form of distortion in data transmission. Phase delay is related to the deay characteristics of the transmission medium. The signal is delayed more at some frequencies than at other. This is referred to as phase frequency distortion or delay distortion. If there were no delay distortion, the curve of phase of the received signal plotted against frequency would be a straighuy line. That is the phase response is directly proportional to the frequency. This is referred as uniform envelop delay. The system with uniform envelop delay is called linear phase system. Any deviation from linear characteristics is referred as envelop delay defined as the slope of such curves and may be measured in microseconds.

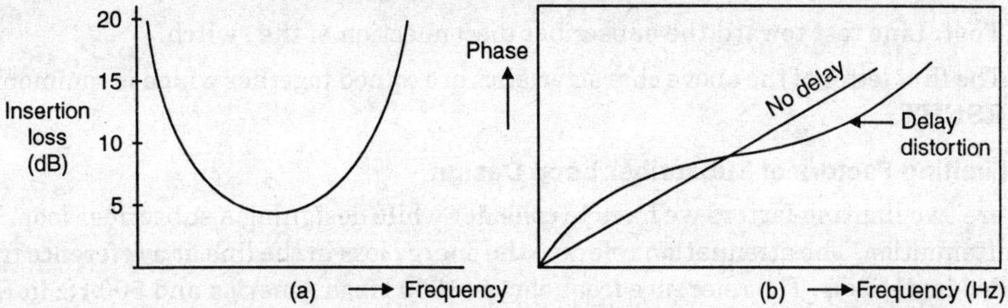

Fig. 3.18. Attenuation distortion and phase distortion.

3.5. SUBSCRIBER LOOP DESIGN

The cables that connect the telephone handsets or other devices to the local switching office or end office is referred as subscriber loop or local loop. Every suscriber has his own pair of wires to the local switching office. Twisted pair local loop is an excellent transmission medium for analog voice signals. But it is limited to low frequency audio signals. The introduction of fiber cable needs a device at subscriber premises to convert electrical energy into light energy and this is the additional cost to the customer. But for high speed data transmission, switched cable TV, videophone, teleconferencing the fiber optic local loop has become essential.

One end of each subscriber loop is terminated on a Main Distribution Frame (MDF) at the exchange. The drop wires (DW) from the telephones are connected to the distribution point (DP) which is located near the subscriber's premises. The distribution points at various locations are connected together by a distribution cables (DC) and terminated to the feeder points (FD). The DC carries 10–500 pair of wires. Many feeder points related to a particular geographical area connected by a branch feeder (BF). From BF, through main feeder, all the subscriber loops are connected to MDF at the end office or local exchange. Typically, the MF carries 100–2000 pair of wires to MDF. For the purpose of flexible interconnection such as transfer from location to other location or within the geographical area, the subscriber pair and exchange pairs are interconnected at the MDF by means of jumpers.

3.5.1. Fundamental Characteristics

The subscriber loop is the most common interface in the network. The fundamental characteristic of this interface are.

Battery. To enable dc signalling and to provide bias current for carbon microphone, a battery of about 48 V is connected to subscriber loop at exchange.

Overvoltage protection. Protection of equipment and personal from lightning strikes and power line induction or shots.

Ringing. Application of a 20 Hz signal at 86 V rms for ringer excitation.

Supervision. Supervise the network by detecting the off hook/on hook and flow/no-flow dc current.

Coding. In the case of digital end office, analog to digital coding and digital to analog decoding functions necessary.

Hybrid. For two wire to four wire conversion, hybrid in necessary.

Test. Line test toward the subscriber disconnection of the switch.

The first letter of the above characteristics are coined together which is commonly known as **BORSCHT.**

3.5.2. Limiting Factors of Subscriber Loop Design

There are two limiting factors we have to consider while designing a subscriber loop. First one is the attenuation. The attenuation refers to the energy loss in the line at a reference frequency, measured in decibels. The reference frequency is 1000 Hz in America and 800 Hz in Europe. If the length of the loop increases the attenuation also increases. The attenuation limit of the subscriber loop is normally 6 dB.

The second limiting factor is voltage drop. If the battery voltage is kept constant with increase in length, the effectiveness of the signalling and conversation will be limited. This is due to IR drop of the line. The IR drop of the line varies with resistances of the battery used in the system, telephone set resistance and the allowable resistance of the subscriber loop.

The maximum allowable resistance in the subscriber loop and the loop resistance limit is calculated as follows. The loop resistance limit is calculated as follows. The loop resistance limit is used to determine the cable length and the cable gauge required.

$$R_m = \frac{V_B}{I_C} \qquad \text{...(3.10)}$$

where R_m = maximum allowable resistance of subscriber loop.

V_B = Battery voltage

I_c = minimum current required for proper operation of carbon microphone

The loop resistance limit is $R_L = R_m - (R_B + R_T)$ ...(3.11)

where R_B = approximate resistance maintained at the battery protect against short circuit in the wire between subscriber and local office

R_T = Telephone set resistance.

Example 3.1. *If the minimum current required for carbon microphone is 23 mA, battery voltage is 50 V, the battery resistance is 400 ohm and the telephone set resistance is 200 ohms, calculate the loop resistance limit.*

Sol.

$$R_m = \frac{50}{23 \times 10^{-3}} = 2200\text{m}\Omega$$

$$R_L = 2200 - (400 + 200)$$

$$R_L = 1600\ \Omega.$$

Example 3.2. *An exchange uses – 48 V battery, a resistance of 300 ohm is placed in series with the battery. If the telephone set resistance is 50 ohm, calculate the loop resistance limit for the minimum current requirement of 23 mA for carbon microphone.*

Sol.
$$R_m = \frac{48}{23 \times 10^{-3}} = 2087\ \Omega$$
$$R_L = 2087 - (300 + 50)$$
$$R_L = 1737\ \Omega.$$

3.5.3. Loop Length

The method of determining subscriber loop length using the signal resistance limit as a basis is called the *basic resistance design*. The maximum subscriber loop length, which is *defined* as the distance from the subscriber to the central office, is expressed as

$$L = \frac{\text{Loop resistance limit}}{\text{dc loop resistance}} = \frac{R_L}{R_{dc}} \quad ...(3.12)$$

The dc loop resistance is measured in ohms per kilometer and expressed as

$$R_{dc} = \frac{21.96}{d^2} \quad ...(3.13)$$

Where 'd' is the diameter of the conductor in millimeters.

Example 3.3. *Calculate dc loop resistance, if the loop resistance limit is 1250 ohm for the loop length of 10 km.*

Sol.
$$L = \frac{R_L}{R_{dc}},\quad R_{dc} = \frac{R_L}{L} = \frac{1250\ \Omega}{10\ \text{km}} = 125\ \Omega/\text{km}.$$

3.5.4. Cable size for the Loop

From the equation 3.13 with the knowledge of dc loop resistance, the diameter of copper wire can be determined. The table shows American Wire Gauge (AWG) versus wire diameter and resistance. From the table 3.4, for the required diameter of the cable, the size of wire gauge can be determined.

Table 3.4. AWG versus wire diameter and resistance

GAUGE NO (AWG)	diameter 'd' (mm)	R_{dc} (Ω/KM)	Attenuation or loss per km (dB/km)
19	0.91	26.39	1.68
22	0.64	52.95	1.35
24	0.51	84.22	1.05
26	0.41	133.90	0.69

The cable sizes of 19, 22, 24 and 26 gauge are the most commonly used cable for different dc resistance of various subscribers. The higher the gauge number the smaller the wire diameter. With 26-gauge wire a loop distance of only about 6.4372 Km (4 miles) is possible. With 19-gauge wire the loop distance might be extended to as much as about 28.96 Km (18 miles).

Example 3.4. *For a 24 gauge loop and a 1250 ohm loop resistance find the loop length.*

Sol. From the table 3.3, for 24 gauge cable, the R_{dc} is 84.22 Ω/km. Hence, the length of the cable

$$L = \frac{R_L}{R_{dc}} = \frac{1250}{84.22} = 14.84 \text{ km.}$$

Maximum permissible loop length. The method of determing the maximum subscriber loop length using the attenuation or loop loss is called the basic transmission design, the maximum loop length is calculated from the formula

$$L_m = \frac{\text{Attenuation limit}}{\text{loss per km}} \quad \text{...(3.14)}$$

Example 3.5. *For 24 gauge loop and a 6 dB loss, find the maximum loop length.*

Sol. For 24 gauge cable, the loss per km is 1.05 dB/km

$$L_m = \frac{6}{1.05} = 5.71 \text{ km.}$$

3.5.5. Inductive Loading

Attenuation limits arise from the ac response of the loop and refers to loop loss in decibels. The amplifiers/loop extenders or inductive loading coil methods are used to admit the loop length to increase beyond the limit. Inductive loading is the process of inserting series inductances (loading coils) into the loop at fixed intervals. These will reduce the transmission loss on subscriber loops. This methods enables the designer to increase the loop length without using higher diameter copper wire.

The attenuation loss in twisted pair cable is because of the capacitance between wire pair. The wires of the local loop have a capacitance of approximately 0.1335 MicroFarad per km regardless of the cable gauge. The longer the cable pairs the higher the capacitance. The capacitance can be reduced by separation of the conductors but practically difficult. Heavyside in 1887 proved that the distortion can be reduced by satisfying the equation RC = LG, where R is the resistance, C is the shunt capacitance, G is the conductance and L is the inductance. All are measured per unit length.

(By inserting loading coils with proper inductance and at appropriate distances, the distortion can be minimized and high-speed transmission can be achieved) Table 3.5 gives the standard letters and its associated spacing. B, D and H are the most, commonly and spacing. To offset the mutual capacitance (0.1335 MicroFarad), load coils having 88 mH inductances are placed at 1.8288 km (6000 ft) intervals on the cable. The frist load coil placed at 0.9144 km (3000 ft) from the local exchange. Then for every 1.8288 km, one loading coil is placed till the subscriber premises. The loading is not necessary, for the loop shorter than 4.83 km. For local loop length, above 18000 ft (5.48 km), loading coils are necessary. Fig. 3.19 (*b*) shows typical load coil spacing.

Table 3.5. Standard letters and spacing of loading

Letter code	A	B	C	D	E	F	H	X	Y
Spacing (ft)	700	3000	929	4500	5575	2787	6000	680	2130
Spacing (km)	.213	.914	.283	1.372	1.7	0.85	1.83	.2107	.699

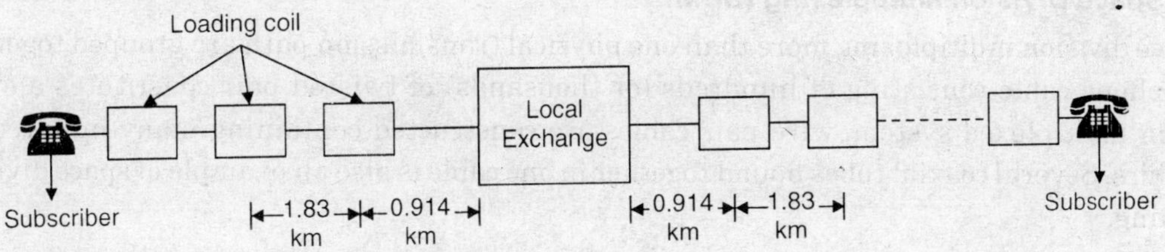

Fig. 3.19. Typical load coil spacing.

Almost all loops are loading with 88mH load coils. It results in a cutoff frequency of around 3800 Hz. Hence 88 mH coil loaded loops effectively pass signals above 3800 Hz. The loading coils are represented for example, as 19H88, 19 indicate the gauge size, H indicates the spacing of the coils and 88 indicate the inductance of the coils. For large volumes of data and high frequency transmission, special facilities and/or conditioned (introduction of amplifiers and equalizer) local loop will be leased. Also, it should be regenerated every 1.6 to 4.8 km depending on speed of transmission and volume of data are to transmitted. The quality of reception at the subscriber end is tested by a rating system standardised by CCITT.

3.6. MULTIPLEXING

Multiplexing defined as any process of sending of a number of separate signals together, over the same cable or bearer, simultaneously and without interference. Thus many speech channels are transmitted togtether as a single channel occupying the bandwidth of the physical facility. Hence the multiplexing is more economical and efficient. In multiplexing, the signals, which may be voice, video or data, are multiplexed together and the resulting signal is transmitted over a system with a suitably high bandwidth. When it is received, it is split up into the separate signals at which it is composed. This process is called demultiplexing.

In telecommunication, the multiplexing means the use of one telecommunication line to handle several channels of voice or data. The best example of multiplexing is our TV cable. We select a particular channel using remote control from the cable, which carries many channels. The individual channels entering and leaving the terminal station is called baseband channels. A link, which carries multiplex signal, is called broad band channel. The combination of multiplexer and demultiplexer at terminal station is called MUX.

The primary use of multiplexing is to save communication line costs. A common application of multiplexing is long distance communication using high-speed point to point links for transferring large quantities of voice signals and data between users.

(There are three methods of transmitting more than one signal over one path. They are space division multiplexing (SDM), frequency division multiplexing (FDM) and time division multiplexing (TDM).)

3.6.1. Space Division Multiplexing (SDM)

In space division multiplexing more than one physical transmission path are grouped together. A telephone cable consisting of hundreds (or thousands) of twisted pair constitutes a space division multiplexed system, wire pair cables are constructed containing many hundreds of wire pairs. Several coaxial tubes bound together in one cable is also an example of space division switching.

A very large number of separate telephone calls can be transmitted together down a coaxial system. Single wire pair commonly carries 12 or 24 voice channels. But one single coaxial tube commonly carries 3600 and the higher capacity one can carry 10800. In India, the telephone cables (5 or 10 lbs per mile) containing copper wires pairs of 20, 40, 60, 80, 100, 1000 and 5 per cent extra used in various towns and cities. For trunk cable lines, 70 – 300 lb copper wire are used. With coaxial cable 10800 two way telephone conversation are possible.

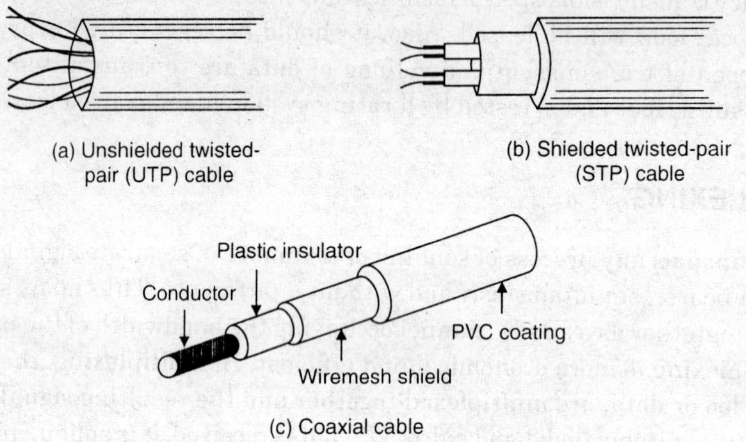

Fig. 3.20. Twisted pair cable and coaxial cable.

The space division switching is not limited to voice frequency circuits alone. Many high capacity transmission systems using either frequency or time division multiplexing can also be space division multiplexed.

3.6.2. Frequency Division Multiplexing (FDM)

FDM is a broad band analog transmission technique in which multiple signals are transmitted over a single cable simultaneously as shown in Fig. 3.21 (*a*). FDM systems divides the available BW of the transmission medium into a number of narrow band or sub channels. The channels are sent over a common path by modulating each channel to different carrier frequency (higher frequency). The signal thus occupies a relatively narrow bandwidth which is a part of a much wider bandwidth transmitted. Each speech channel occupies 4 kHz of the available bandwidth. Fig. 3.21 (*b*) shows FDM in telephone transmission.

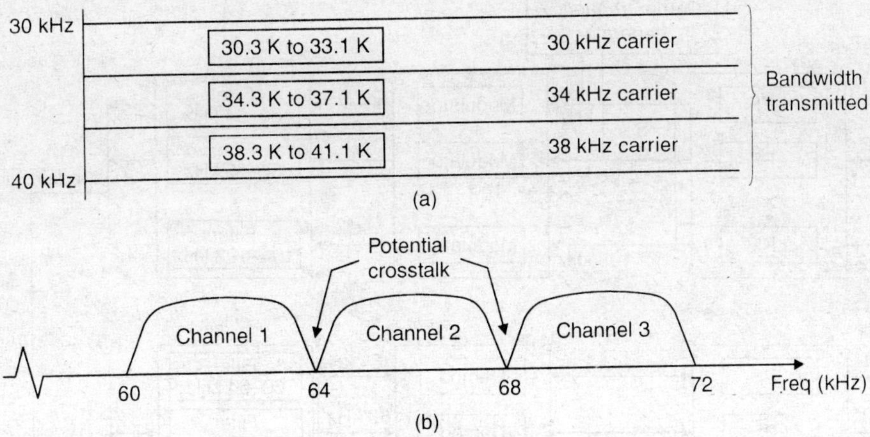

Fig. 3.21. (*a*) Telephone multiplexing (*b*) FDM in telephone transmission.

These modulated carriers are all amplified and transmitted together over the channel. Analog FDM used extensively in point to point microwave radios, coaxial cable wire line system and fiber optic transmission (Wavelength division multiplexing).

FDM Hierarchy. To standardize the equipment in the broad band transmission system. CCITT recommended the following FDM hierarchy.

Table 3.6. FDM hierarchy recommended by CCITT

Multiplex level	Number of voice circuits	Formation	Frequency band (kHz)
Voice channel	1	—	0–4
Group	12	12 voice circuits	60–108
Super group	60	5 group (12 × 5 = 60)	312–552
Master group	600	10 super group (60 × 10 = 600)	564–3084
Jumbo group	3600	6 master group	564–17548
Jumbo group mix	10,800	3 Jumbo groups	3000–60000

All multiplex equipment in the FDM hierarchy uses SSB modulation. Each level of the hierarchy is implemented using a set of standard FDM modules. The multiplex equipment is independent of particular transmission media.

FDM principle. FDM for 12 telephone channels (group multiplex level) is considered for example. The signals pass through 12 low pass filters (LPF) to remove any high frequency components. The LPF outputs are modulated on 12 separate carrier signals separated by 4 kHz. The modulations in all FDM hierarchy is single side band (SSB) modulation. The output of each of the 12 modulations must be filtered (band pass filter) to avoid the interference with each other. BPF are used to restrict each signal to the allocated 4 kHz band. The principle of operation is shown in Fig. 3.22.

When the signal is received, converse process takes place as shown. In fact, the modulators and demodulators are combined into 12 single units to permit two way transmissions.

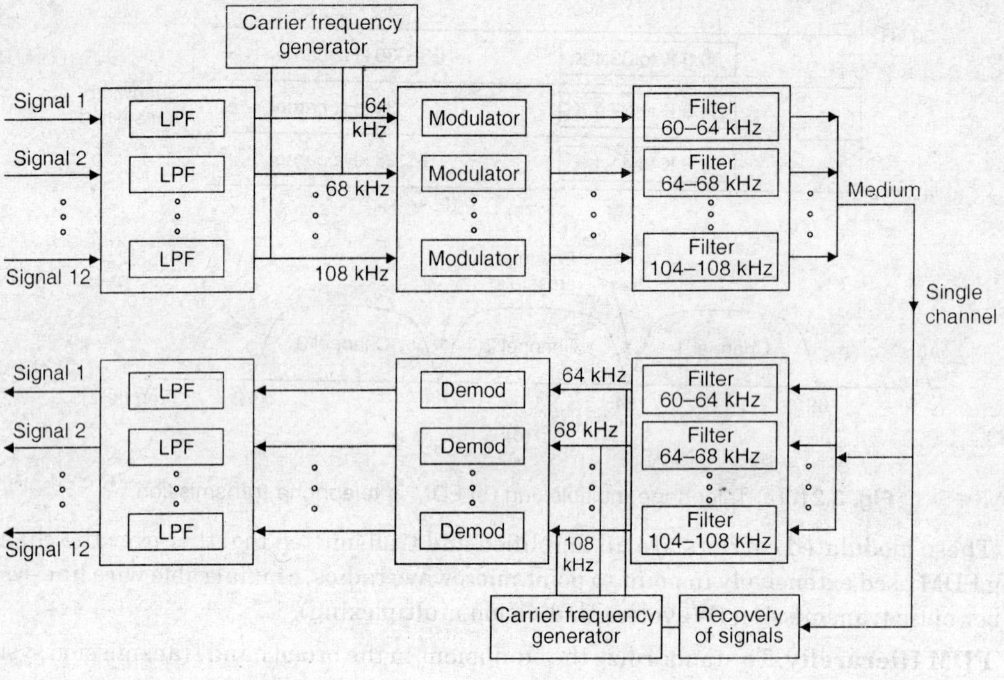

Fig. 3.22. Principle of operation of FDM.

3.6.3. Time Division Multiplexing (TDM)

In 1970's TDM was introduced and now favored over FDM. TDM is the sharing of a common transmission medium in time. In TDM, the time available is divided into small slots, and each of them occupied by a piece of one of the signals to be sent. Thus the multiplexing device should scan the input signal in round-robin fashion. TDM is a base band technology in which individual channels of data or voice are interleaved into a stream of framed bits across a communication channel. Fig. 3.23 shows the Principle of TDM.

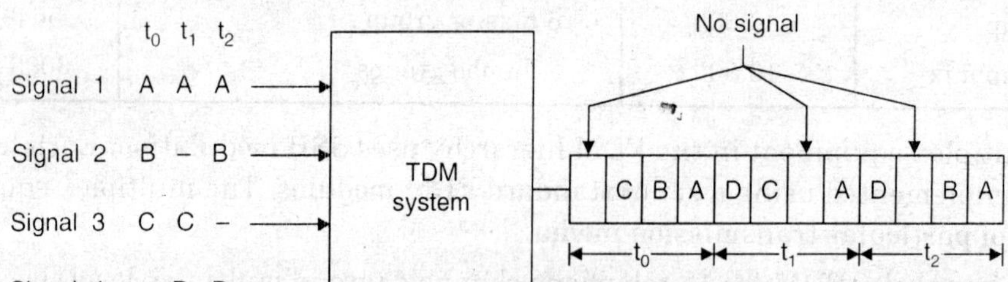

Fig. 3.23. Principle of TDM.

If analog signals are to be multiplexed, the signals should be sampled. The analog signals are digitized by CODEC (coder/decoder) device. It produces 7 or 8 bit number. Sampling is 8000 per sec (125 mirco sec/sample) to capture all the information from the 4 kHz telephone channel bandwidth. Pulse code modulation (PCM) technique is used in TDM. T1 carrier is the widespread method (in North America and Japan). T1 consists of 24 voice channels multiplexed together. Each channel has 8 bits, 7 for information and 1 for signalling. The output frame of 125 micro sec consists of 24 × 8 = 192 bits plus one extra bit for framing.

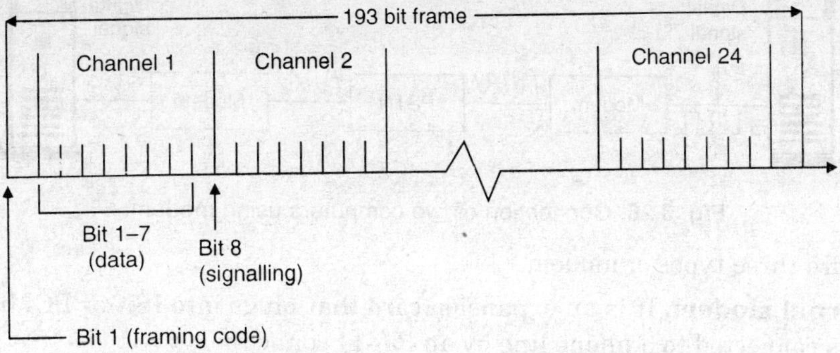

Fig. 3.24. The T1 carrier format.

A PAM or PPM can be employed for TDM. These methods are now obsolete as the transmitted pulses dispersed due to attenuation and delay distortion. They spread in time and interfere with the pulses of adjacent channels, thus causing interchannel cross talk. The PCM technique used in TDM eliminates this problem. Typical PCM system for TDM transmission is shown.

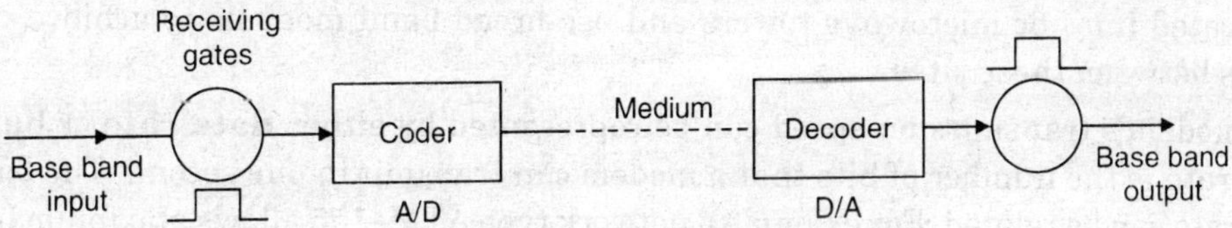

Fig. 3.25. TDM transmission.

3.6.4. Digital Transmission

Since 1980's digital transmission is slowly replacing the analog switching and transmission. In digital transmission system, the message signals are converted into digital form. There are two basic modes of digital transmission. They are asynchronous transmission and synchronous transmission. Both techniques are explained in detail in transmission networks chapter. The digital transmission produces a rugged signal with a high degree of immunity to transmission distortion. The quantization noise also associated with the signal, which is the fundamental limitation on waveform reconstruction. A large bandwidth of coded PCM is used to reduce the quantization noise. The PCM loading also discussed in the chapter 10.

3.7. MODEMS

Modems (modulator/demodulator) are data communication devices that convert digital signals to analog signals and vice versa. Modem allow digital transmission over analog telephone lines. Modems enable users to establish a link between computers using a telephone line.

Modems enable computers to access the internet by cable TV lines. A modem is also known as Data Circuit Terminating Equipment (DCE) which is used to connect a computer or data terminal to a network. Fig. 3.26 shows the connection of two computers using modems.

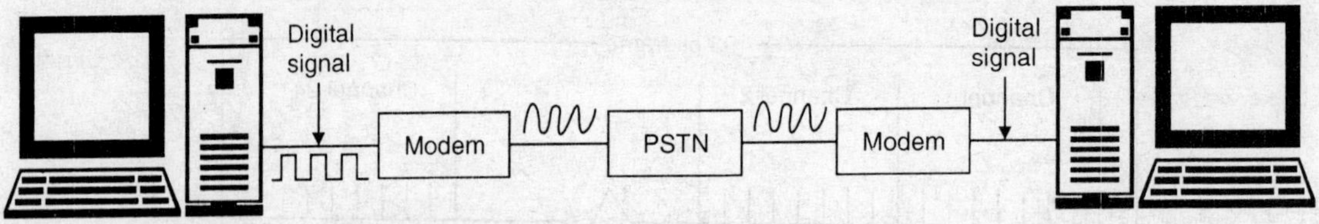

Fig. 3.26. Connection of two computers using modem.

There are three types of modem.

1. **Internal modem.** It is an expansion card that plugs into ISA or PCI bus inside the computer. It is connected to a phone line by an RJ–11 connection.

2. **External modem.** It is housed in a separated casing and typically uses a DB–9 connector to attach to one of the computers serial ports.

3. **Laptop modem.** The modems in laptop and note book computer consists of a PC card that houses entire circuitry for the modem.

The above three are related to the consumer voice grade modems. These modems employ communication techniques according to the ITU V series standards. There are modems for non-telephone system connection referred as broad band modems. A company may setup its own dedicated lines or microwave towers and use broad band modems to achieve very high data rates between those sites.

A modem's transmission speed can be represented by either **data rate** or **baud rate**. The data rate is the number of bits that a modem can transmit in one second. The bandwidth and data rate can be related. For example a network type ATM–155 allows maximum frequency of 100 MHz and equivalent data rate of 155 Mbits/sec. The relations vary for different network. The baud rate name comes from the French man Baudot, who developed an encoding scheme for French telegraph systems in 1877. The baud rate is the number of signals that a modem can transmit in one second. Otherwise, baud represents the number of times the state of communication line changes per second.

Modulation methods. The amplitude of the signal varies with frequency over telephone lines. It is well known that the signals with frequencies 300 to 3400 Hz are strong at reception. Thus the data sent over analog telephone channels must fit into this bandwidth. The modem uses some modulation technique to fit the signal into this bandwidth. A carrier frequency is modulated by the data to be sent so those significant signal components are inside the speech bandwidth.

The types of modulation techniques used to convert digital signals to analog signals are Amplitude Shifts Keying (ASK), Frequency Shift Keying (FSK) and Phase Shift Keying (PSK). Most advanced modems use a combination of modulation techniques to transmit multiple bits per baud and increased transmission speed. The hybrid modulation technique called a Quadrature Amplitude Modulation (QAM) is the combination of PSK and ASK.

Constellation Patterns. The original signal can be represented as different signals by phase shifting it at various degrees. Each signal can be represented by binary number. The relation between phase and the binary representation of each phase can be plotted on a coordinate system called constellation diagram. Fig. 3.27 (*a*), (*b*) and (*c*) shows the constellation diagram using 90-degree phase shift, 45-degree phase shift and constellation diagram for QAM.

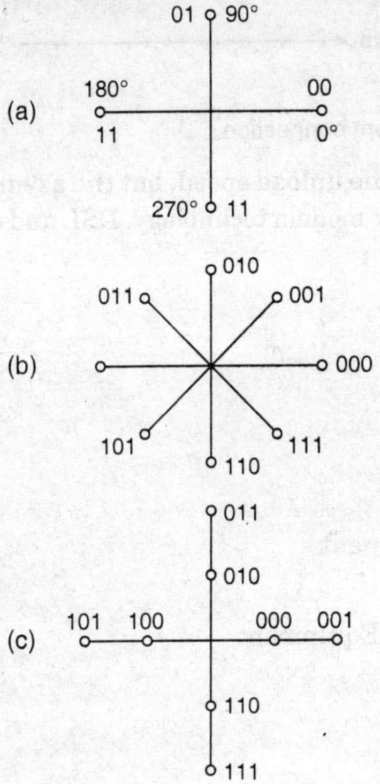

Phase shift	Binary value
No shift	00
90°	01
180°	10
270°	11

0°	000
45°	001
90°	010
135°	011
180°	100
225°	101
270°	110
315°	111

No	000	A1
No	001	A2
90°	010	A1
90°	011	A2
180°	100	A1
180°		

Fig. 3.27

The modem's speed using a 90 degree phase shift is 2 × 400 which is equal to 8 kbps. The signal shifted by 45 degrees increases the speed to 3 × 400 which is equal to 12 kbps.

V. 90 (56 Kbps) Modem. ITU – T developed many version of modem named modem standards. The standards ranges from V. 22 with speed of 1200 bits/sec, full duplex, PSK encoding to V. 90 with 56 kbps download speed and 33.6 kbps upload speed and V. 92 which is the upgraded version of V. 90 standard with upload speed of 47 kbps. In this section, the version V. 90 modem standard created in 1998 by ITU is considered for discussion.

The 56 kbps modem was designed for a one end digital connection from the server to the PSTN. V. 90 have become a world wide standard with 56 kbps modem technology, the transmission rate is 33.6 kbps and receiving rate is 56 kbps. This is due to the fact that the upstream channel (user to server) requires ADC at local exhchange.

Fig. 3.28 explains the 56 kbps modem connection. This conversion produces noise and reduces the speed of the modem to 33.6 kbps.

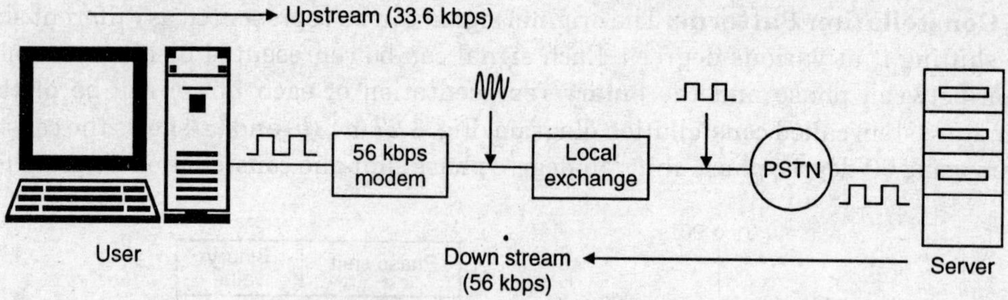

Fig. 3.28. 56 kbps modem connection.

The newer version V. 92 standard improves the upload speed, but the asymmetry is still there. The Digital Subscriber Line (DSL) is the latest modem technology. DSL and other modem technology are discussed in the later chapter.

ACRONYMS

AGC	—	Automatic Gain Control
ASK	—	Amplitude Shift Keying
AWG	—	American Wire Gauge
CPE	—	Customer Premise Equipment
DC	—	Distribution Cable
DCE	—	Data Circuit terminating Equipment
DP	—	Distribution Point
DSL	—	Digital Subscriber Line
DSO	—	Digital Signal, Level Zero
DTMF	—	Dual Tone Multiple Frequency
DW	—	Drop Wires
ELFEXT	—	Equal Level FEXT
FD	—	Feeder Point
FDM	—	Frequency Division Multiplexing
FEXT	—	Far End Cross Talk
FSK	—	Frequency Shift Keying
LSSGR	—	Local Switching System Generic Requirements
MDF	—	Main Distribution Frame
MODEM	—	Modulator/Demodulator
NEXT	—	Near End Cross Talk
PAM	—	Pulse Amplitude Modulation
PCM	—	Pulse Code Modulation
PPM	—	Pulse Position Modulation

PSK	—	Phase Shift Keying
QAM	—	Quadrature Amplitude Modulation
SDM	—	Space Division Multiplexing
TDM	—	Time Division Multiplexing

RELATED WEBSITES

Networking updates	—*http://www.Linktionary.com*
Multiplexing TCP	—*http://www.aciri.org/floyd/*
Bit pipe (search for multiplexing)	—*http://www.bitpipe.com/*
Data communication	—*http://telecom.tbi.net/transmission.htm.*
Network magazine	—*www.Network magazinej.com*
Data communication	—*http://www.icc.org/*
V.g0/X2/k56	—*http://www.hal.pc.org/*

REVIEW QUESTIONS

1. List the end instruments and its use.
2. What are the primary functions of transmission systems ?
3. Define local loop and trunks.
4. What is the use of multiplexing ? List the variuos multiplexing techniques.
5. Distinguish simplex, half duplex and full duplex transmission
6. What is known as singing ? How singing can be reduced ?
7. List various impairments which affects the signals.
8. Draw echo canceller circuit and explain.
9. Define noise related to telecommunication.
10. What are the various sources of impulse noise ?
11. What are called cross talk ?
12. Explain briefly NEXT and FEXT.
13. Explain the various types of distortion.
14. What is BORSCHT ?
15. Consider a subscriber loop of 12 km long. The loop resistance is 1600 ohm. Calculate the d.c. loop resistance and determine the cable gauge for the loop.
16. Explain the concept of inductive loading.
17. What is the use of modem ?
18. What are the various types of modems ?

4

Evolution of Telecommunication Switching System

4

Evolution of Telecommunication Switching Systems

4.1. INTRODUCTION

Telecommunication system is an important and integral part of modern society. In addition to public switched telephone network (PSTN), it plays vital role in radio and television networks, internet and Asynchronous transfer mode (ATM) networks. The switching system provides various services to the subscribers. The switching system is a collection of switching elements arranged and controlled in such a way as to setup a communication path between any two distant points. This chapter demonstrates the switching systems of manual exchanges to the electronic switching systems.

The process of transferring message from one place to another (or line to line) is called switching related to outside the switching plant or systems. There are three types of switching namely a circuit switching, message switching and packet switching. In telecommunication switching, the circuit switching and message switchings are used. The switching technique used in computer communication network or data transfer is packet switching. The circuit switching and message switching are explained in this chapter. The packet switching is dealt in the latter chapter.

4.2. EVOLUTION OF PSTN

Telecommunication is the communication of voice or data over long distances using public switched telephone network (PSTN). PSTN consists of transmission component, switching components and facilities for maintaining equipment, billing system and other internal components. PSTN also referred to as plain old telephone system (POTS). The switching technique used in PSTN is circuit switching in general.

To setup connection between subscribers, the PSTN consists of the transmission systems, switching system and signalling systems. The transmission system is dealt in the chapter 3. This chapter explains various switching systems and the signalling system is explained exclusively in the chapter 7. The PSTN consists of the following hierarchy. Local networks which connect subscribers and local exchanges. Function networks, which interconnect a group of local exchanges serving an area and a trunk exchange. Third one is the trunk network which provides long distance connections nationally and internationally.

4.2.1. Elemental Telecommunication Systems

A simplest telephone communication system between two subscriber is shown in Fig. 4.1. The telephone set which contains transmitter, receiver and base unit circuits were discussed in chapter 3. The central battery and its use in telecommunication also discussed in chapter 3. In the circuit of Fig. 4.1 shown, there is a quiescent current flowing even in the asbence of sound. This quiescent current is necessary for faithful sound reproduction. The inductor L offers no dc resistance but is nominally opposes the voice frequencies.

The audio signal transmitted from calling subscriber not only heared at called subscriber receiver, but also at senders receiver. The audio signal heared at the generating end is called.

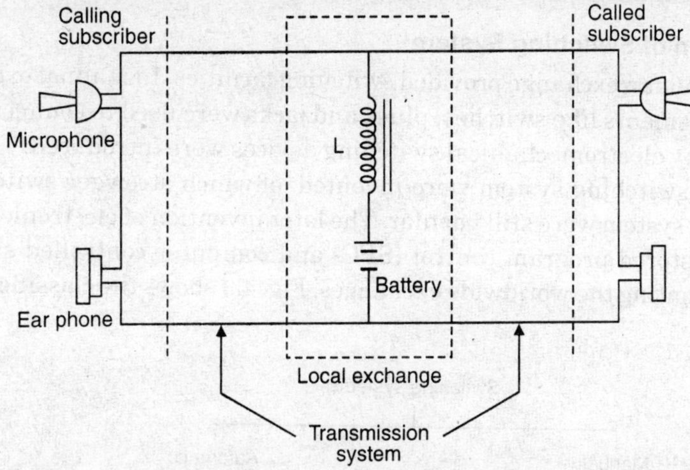

Fig. 4.1. Basic telecommunication system.

Sidetone. Certain amount of sidetone is necessary. When a calling subscriber does not hear his own audio unconsciously, he may raise his voice level. But the sidetone sound should be within a particular limit. The Fig. 4.2. shown is a modern telecommunication system which provides facility for reducing the sidetone.

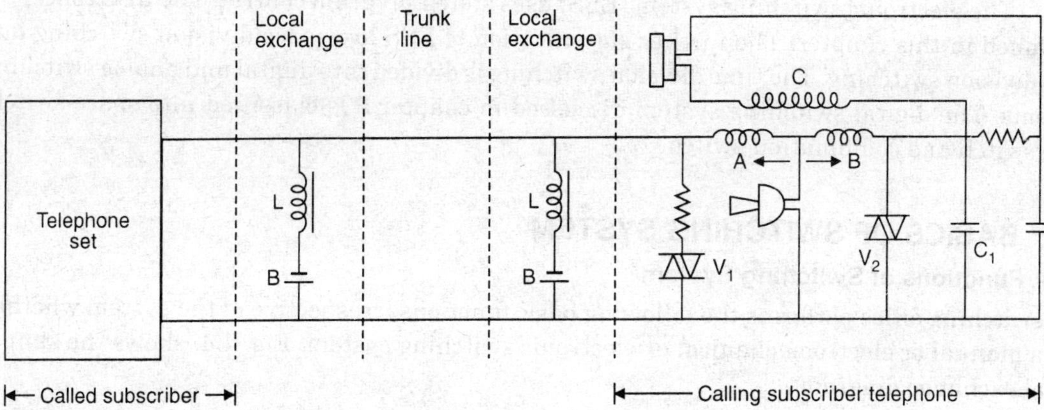

Fig. 4.2. Modern telecommunication system.

The device V_1 and V_2 are varistor. Varistor is a semiconductor device whose resistance is a function of the current flowing through it. The resistance decreases with increasing current. The voice signal generated by transmitter of calling subscriber passes through coils A and B in opposite direction. The impedance offered by V_2 with C_1 and C_2 (simulated) approximates the impedance of V_1. Thus the current in A and B are opposite but with equal magnitude. Thus, the side tone is kept considerably small and within the acceptable limits.

The voice signal from the called subscriber flows through the coil A and B in same direction. The current in coil AB is now coupled to the secondary coil C and transmitted to earphone of calling subscriber. Thus the voice signal is received without attenuation. On shorter loops, the varistor exhibits a high resistance on long loops, it has a low resistance.

4.2.2. Classification of Switching System

In early days, the human exchange provided switching facilities. In manual exchanges, a human operator and the elements like switches, plugs and jacks were used to connect two subscribers. Around 1890's many electromechanical switching devices were introduced. Till 1940, different electromechanical switching system were invented, of which strowger switching system and cross bar switching system were still popular. The later invention of electronic switching system (ESS) which uses stored program control (SPC) and computer controlled switching systems are presently dominating the worldwide exchanges. Fig. 4.3 shows the classification of switching system.

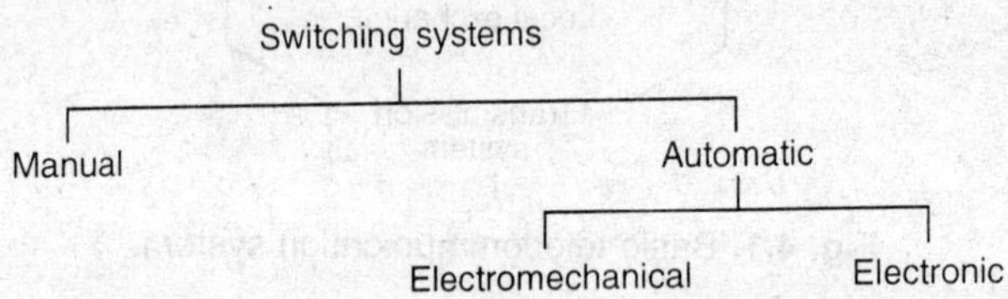

Fig. 4.3. Classification of switching system.

Under electromechanical switching system, the popular systems of strowger and crossbar switching system are discussed in this chapter.

The electronic switching system (ESS) uses stored program control. The SPC concept is explained in this chapter. The further classification of ESS are space division switching and time division switching. The time division switching is divided into digital and analog switching systems. The digital switching system discussed in chapter 5 is classified into space switch, time switch and combination switch.

4.3. BASICS OF SWITCHING SYSTEM

4.3.1. Functions of Switching System

The switching office performs the following basic functions irrespective of the system whether it is a manual or electromechanical or electronic switching system. Fig. 4.4. shows the simple signal exchange diagram

1. **Identity.** The local switching center must react to a calling signal from calling subscriber and must be able to receive information to identify the required destination terminal (seize.)

2. **Addressing.** The switching system must be able to identify the called subscriber from the input information (train of pulses or multiple frequency depends on the dialing facility).

The address may be in same local centre or some other exchange. If the terminal or trunk group is busy, a suitable signal must be returned to the calling subscriber. If there are more than one free circuit, particular one will be selected.

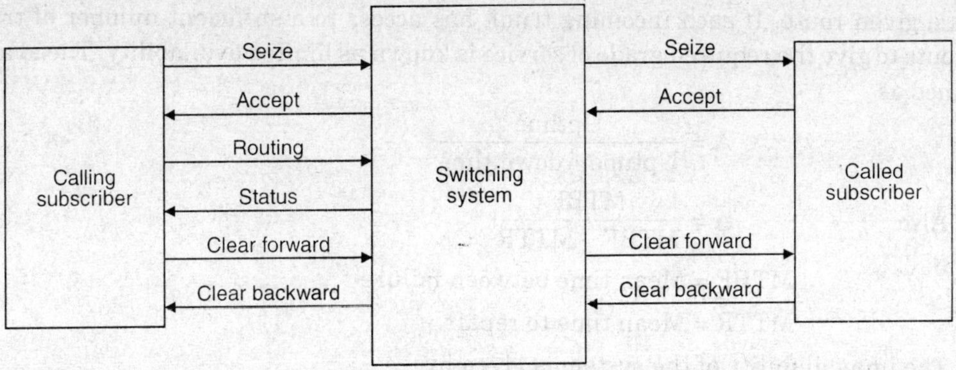

Fig. 4.4. Simple signal exchange diagram.

3. **Finding and pathsetup.** Once the calling subscriber destination is identified and the called subscriber is available, an accept signal is passed to the switching system and calling subscriber. Based on the availability, suitable path will be selected.

4. **Busy testing.** If the number dialled by the calling subscriber is wrong or the called subscriber is busy (not attending the phone) or the terminal may be free (lifting the phone) but no response (not willing to talk or children handling), a switching system has to pass a corresponding voice message or busy tone after waiting for some time (status).

5. **Supervision.** Once the path is setup between calling and called subscriber, it should be supervised in order to detect answer and clear down conditions and recording billing information.

6. **Clear down.** When the established call is completed, the path setup should be disconnected. If the calling subscriber keeps the phone down first, the signal called clear forward is passed to the switching system. If the called subscriber keeps the phone down first, a signal called clear backward signal is passed to the switching system. By clear signal, the switching system must disconnect the path setup between calling and called subscriber.

7. **Billing.** A switching system should have a mechanism to meter to count the number of units made during the conversation. The cumulative number of units made for a particular duration by the calling subscriber is calculated. This information and if any should be sent to the calling subscriber.

4.3.1. Requirements of Switching System

All practical switching system should satisfy the following requirements for the economic use of the equipments of the system and to provide efficient service to the subscribers. Depends on the place (Rural or town, big town, city or big cities). The local exchange located and the service provided to the subscriber may vary. Some important requirements are discussed briefly.

High availability. The telephone system must be very reliable. System reliability can be expressed mathematically as the ratio of uptime to sum of the uptime and down time. The uptime is the total time that the system is operating satisfactorily and the down time is the total time that is not. In telephone switching networks, the availability or full accessibility is possible if all of the lines are equally accessible to all incoming calls.

The full accessibility is also defined as the capacity or number of outlets of a switch to access a given route. If each incoming trunk has access to a sufficient number of trunks on each route to give the required grade of service is known as limited availability. The availability is defined as

$$A = \frac{\text{Uptime}}{\text{Uptime} + \text{down time}} \quad \text{...(4.1)}$$

Also

$$A = \frac{\text{MTBF}}{\text{MTBF} + \text{MTTR}} \quad \text{...(4.2)}$$

where, MTBF = Mean time between failure

MTTR = Mean time to repair.

The unavailability of the system is given by

$$U = 1 - A < \frac{\text{MTTR}}{\text{MTBF} + \text{MTTR}} \quad \text{...(4.3)}$$

High speed. The switching speed should be high enough to make use of the switching system efficiently. The speed of switching depends on how quickly the control signals are transmitted. For instance, the seize signal from the calling terminal must be identified quickly by the system to realise the need of path setup by the subscriber. The common control should be used effectively to identify the called terminal or the free trunks to setup a path. Thus the switching system must have the facility of quick access of the switching equipment and networks.

Low down time. The down time is the total time the switching system is not operating satisfactorily. The down time is low enough to have high availability. The unavailability of switching system may be due to failure of equipments, troubles in transmission media, human errors in switching etc.

Good facilities. A switching system must have various facilities to serve the subscriber. For example wake up calls, address identification on phone number or phone number identification on address, recording facilities, quick service for the emergency numbers, good accessibility etc. Also it should have good servicing facilities in case of repair of equipments, skilled technicians, standby systems, etc. Good facilities is possible in any switching system whether it is at rural or town or in cities, if that exchange is not overloaded.

High security. To ensure satisfied or correct operation (*i.e.* providing path and supervising the entire calls to pass necessary control signals) provision should be provided in the switching system. Duplicated common control circuits, registers, processors and standby systems are used provide high security.

4.3.2. Basic Elements of Switching System

There are three classes of switching system based on the division of information in space, time and frequency. They are (*i*) Space division switch (*ii*) Time division switch and (*iii*) Frequency division switch. The space division provides fixed path for the entire duration of a call. Simplicity, unlimited bandwidth, cross talk limitations are the advantages of space division switches. But these space switches are slow to operate, bulky, and involves large amount of wiring.

In time division switching all inlets and outlet one connected to a common switch mechanism. The switch is connected to the required inlet and outlet for short durations. Each input is sampled to change the connecting pattern. Thus switch is fast and compact. This technique may only be used where the signal is not affected by the sampling process. Time division switches of analog signals have limited applications. Thus time division switches have more practical value only when the signal is already in digital form.

In frequency division switching, the incoming signal is modulated onto a difficult carrier frequency. Switching is achieved if each outlet is provided with a demolulator which can have its carrier frequency changed. Other than radio communication, until recently, there was no practical applications with this switching. Frequency division switching is now finding applications in demand assigned satellite communication links.

The basic elements required in a switching centre are (*i*) Switches (*ii*) a means of receiving signal from terminals and other switching centres and (*iii*) a control system which is required to (*a*) perform logical operations (*b*) store information and (*c*) provide an interface between the control and the switching and information elements. The switching elements used in space division and time division switching are (*a*) conventional relays (*b*) read relays (*c*) uniselectors (*d*) two motion selector (*e*) arbiters (*f*) junctors etc. In this section the functions of arbiters is described briefly. The remaining elements are explained at appropriate places.

Arbiters. A common control performs a specific call processing function. It serves different calls on a time division basis. Since a number of trunks may request the use of common control at the same time, contention can arise. A circuit in a common control used to resolve this contention is called a one only selector, an allotter or an arbiter. The arbitor has inputs from N calling control units and can be in one of two states, busy or free. Its operation is as follows.

(*i*) If the arbiter is in the free state, a seize signal from any one of its inputs puts it into busy state. The arbiter then sends a seize signal to its associated main control unit and returns an accept signal to the calling control unit which has seized it.

(*ii*) If the arbiter is in the busy state, a further seize signal on any of its inputs has no effect.

(*iii*) If two or more seize signals are received simultaneously, only one seize signal is accepted.

(*iv*) Once seized, arbiter behaves as a simple electrical path, to exchange, subsequent signals between the calling and associated called control units.

(*v*) Till the reception of release signal from the main control unit, the arbiter remains in busy state.

(*vi*) For the simultaneous reception of seize signals, a arbiter may be designed to perform a queueing function. For this purpose, an arbiter must include a mechanism for storing the order of arrival of the seize signals.

To seize the simultaneous reception of signals, the arbitration may be done serially or parallely. The serial arbitration is done by scanner. The scanner scans the new calling condition and once detected, the path is provided and the scanner restart the look for calling conditions. Thus if two or more sources call simultaneously only one will be processed at a time.

In electromechanical systems and more advanced electronic systems parallel arbiters are required. Electromechanical can not use scanner, becasue of mechanical wear and resulting noise. In parallel arbitration, if a calling condition occurs, the states of all the conditions are sampled and stored. The circuit then selects only one on some priority basis. Then the arbiter is immediately reactivated. It is also referred as priority encoder or nonpreemptive interupt mechanism.

4.3.3. Simple Human Exchange

In the manual exchange (until 1892), the control was provided by a human operator and the elements of the switch assemblies are plugs and jacks. All the local exchange hardwares are duplicated for each subscriber except the ringing generator, operator's head set and the battery.

If a subscriber A initiates a call to the subscriber B, A lifts the telephone handset from the cradle. This action, closes the subscribers loop which includes transmitter and receiver of the handset. The closing circuit causes a dc current (from battery) to flow through line relay and illuminates the lamp of subscriber A. By seeing the light, the human operator, closes the speak key and ask the subscriber A "number please". By knowing the called subscriber is B, the operator throws ring key B to the ringing generator.

The ringing generator provides a dc current to alert the subscriber B. If B does not pickup the phone after reasonable time, the operator reports to A that the call cannot be connected. If B lefts his phone from its cradle, the lamp of B glows, The operator then connect jack A to jack B and then say to A go ahead please. Both lights of A and B are 'on' till their conversation. If any one or both lamps goes off, the operator will disconnect the jacks.

Limitations of manual exchanges :

1. Language dependent. The operation of a human exchange is language dependent as the subscriber needs to communicate with the operator. In multilingual areas (big towns, cities and tourist spots) this language dependency poses severe problems.

2. Lack of privacy. As a human operator is involving in connecting two subscribers, he or she may be willing to hear the conversation of two VIP's or record the message. So in human exchanges, privacy is not possible.

3. Switching delay. Before setting a path between two subscriber, the operator has to monitor various signalling and if the operator is not active, the delay in switching will be high. Normally it takes minutes to setup a call or release a call.

4. Limited service. An exchange can provide service only to minimum number of subscriber. If the subscriber rate increases, overload and thus congestion are not unexpected. To avoid congestion, more handware should be duplicated and more human operator is necessary. These all will results in large overhead for the exchange.

4.4. THE STROWGER STEP BY STEP SWITCHING SYSTEM

Several electromechanical switching system were developed around 1880–1890 to eliminate the limitations of manual exchanges and to establish automatic exchanges to improve the speed and carry more leads (subscribers). Among those electromechanical automatic switches, strowger's step by step switching system was the most pupular and widely used and even now in some part of the world, it is in use. The history of this switching system is described briefly in the following paragraph.

The first automatic electromechanical switch was developed by connolly and Mcig the in 1879. But Almon B. Strowger, an undertaker in Kensas city, USA was the first to put it to effective uşe. Strowger developed a electromechanical system with electromagnets and pawls. Along with his nephew Walters, Strowger produced a working model in 1888 (US patent No. 447918, 10/6/1891). In this system, a moving wiper (with contacts in the end) moved upto and around a bank of many other contacts, making a connection with any one of them. Together with Joseph B. Harris and Moses A. Meyer, strowger formed his company "Strowger Automatic Telephone Exchange" in october 1891.

The reasons for survival of this system even in some part of the world are its (*a*) high system availability (*b*) comprehensibility and (*c*) cheapness and simplicity.

4.4.1. Basic Elements of Strowger Switching System

There are two types of basic elements which performs most of the functions of the strowger switching system. They are (*a*) Uniselectors and (*b*) Two motion selectors.

Uniselectors. A uniselector is a one which has a single rotary switch with a bank of contacts. Depending upon the number of switching contacts, uniselectors are identified as 10 outlet or 24 outlet uniselectors. A single 10 outlet or 24 outlet uniselector can be used as a switching element for 10 or 24 subscribers. Several uniselectors can be graded together so that multiple incoming circuits can be connected to multiple outgoing circuits. Fig. 4.5 shows the simple arrangement of uniselectors.

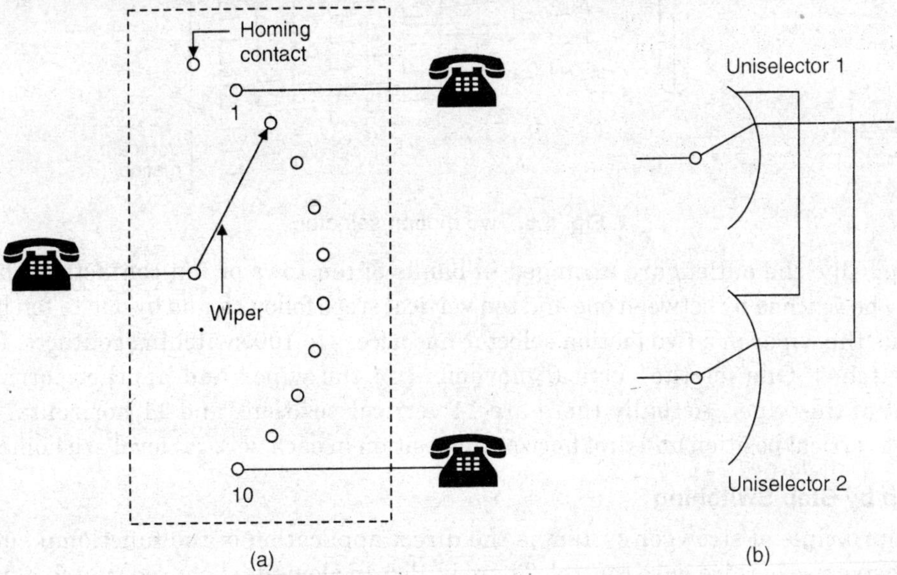

Fig. 4.5. (*a*) 10 contact uniselector, (*b*) graded uniselectors.

The contact arm (wiper) moves across a fixed set of switch contacts. In the case single uniselector, each contact is connected to an outgoing channel, so a caller can choose to connect to any of 10 different subscribers by dialling any digit from 1 to 10. As this selector moves in just one plane, this kind of automated selector is known as uniselector.

An uniselector is operated by and is wiper movement performed by a drive mechanism of a rotary switch. This mechanism contains an armature, electromagnet, Pawl, and Ratched wheel. The wiper is attached to the ratchet wheel. When the line relay detects a calling signal, the magnet is energised and operates the ratchet wheel, pawl and its associated wiper. When the electromagnet is deenergised the armature is released and returns to its rest position. Thus, if the electromagnet is energised and deenergised, (for example 3 times by applying 3 pulses), the wiper moves by three contacts.

Two motion selectors. A two motion selector is a selector in which a set of wipers is moved in two different planes by means of separate mechanisms. By mounting several arcs of outlets on top of each other, the number of outlets can be increased significantly, but the wipers are then required to move both horizontally to select a bank and then vertically to move around that bank to the required outlet. Such a selector is known as a two motion selector. Fig. 4.6 shows a typical two motion selector arrangement.

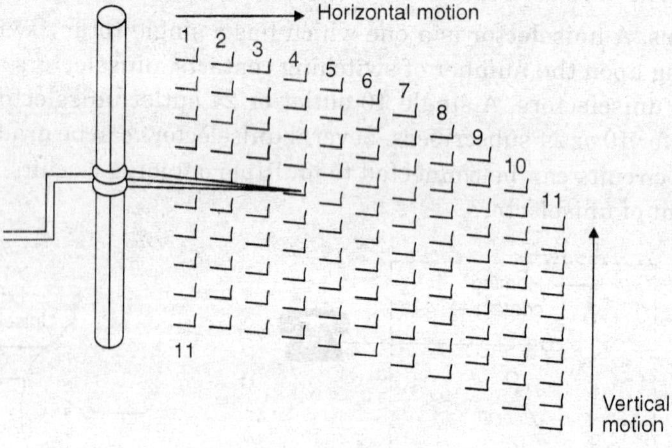

Fig. 4.6. Two motion selector.

Typically, the outlets are arranged in banks of ten rows or ten contacts each. A given outlet may be reached by between one and ten vertical steps followed and by one to ten horizontal steps. Thus the wiper in a two motion selector has access to 100 switching contacts. It has two rotary switches. One for the vertical movement of the wiper and another for horizontal movement of the wiper. Actually there are 11 vertical positions and 11 horizontal contacts. The lowest vertical position and first horizontal position in each vertical level are home position.

4.4.2. Step by Step Switching

The basic principle of strowger system is the direct application of the functional subdivision with extensive use of third wire control. There is also an element of shared switch network but without any common control. In general, the strowger switching system consists of subscriber's line circuit, line finder & alloter circuit, Group selector and final selector. Fig. 4.7 shows the block diagram of strowger switching which explains the process by which the switching system connects a calling subscriber and called subscriber.

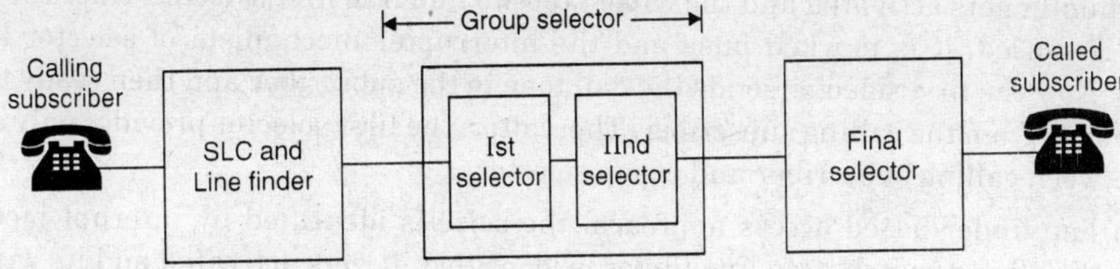

Fig. 4.7. Block diagram of strowger switching system.

Subscriber line circuit (SLC). Every subscriber is connected to his local exchange by one pair of wires. This single pair carries the voice in both directions and the ring current to ring the bell when a call is received. At the exchange, every subscriber line terminates into its own subscriber line circuit (SLC). This consists of a pair of relays dedicated to that subscriber. If there are 1000 subscriber on that exchange, then there are 1000 SLCs. Remaining switching circuits are shared by all the subscribers. When the subscriber lifts his handset, current starts to flow on the line. This is detected by the SLC.

Line Finder & Alloter. As there are many subscribers, but only a few selectors, there has to be a method for finding a free selector and to connect the calling subscriber to that free

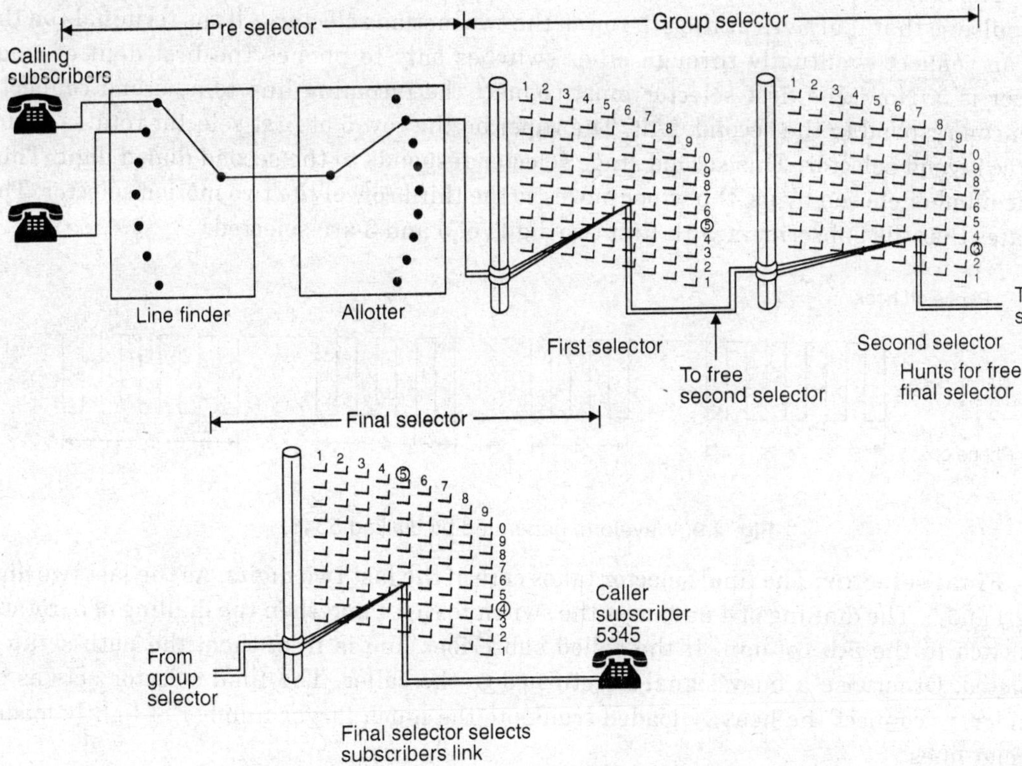

Fig. 4.8. Routing of a local call in strowger switching system.

selector. To find a free selector, alloter switch is used for connecting calling subscriber and selector line, selector hunter based access or line finder based access can be used. In selector hunter based access, when a subscriber lifts his handset, the interrupter mechanism in his selector hunter gets activated and the wiper steps to find free first selector. Once the free first selector is sensed, it is marked busy and the interrupter mechanism of selector hunter is disabled. Now the first selector sends the dial tone to the subscriber and then ready to receive dialled pulses from the calling subscriber. Thereafter, the first selector provides only electrical paths between calling subscriber and group selector.

In line finder based access approach, the seize is identified by interupt mechanism. Through the alloter switch, free line finder is identified. It gets activated and its wiper steps forward to reach the subscriber contact. Now the corresponding first selector sends the accept signal as dial tone. Thereafter, it acts as a simple electrical path between calling subscriber and group selector.

Fig. 4.8 shows a simplified routing of local call in a strower switching system. Although the line finder shown is looking like a uniselector, it is in fact normally a two motion selector.

Group Selector. Depends on the subscriber number, the group selector may comprise one or two selectros, generally referred as first and second selectors. For 3 digit number, only one selector is required. For a 4 digit number, two selectors are required. Let the called subscriber number is 5345. When the subscriber dial the first number 5, the voltage level corresponding to '5' is represented by the sequence of 5 negative pulses as shown in Fig. 4.9.

The response of the first selector to these 5 pulses is to advance vertically one step for each pulse so that it arrives at the 5th row of the two motion selector. All the terminals on this row can connect eventually through other switches only to phones the first digit of whole number is 5. Now this first selector must connect the incoming line to a second connector which can respond to the second digit. The wiper on the row 5 of first selector rotates to find the free second selector. This second stage selector responds to the second dialed digit. Thus, for the number chosen by us, the wiper moves to the third row of the two motion selector. This indicates that the subscribers with first two digits of 5 and 3 are selected.

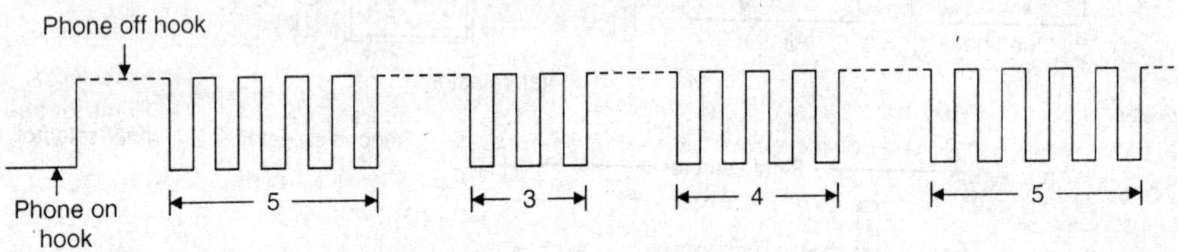

Fig. 4.9. Waveform generated by dialling 5345.

Final selector. The final selector takes care of the last two digits. As the last two digit being 4 and 5. The dialling of 4 advances the switch to row 4 and then the dialling of 5, rotates the switch to the 5th column. If the called subscriber line is free, then, the path setup is completed. Otherwise a busy signal is returned to the caller. The final selector acts as an expander, to connect the heavily loaded trunks to the much larger number of lightly loaded customer lines.

As the path setup between calling and called subscriber is in response to the digits dialled, the system is called the step by step system. It is also referred as a direct controlled switching system because each switching stage will be under direct control of the orginating telephone's dial. As the strowger system provides dedicated path for the subscribers during conversation, it may be referred to as space division technology. In practice, with 4-digit numbering scheme, this switching system provides access to fewer than 10000 subscribers.

Disadvantages. The step by step system has the advantage of being inexpensive for small system and highly reliable due to the distributed nature of equipment. However, the system has several draw backs. Some of them are

1. As this switching involves heavy mechanical displacements, regular maintenance by the skilled technicians are necessary.

2. It is not feasible to select an alternate route for interoffice calls, if all the trunks are busy as the switching is by step through various selectors.

3. Step by step switching is limited to dial pulses. For touchtone telephones, special devices has to be introduced between line finder and first selector to convert the tones into dial pulses.

4. If calling rate is high, heavy operation is performed by the system and the life time of the system is less.

5. The last two digits of the called line numbers are specifically determined by their location on the connector. Congestion could arise when the switching system is heavily loaded.

6. The capacity of switching system reduces if codes of different numbers are alloted to various subscribers, such as fire service, police ambulance, fault regorts, directory enquiry, operator assistance etc. In certain cases, the exchange capacity may be reduced from 10000 to even 6000 customer lines.

7. The strowger system can accept only 7 to 9 pulses in 1 second. Hence if we dial fast, the system can not give correct perfomance.

4.5. CROSSBAR EXCHANGE

In the late 1930's and throughout 1940's, AT & T introduced various versions of the crossbar switches. This crossbar switch basically consists line link frames trunk, link frames and common control equipments. With crossbar switchies and common control equipments, the crossbar exchange achieves full access and nonblocking capabilities. Active elements called crosspoints are placed between input and output lines. In common control switching systems, the switching and the control operations are separated. This permits a particular group of common control circuits to route connections through the switching network for many calls at the same time on a shared basis. The unique features of the crossbar switches are

(*i*) Common control allows the customer and the switch to share the common equipments used to process the call.

(*ii*) Wire logic computer allows specific routine functions of call processing to be handwired into the switch.

(*iii*) Flexible concentration ratios allows the system designer to select the appropriate ratio for a specific switch based on customer mix in a specific location.

(*iv*) Crossbar switches are easier to maintain because the switch have significantly fewer moving parts than strowger switching system.

Basic principle. The fundamental concept of crossbar switching is that it uses common control networks. The common control networks enables the exchange to perform event monitoring, call processing, charging, operation and maintenance. The common control also facilitates uniform numbering of subscribers in a multiexchange area like big cities and routing of calls from one exchange to another via some intermediate exchanges. The common control method of switching overcomes the disadvantages of step-by-step switching. The common control makes no call processing until it receives entire number. It receives all the number, stores, and then establishes connection.

Crossbar switching matrix. The basic crossbar matrix requires atleast M × N sets of contacts and M + N or less activactors to select one of the contacts. Fig. 4.10 illustrates the 3 × 4 crossbar switching. It contains an array of horizontal and vertical wires (shown as a solid line). Both wires are connected to initially separated contact points of switches. Horizontal and vertical bars (shown as dotted lines) are mechanically connected to these contact points and attached to the electromagnets.

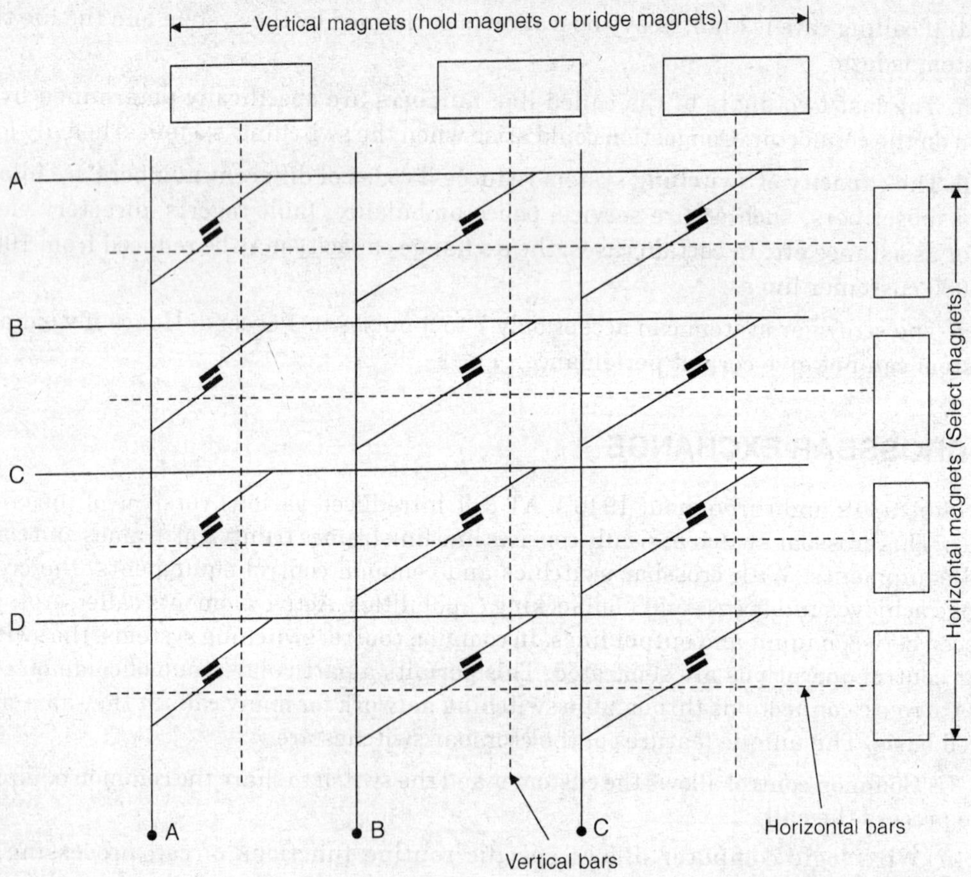

Fig. 4.10. 3 × 4 crossbar switch.

When both horizontal and vertical bars connected to the electromagnet are activated, the contact of the intersection of the two bars will close together. Thus the contact is made and continues to hold. When the electromagnets are deenergized both horizontal bar and vertical bars are released from the contact. In order to prevent the catching of different crosspoint in the same circuit, a procedure is followed to establish a connection. Accordingly, horizontal bar is energised first and then vertical bar is energised to make contact or in reverse. But while removing contact horizontal bar is deenergized first and then the vertical bar is deenergized.

The crossbar switch is known as a non-blocking crossbar configuration. It requires N^2 switching elements for N subscribers. Thus for 100 subscribers, 10000 crosspoint switches are required. Hence, crossbar is economic only for small private exchanges requiring small switches.

For connecting and releasing the subscriber, the select magnet and bridge magnet should be energised and deenergised respectively. External switch must decide which magnet to operate. This is called marker. A marker can control many swtiches and serve many registers. Thus, even a large exchange needs few markers. In Ericsson ARF system, groups of 1000 subscribers are served by a line switch network controlled by the two markers.

Diagonal crosspoint matrix. A diagonal matrix for 5 subscriber is shown in Fig. 4.11. The number of crosspoints are reduced to $N(N-1)/2$, where N is the number of subscribers. It is also called triangular matrix or two way matrix.

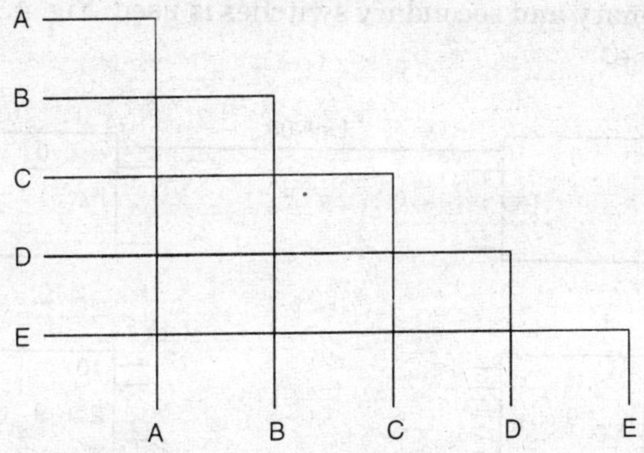

Fig. 4.11. Diagonal crosspoint matrix.

The diagonal crosspoint matrix is fully connected. When subscriber *c* initiates a call, his horizontal bar is energised first and then the appropriate bar. The diagonal crosspoint mattrix is nonblocking configuration. The difficulty is that the failure of a single switch will make some subscribers inaccessible.

Blocking Configurations. By blocking configuration the crosspoint switches required can be reduced significantly. Fig. 4.12. shows the two stage matrices.

Fig. 4.12. shows that there are now four paths between input and output in Fig. 4.12 (*a*) and four paths between any of the stations in Fig. 4.12 (*b*). For the rectangular matrices with N inputs and N outputs, the number of switches is now $2N^2$ compared to N^2 for the single stage. Here, the random failure of a limited number of switches will not preclude connections.

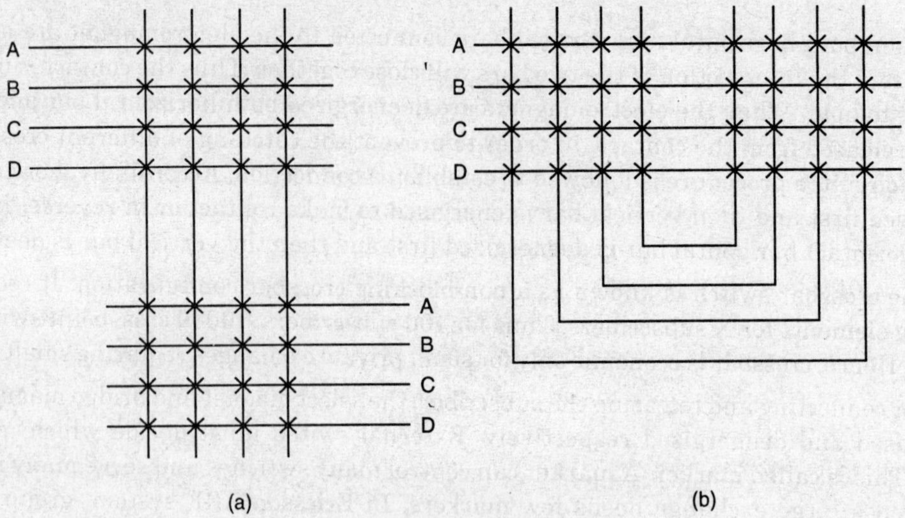

Fig. 4.12. Two stage matrics.

Multistage switching. The single stage structures are called non-blocking and the multistage crosspoints are called blocking. Many of the limitations of the single stage matrix can be remedied by using a multistage structure. In order to produce longer switches a two stage link system of primary and secondary switches is used. Fig. 4.13 shows a two stage link network called line frame.

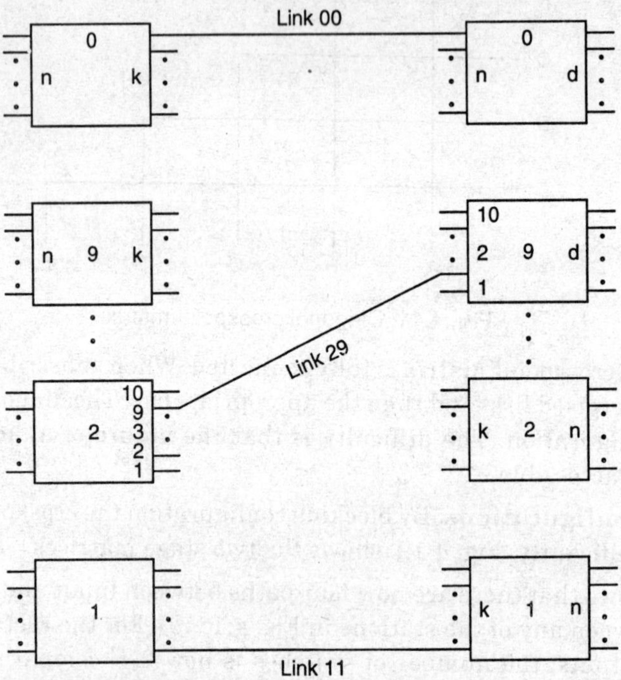

Fig. 4.13. Two stage link network.

The Fig. 4.13 shows twenty switches of size 10 × 10 used to connect 100 incoming trunks to 100 outgoing trunks. The links between primary and secondary are arranged systematically. The link 29 connects the outlet of 9 of primary switch 2 and inlet of 2 of secondary switch 9.

The marker sets up a connection between incoming and outgoing trunk only when both are found to be free. This is called conditional selection.

4.5.1. AT & T No. 5 Crossbar System

This system was developed by the Bell Telephone Laboratories and brought into service in 1948. This system is especially suitable for isolated small cities and for residential areas on the fringes of large cities. When the percentage of calls connected to subscribers in the same office is relatively high, No. 5 crossbar system is useful. Improvements and added features have widened the applications of the No. 5 equipment. It is presently being used in almost all areas including metropolitan business exchanges and rural centres of about 2000 lines or more.

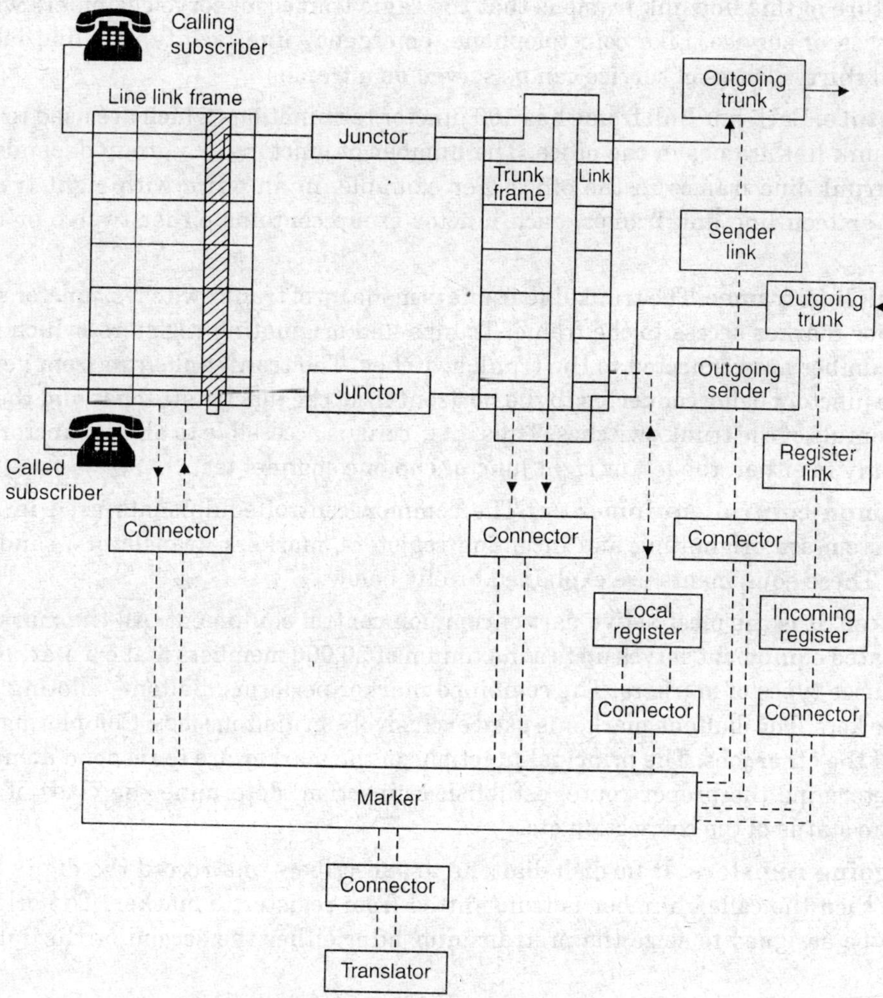

Fig. 4.14. No. 5 crossbar switching structure.

The basic stucture of the No. 5 crossbar system is shown in Fig. 4.14. It consists of line link frame, trunk link frame and common control equipment. The No. 5 system employs a single switching train to handle incoming, outgoing or switch through. Also, the connection of the subscriber to the dial register circuit is also made through this switching train. Subscriber lines and incoming trunks terminate on the line link frames, trunks and orginating registers termination on the trunk link frames. The line link and trunk link frames are interconected by "junctors" which give each link frame access to all of the trunk link frames.

Line link frames. The basic line link frame is a 2 bay frame work with each bay mounting ten 200 point, 3 wire switches. The ten switches on one bay are used as combined line and junctor switches and provide terminations for 100 junctors and 100 lines. The ten switches on the other bay are line switches which provide terminations for 190 additional lines and ten "no test" verticals used to obtain access to busy lines. Thus a line link frames provides 290 line terminations and 100 junctor terminations.

A feature of this line link frame is that the same frame can serve customers who have various classes of service. Like coin telephone, emergency line facility, hot line facility, a maximum of thirty classes of service can be served on a frame.

Junctors. Each line link frame has 100 junctor terminations which are used to connect to all the trunk link frames in the office. The number of junctors in a group depends on the number of trunk line frames in the office. For example, in an office with eight trunk link frames and sixteen line link frames, each junctor group contains either twelve or thirteen junctors.

Trunk link frames. The trunk link frame is made up of trunk switches, junctor switches and relays for marker access to the frame. Trunks and originating registers, which register the called number are connected to the trunk switches. The trunk links run from vertical to vertical, the junctors being connected to the horizontals of the junctor switches and the trunks to the horizontals of the trunk switches. Thus, the trunk is accessible to all the junctors on the frame but only to either, the left or right junctors on one channel test.

Common control equipments : The common control equipments used in AT & T crossbar system are originating and incoming registers, markers, translators, senders and connectors. These equipments are explained briefly below.

Marker. It is the most active part of common control equipment. All the markers and their associated equipment serves up to a maximum of 20,000 members make a marker group. There are three types of markers. The combined marker performs dialtone, alloting the jobs between markers. The dialtone marker is used exclusively on dialtone jobs. Completing marker performs all the other jobs. The principal functions of the marker are to respond demands for dialtone, determine the proper route, establish connection, determine the class of service, recognize the status of the connection etc.

Outgoing registers. It furnish dialtone to subscribers and record the digits that are dialed, and then the called number is transmitted from register to marker. The originating circuit may be assigned to seize the pretranslator after either the second or third digit has been dialed.

Pretranslator. The pretranslator determines, how many digits the register should expect before seizing a marker. A pretranslator can be placed in the outgoing register. For complex numbering plans, a separate pretranslator circuit is provided. The pretranslator determines how many more digits should be dialed and instructs the register, the time it must wait to make connection to marker.

Outgoing Sender. The marker transfers the required digits of the called number to a sender which is connected to an outgoing trunk. The sender type is based on the switching system used. In crossbar No. 5 office, four different types of outgoing senders provided are Dial pulse (DP), Multi frequency (MF), Reventive pulse (RP) Panel call indicator (PCI) This four types of outgoing senders and intermarker group senders may be located on one sender link frame.

Connectors. It is used for interconnecting two equipment elements for short interval of time. If more than one type of equipment originate action toward another type, the connector is named according to both the originating and terminating action :

Important features of No. 5 crossbar system :

(*i*) The use of precious metal, non-sliding contacts results in noise free conversations.

(*ii*) Various options of charging methods such as AMA or message resister and coin.

(*iii*) Provision of toll and tandem switching features with the same common control equipment as is used for local traffic.

(*iv*) Good trouble shooting procedures and additional features like eleven digit capacity, alternate routing, code conversion, marker pulse conversion etc.

(*v*) Improved distribution of usage over various equipment units by means of rotating sequence and memory circuits.

4.6. SPC EXCHANGE

In last two sections, the strowger's step by step switching system and crossbar switching system were studied. In each case, electromechanical components were used for both switching and control elements. In 1965, Bell system installed the first computer controlled switching system which uses a stored program digital computer for its control functions. The SPC concepts permitts the features like abbreviated dialing, call forwarding, call waiting etc. The SPC provides significant advantages to end users. The SPC enables easier number changes, automated call tracing message unit accounting (for billing) etc.

4.6.1. Basic of SPC

In SPC, a programe or a set of instructions are stored in its memory and executed automatically one by one by the processor. Carrying out the exchange control functions through programs stored in the memory of a computer led to the name stored program control. A computer can be programmed to test the conditions of the inputs and last states and decide on new outputs and states. The decisions are expressed as programs which can be rewritten to modify or extend the functions of control system. All switching systems manufactured for use as public switching systems now use computers and software programming to control the switching of calls.

Using SPC, 20 mA transmitter (old transmitter need 23 mA) with 52 V battery feed and longer subscriber loop can be achieved.

Basic view of SPC telephony switch. Fig. 4.15 shows a basic control structure of a SPC telephony exchange.

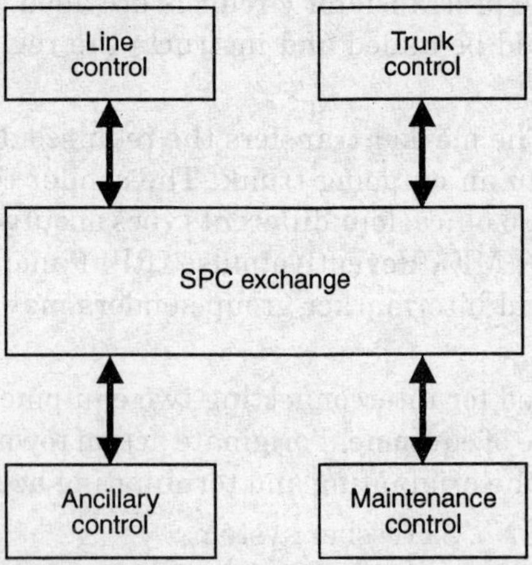

Fig. 4.15. Basic control structure of SPC.

The SPC uses processors designed to meet the various requirements of the exchange. More than one processors are used for the reliability. Normally these processors are duplicated. Also the SPC system uses distributed software and hardware architectures. To carry over the maintenance functions of the switching system, a separate processor is used.

Using the above setup, the SPC performs trunk routing to other control or tandem offices. Special features and functions are also enabled with sophisticated equipments and in compact form.

There are two types in SPC exchanges, namely centralised SPC and distributed SPC. In the following sections, both types are described.

4.6.2. Centralised SPC

Early electronic switching systems are centralised SPC exchanges and used a single processor to perform the exchange functions. Presently centralised exchanges uses dual processor for

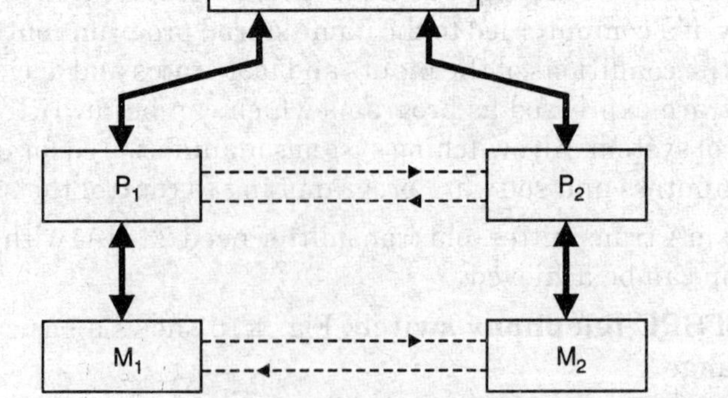

Fig. 4.16. Centralised SPC.

high reliability. All the control equipments are replaced by the processors. A dual processor architecture may be configured to operate in (*a*) standby mode (*b*) synchronous duplex mode and (*c*) Load sharing mode.

Standby mode. In this mode, any one of the processors will be active and the rest is standby. The standby processor is brought online only when the active processor fail. This mode of exchange uses a secondary storage common to both processors. The active processor copies the status of the system periodically and stores in its secondary storage. In this mode the processors are not connected directly. In secondary storage, programs and instructions related to the control functions, routine programs and other required informations are stored.

Synchronous duplex mode. In this mode, the processors P_1 and P_2 are connected together to exchange instructions and controls. Instead of a secondary storage common to P_1 and P_2, separate memory M_1 and M_2 are used. These processors are coupled to exchange stored data. This mode of operation also uses a comparator in between P_1 and P_2. The comparator compares the result of the processors. During normal operation, both processors receives all the information from the exchange and receives related data from their memories. Although only one processor actually controls the exchange and remaining is in synchronism with first one. If a mismatch occurs, the fault is identified by the comparator, and the faulty processor is identified by operating both individually. After the rectification of fault, the processor is brought into service.

Load sharing mode. In this mode, the comparator is removed and alternatively an exclusion device (ED) is used. The processors calls for ED to share the resources, so that both the processors do not seek the same resource at the same time. In this mode, both the processor are active simultaneously and share the resources of exchange and the load dynamically. If one processor fails, with the help of ED, the other processor takes over the entire load of the exchange. Under normal operation, each processor handles one half of the calls on a statistical basis. However the exchange operator can vary the processor load for maintenance purpose.

4.6.3. Availability

Single processor. Availability $A = \dfrac{\text{MTBF}}{\text{MTBF} + \text{MTTR}}$ ...(4.4)

where MTBF = Mean time between failures

MTTR = Mean time to repair

Unavailability = 1 – A

$$U = 1 - \frac{\text{MTBF}}{\text{MTBF} + \text{MTTR}} \;;\; U = \frac{\text{MTTR}}{\text{MTBF} + \text{MTTR}} \qquad \text{...(4.5)}$$

If MTBF >> MTTR, $U = \dfrac{\text{MTTR}}{\text{MTBF}}$ ...(4.6)

Dual Processor. A dual processor system is said to have failed only when both processor fails and the total system is unavailable. The MTBF of dual processor is given by

$$(\text{MTBF})_D = \frac{(\text{MTBF})^2}{2\text{MTTR}} \qquad \text{...(4.7)}$$

where $(MTBF)_D$ = MTBF of dual processor

MTBF = MTBF of single processor

Availability $$A_D = \frac{(MTBF)_D}{MTTR + (MTBF)_D} \quad ...(4.8)$$

Substituting $(MTBF)_D$ in the above equation, we have

$$A_D = \frac{(MTBF)^2 / 2MTTR}{MTTR + \dfrac{(MTBF)^2}{2MTTR}}$$

$$A_D = \frac{(MTBF)^2}{(MTBF)^2 + 2(MTTR)^2} \quad ...(4.9)$$

Unavailability $$U_D = 1 - A_D = 1 - \frac{(MTBF)^2}{(MTBF)^2 + 2(MTTR)^2}$$

$$= \frac{2(MTTR)^2}{(MTBF)^2 + 2(MTTR)^2} \quad ...(4.10)$$

If MTBF >> MTTR, $$U_D = \frac{2(MTTR)^2}{(MTBF)^2} \quad ...(4.11)$$

Example 4.1. *Given that MTBF = 2000 hrs and MTTR = 4 hrs. Calculate the unavailability for single and dual processor systems for 10 years and 30 years.*

Sol. Given : MTBF = 2000 hrs

MTTR = 4 hrs.

Unavailability of single processor

$$U = \frac{MTTR}{MTBF} = \frac{4}{2000} = 2 \times 10^{-3}$$

for 10 years, $U = 24 \text{ hrs} \times 365 \text{ days} \times 10 \text{ years} \times 2 \times 10^{-3}$

$U = 175.2$ hrs.

for 30 years, $U = 24 \text{ hrs} \times 365 \text{ days} \times 30 \text{ yrs} \times 2 \times 10^{-3} = 525.6$ hrs.

Unavailiability of dual processors

$$U_D = \frac{2(MTTR)^2}{(MTBF)^2} = \frac{2(4)^2}{2000^2} = 8 \times 10^{-6}$$

for 10 years, $U = 24 \times 365 \times 10 \times 8 \times 10^{-6} = 0.7008$ hrs = 4.2 minutes

for 30 years $U = 2.1$ hrs.

4.6.4. Distributed SPC

The introduction of distributed SPC enabled customers to be provided with a wider range of services than those available with centralised and electromechanical switching system. Instead of all processing being performed by a one or two processor in centralised switching, functions are delegated to separate small processors (referred as regional processors). But central processors is still required to direct the regional processors and to perform more complex tasks.

The distributed SPC offers better availability and reliability than the centralised SPC. Entire exchange control functions may be decomposed either horizontally or vertically for distributed processing.

In vertical decomposition, the exchange environment is divided into several blocks and each block is assigned to a processor that performs all control functions related to that block of equipments. In horizontal decomposition, each processor performs one or some of the exchange control functions. Figure shows the distributed control where switching equipment is divided into parts, each of which have its own processor.

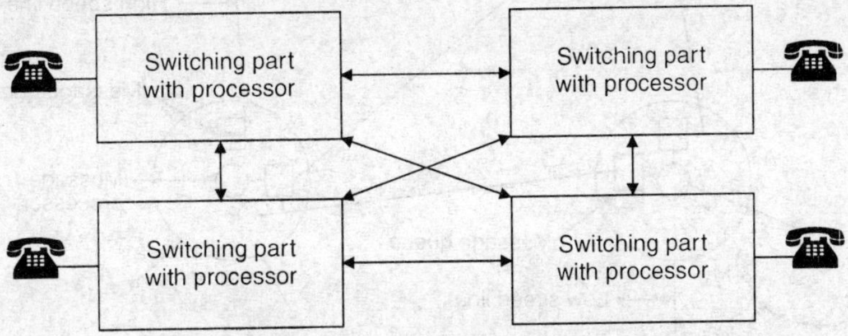

Fig. 4.17. Distributed SPC.

4.7. SWITCHING TECHNIQUES

This section describes various techniques used to establish connections between users exchanges. Switches are hardware and/or software devices used to connect two or more users temporarily. Message switching, circuit switching and packet switching are the most important switching methods.

The terminals of the message switching systems are usually teleprinters. In this switching, delays are incurred but no calls are lost as each messages are queued for each link. Thus much higher link utilisation is achieved. The reason for the delay is that the system is designed to maximise the utilisation of transmission links by queueing message awaiting the use of a line. This switching is also called store and forward switching.

The circuit switching sets up connection between the telephone, telex networks etc. which interchange informations directly. If a subscriber or system to which connection to be made as engaged with other connection, path setup cannot be made. Thus circuit switching is also referred as lost call system.

The modified form of message switching is called packet switching. Packet switching system carries data from a terminal or computer as a short packets of information to the required destination. This system is midway between message switching and circuit switching. The packet switching is explained in latest chapter 11.

4.7.1. Message Switching

In message switching, the messages are stored and relayed from secondary storage. So, message switching is best known by the term store and forward. In message switching, there is no direct link between the sender and the receiver. A message delivered to the destination is rerouted along any path before it reaches the destination. It was common in 1960's and 1970's. Typical message switching network is shown in Fig. 4.18.

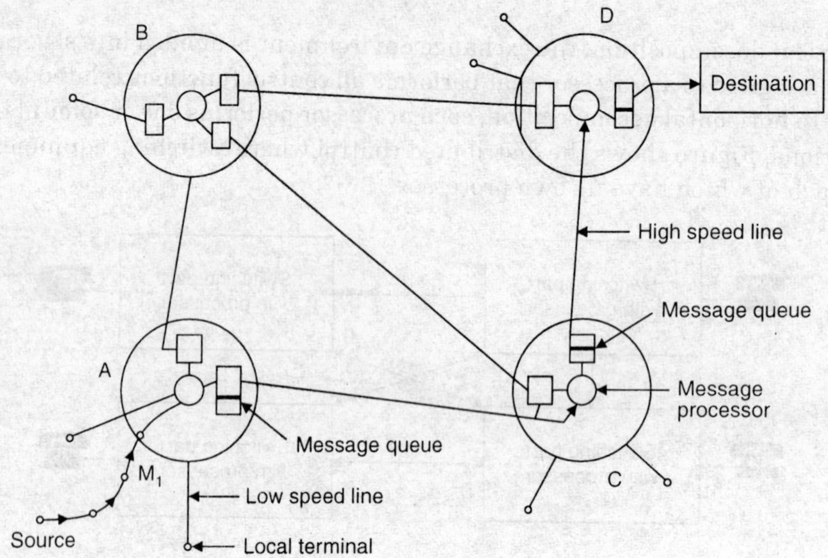

Fig. 4.18. Message switching network.

Message switching offers the possibility of greatly imporved economy. The working of message switching is as follows. Source sends message M_1 to the destination. Suppose that the transmission path seleted is A—C—D. In message switching no complete connection is required. Thus the each message includes a header contains the destination address, routing information and priority information (for special cases).

The node A will transmit message M_1 to switching centre C, where it will be stored in a buffer. The processor in each node maintains message queues for each outgoing link. These queues are normally serviced on a first come, first served basis. If the header have priority information, the message will be passed earlier. Then, after storage and possible delay, the message will go from node C to D and then to the destination. To store the messages, large capacity storage media is necessary at each node. In increases the node cost high. Also, most message switching networks could deliver a message on a delayed basis if destination node is busy or otherwise unable to accept traffic. Also, the message switching was used in unintelligent devices such as telegraphy. Since these devices has been replaced by the high speed intelligent devices, this switching is unpopular for direct communications.

In message switching network, the transmission links are never idle. Utilization of transmission link of a message switching network is directly related to the actual flow of information. With increased store and forward queing delays, utilization efficiency can be increased.

4.7.2. Circuit Switching

Circuit switching creates a direct physical connection between two devices such as phones or computers. In order to setup a direct connection over many links it is necessary that each link to be simutaneously free. This implies that the average utilization of the links must be low if the probability of demand for connection is more. It is therefore used in voice networks mainly and not in networks designed for data transfer. A circuit switch is a device with n inputs and m outputs that creates a temporary connection between source and destination. The inputs n and outputs m need not be equal. In order to transmit information, a circuit switched network finds a route along which it has free circuits. The network connects the circuits together and reserves them for the transmission. Fig. 4.19 illustrates the circuit switching.

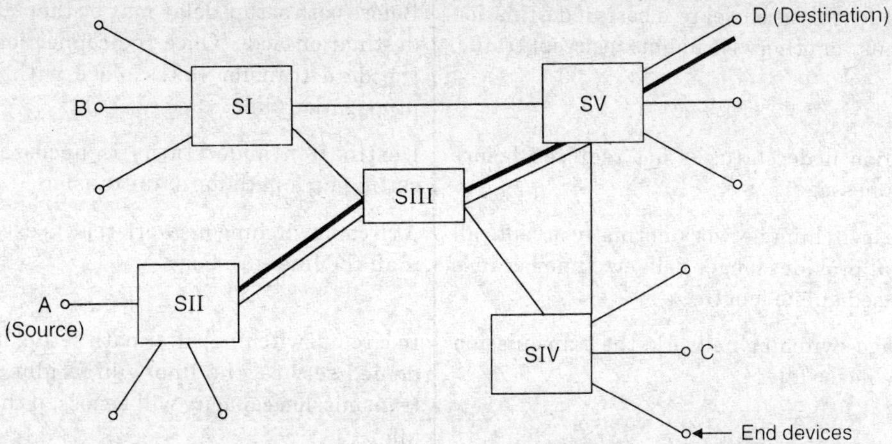

S1-SN → Switches, —Path between switches, ▬Path setup between end device A&D,

Fig. 4.19. Circuit switching network.

In the Fig. 4.19 shown, the end device A is connected to the end device D through switches II, III and V. Circuit switching inovles three phases. First, the source, request the network for the route. The network assigns a route. Second, the data transfer occurs, the duration of the data transfer is called holding time. Third, once the data transfer is competed, the path setup is disconnected. By moving the level of switches, any end devices can be connected to any other end devices. Circuit switching is usually accomplished by TDM.

As the data transfer takes place in three phases, the time taken for the data transfer (T) is expressed as

$$T = T_p + T_d + T_r \qquad ...(4.12)$$

where T_p = Path setup time $(N - 1)\, T_{rs}$ ...(4.13)

T_d = data transfer time = M/R ...(4.14)

T_{rs} = average route selection time

T_r = data release time = NT_n ...(4.15)

N = Number of switches in the path

M = Message length in bits

R = data rate in bits per sec

T_h = house keeping entries time.

Thus $$T = (N - 1)\, T_{rs} + M/R = NT_h \qquad ...(4.16)$$

The propagation time is not considered as it is comparatively very small. In our case N = 3, If T_{rs} = 2 sec, T_n = 2 sec, R = 2400 bps and the message is 300 bytes long, the time for the data transfer is T = (3 – 1) × 2 + 300 × 8/2400 + 3 × 0.2 = 4 + 1 + 0.6 = **5.6 sec.**

Table 4.1. Comparison of message and circuit switching :

Message switching	Cirtuit switching
The source and destination do not interact in real time	The source and destination are connected temporarily during data transfer.
Message delivery is on delayed basis if destination node is busy or otherwise unable to accept traffic.	Before path setup, delay may be there due to busy destination node. Once the connection is made, the data transfer takes place with negligible propagation time.
Destination node status is not required before sending message.	Destination node status is necessary before setting up a path for data transfer.
Message switching network normally accepts all traffic but provides longer delivery time because of increased queue length.	A circuit switching network rejects excess traffic, if all the lines are busy.
In message switching network, the transmission links are never idle.	In circuit switching, after path setup, if the users denied service, the line will be idle. Thus, the transmission capacity will be less, if the lines are idle.

4.8. RELAYS

A switching system has to perform many functions including switching, receiving signals from terminals and other switching centres, transmit signals to terminals and other switching centres and control system to operate switches. The control system used to perform logical operations, store information, provide an interface between the control and the switching and information elements. All the above functions can be provided by simple relays. With the advent of latest electronic components and equipments the use of the relays are reduced, but still majority of the telephone terminals throughout the word are still controlled by relay based systems.

Conventional relays are simple devices used for switching. To perform switching relay acts as a crosspoint. This crosspoint is latched by electrical, magnetic or mechanical means. Reed relay is a common electromechanical switch used in modern switching equipments. In the following section, the reed relay device, its uses in telecommunication switching systems are described.

4.8.1. Reed Relays

Reed relay offers several advantages over conventional relays. One of the main feature is the speed. The fastest switching reed relay is the 9800 series, with a typical actuate time of 100 microseconds. Release time is approximately 50 microseconds. Actual time is defined as the period from coil energization until the contact is closed and has stopped bouncing. Reed relay structure achieves a degree of mechnical simplification and operation with a minimum of moving parts. Hence this structure have increased mechanical efficiency and a significant increase in operating speed over the corresponding properties of the conventional telephone relay.

Fig. 4.20 shows the normally open, single contact pair, dry reed relay. A number of such glass enclosed pairs may be included within the same control winding or coil and linked by the coil flux to operate in a manner similar to the multicontact, reed arrangement.

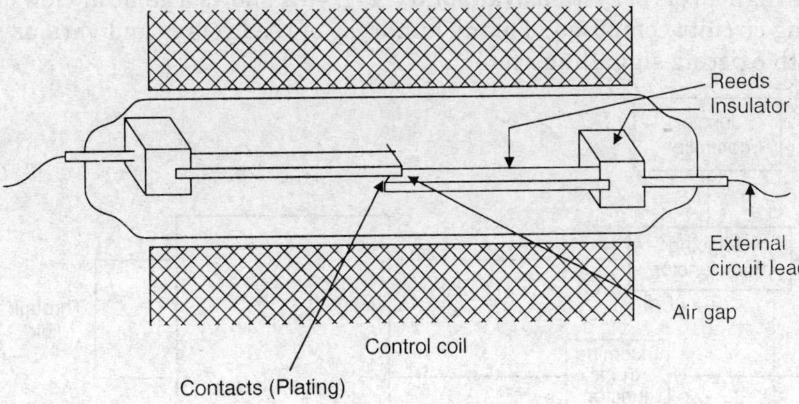

Fig. 4.20. Basic reed relay arrangement.

In basic reed relay, the contacts are plated on the ends of overlapping centilevered magnetic strips or reeds which serve as contact springs. The reeds, contacts and air gap are enclosed in a glass envelop to obtain the benefits of a relatively corrosion free atmosphere for contact operation. The enclosed reeds are wound by the control winding for the magnetic flux. The reeds are magnetized by the flux and are pulled together to close the air gap at the plated contacts. This closes the external circuit path connected to the activated reeds. When a demagnetising current is applied to one or the other of the coils, the contacts open.

Several varieties of the reed relay were developed. Mercury wetted reed relay and dry reed relay are quite popular. The operate and release values (speed) for reed relay operation depend on the structural details of relay such as the gapwidth and overlap, reed dimensions, electrical and magnetic properties of reed relay material.

An alternative to electrical latching is a magnetic latching reed relay called a Ferreed. In this structure, there is one permanent magnet and one magnet whose direction of magnetisation can be reversed by means of current pulse through its windings. When both the magnets are magnetised in the same direction, the contacts operate. The disadvantage is the large current requirement.

Reed relays serve in many different applications requiring low and stable contact resistance, low capacitance, high insulation resistance, long life and small size. They are fitted with coaxial shielding for high frequency applications and process control equipments. Also their low cost and versatility makes them suitable for many security and general purpose applications.

In telecommunication switching systems, relays make a convenient interface elements, used as logic and memory elements for combinational circuits. Relays can be used to provide timing elements by connecting capacitors in parallel with their coil. In can also be used as a translators.

4.8.2. Reed Relay Systems

TEX 2 is a reed relay exchange system (200 to 4000 lines) in common use in United Kingdom. It uses 5 × 5 and 5 × 4 crosspoints. Each crosspoint consists of a 4 reed inserted in a single unit. The TEX 2 is of minimal cost but too inflexible for the introduction of linked numbering within numbering plan areas.

TXE 4 is a reed relay system brought into service in 1976. The capacity of the system is 2000 to 40000 connections or 5000 erlangs. It uses a single general purpose switching network.

TXE 4 achieves high levels of system availability. Fig. 4.21 shows a general view of TXE 4. The actual switching circuits composed of many hundreds of reed relays and various add on units to interface with existing signalling systems.

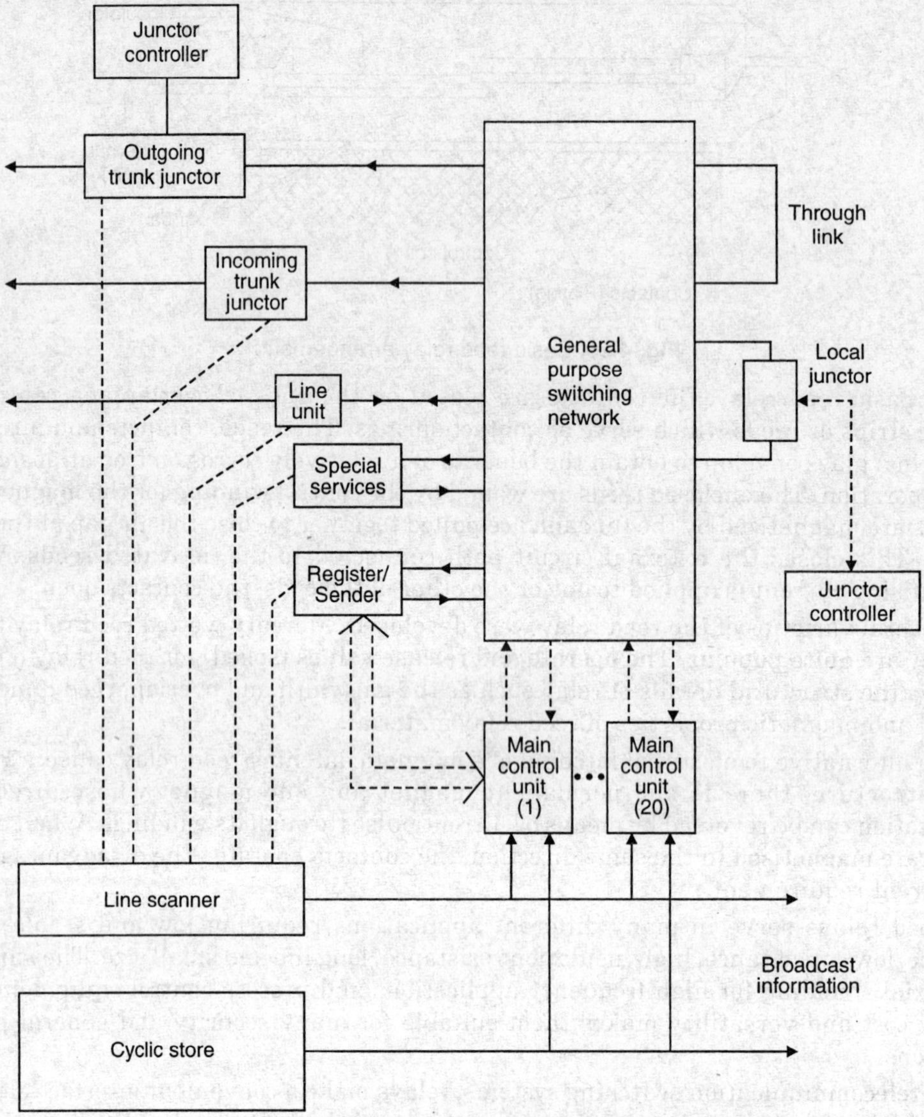

Fig. 4.21. General view of TXE 4.

All local lines, incoming and outgoing trunks, junctors, registers, senders and any special equipments such as coin box controller are connected one side (in figure left handside) of the switching network. The other side of the network shown are internal junctors. The functions of internal junctors are to provide a metallic path (through links) and to provide transmission bridge (local junctors).

The use of single network permits the interconnection of any devices on the traffic side of the switch and provides control of the system. This technique is called serial trunking. In serial switching, for one user to another user, seven stages of links are used is shown in Fig. 4.22.

Each through link or local junctor can be connected to any terminal on the traffic side by three stages of switches on one part and four stages on the other part.

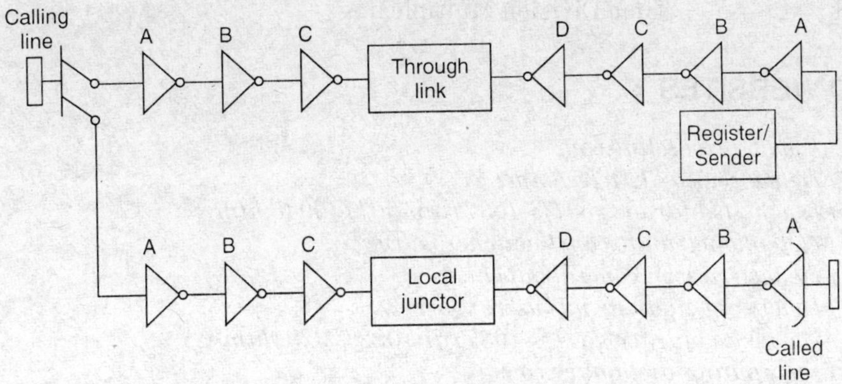

Fig. 4.22. Serial trunking of local call in TXE-4.

The local junctor provides a ringing current, control, clean supervision and path release. For a local call a parallel path *n* established from the calling line to the called line *via* local junctor. Once the outgoing route has been determined, the sender part of the register sender is connected to an outgoing trunk junctors and conversation commences. The connection is then under the control of the outgoing trunk junctor. A, B, C and D are subunits.

The main control unit (up to 20) has number of register/senders (typically 36 in each). The cyclic and line scanner periodically broadcast all the fixed information relating to the system.

The first generation TEX 4 uses a diamond ring arrangement for the cyclic store. The information broadcasted by cyclic store includes the association between directory and equipment numbers between access codes and specific outgoing trunk junctors and so on. The line scanner broadcasts the state of line of the various high traffic items. The rate of broadcast ranges from minimum scanning rate of 156 ms to 36 ms for local calls.

The operation is as follows. Suppose that a subscriber initiates a call. With the reception of dialled pulses, within 156 m sec, the state of the associated line unit is broadcast with the information from the cyclic store also. The main control is designed in such a way that, at any time, it looks for next call. Once the calling condition is detected it instructs the switch network to connect the calling line to one of the registers controlled by the main control unit. The register detects when a significant number of digits has been dialled and itself calls in the main control unit by indicating a change of state to the scanner. The main control unit then selects the junctor and instructs the switch network to setup the required path.

ACRONYMS

ATM	—	Asynchronous transfer mode
DP	—	Dial pulse
ESS	—	Electronic switching systems
MF	—	Multi frequency
PCI	—	Panel call indicator
POTS	—	Plain old telephone systems
PSTN	—	Public switched telephone network

RP	—	Revertive pulse
SLC	—	Subscriber line circuit
SPC	—	Stored program control
TDM	—	Time Division Multiplexing

RELATED WEBSITES

http://www.privateline.com
http://www.seg.co.uk/telecomm/.
http://www.its.bldrdoc.gov/fs-1037/div-001/_0046.htm
http://telephonetribute.com/switches_survey
http://www.ericsson.com/about/history
http://www.erg_abdn.ac.uk/users/gorry
http://www.its.bldrdoc.gov/fs-1037/div-022/_3291htm
http://computing-dictionary.com
http://jhunixhcf.jhu-edu
http://www.webopedia-com/term/c/circuit
http://www.crcmsession.com/article/files
http://www.cotorelay.com
http://www.kellysearch.com/gb-product
http://catalogs.indiamart.com/category/telecom-products.html

REVIEW QUESTIONS

1. Describe the telecommunication systems.
2. How the switching systems can be classified ?
3. Explain the functions of a switching system with signal exchange diagram.
4. Define availability.
5. List the requirements of a effective switching system.
6. What is the use of arbiter ?
7. What are the limitations of manual exchanges ?
8. Write short notes on (*a*) uniselector and (*b*) two motion selector.
9. With a block diagram, explain the functions of a step by step switching system.
10. Explain the routing of a local call with a suitable example and diagram.
11. List the disadvantages of strowger switching system.
12. What are the unique features of cross bar switches ?
13. What is the basic principle of cross bar ? With necessary diagrams explain the operation.
14. Explain AT & T No. 5 cross bar system with neat diagram and list its features.
15. What are the features of SPC ?
16. Explain various modes of SPC and compare their availability.
17. Given that MTBF = 2200 hrs and MTTR = 6 hrs. Calculate the unavailability of (*a*) single processor and (*b*) dual processor for 12 years and 24 years.
18. Compare the centralised SPC and distributed SPC.
19. Explain briefly the basic concepts of message and circuit switching.
20. Tabulate the difference between circuit and message switching.
21. Explain reed relay system with neat diagram.
22. Explain TXE 4 switching system.

5

Digital Switching Systems

5

Digital Switching Systems

5.1. INTRODUCTION

Various facilities of digital switching and transmission is the reason why the analog switching is slowly getting replaced by digital switching. The incorporation of digital switching and transmission technique into telecommunications altered the whole telecommunication industries setup. The reliability of digital switching system is becoming increasingly important for users of telephone services. Voice and/or data can be represented using digital signals efficiently than analog signals.

A switching system is called digital when the input to and output from the switching system can directly support digital signal. Many basic elements of the digital switching system and its operation are very similar to the stored program control (SPC) switching system. The cost of an analogue switch is roughly proportional to the number of cross points, but the cost relationship in digital switching is different.

The functions of the digital switching network is to connect pairs of channels. So that information arriving at the switching centre in a particular channel on one PCM multiplex system can be passed to some other channel on an outgoing PCM multiplex systems. To achieve this switching, two processes referred to as time switching and space switching are used. The principles of these two switching process are described in this chapter.

In digital data communication (analog or digital signal), a fundamental requirement is that the receiver should know the starting time and duration of each bit that it receives. To meet this requirement asynchronous and synchronous transmission are used. These two transmission techniques are described in this chapter.

5.2. EVOLUTION OF DIGITAL SWITCHING SYSTEM

The early version of electronic switching system is the stored program control (SPC). The SPC systems have temporary memory for storing transient call information and to carry programming information. The SPC performs line control, trunk control, ancillary control, maintenance control etc. The instructions required for performing these operations are resided in a single processor. For reliability or high availability, the processor may be duplicated. Thus SPC uses a centralised software and hardware architectures.

A modern digital switching system employs a number of processors that uses distributed softwate and hardware architectures. The digital switching system also referred as Electronic Switching System–III generation is purely electronic in operation, the switching process is by time division/digital transmission, the type of control is stored program common control and the network uses pulse code modulation. Fig. 5.1 shows the evolution of digital switching system and is self explanatory.

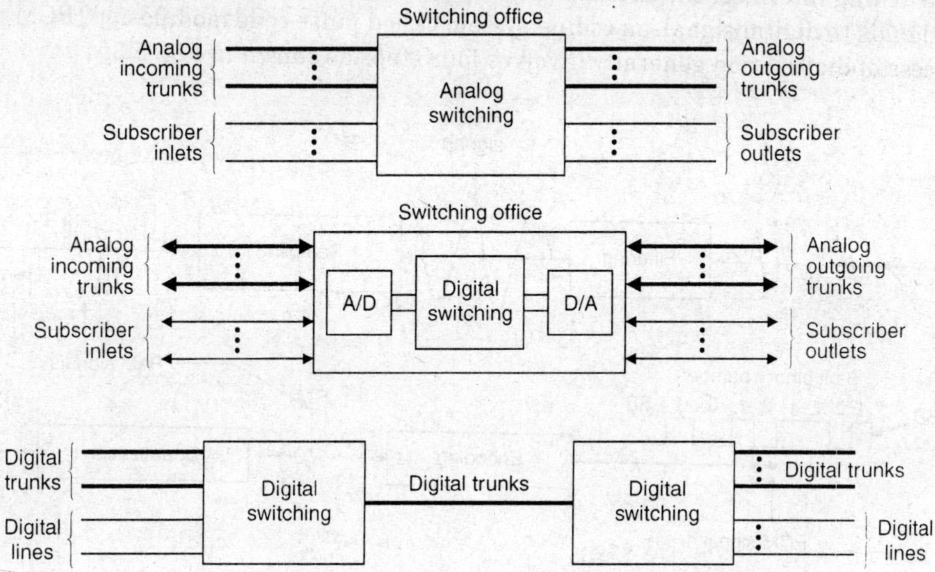

Fig. 5.1. Evolution of digital switching.

5.2.1. Digital Signal

A digital signal is a discrete signal. It is depicted as discrete variable (on/off) against the analog signal which is continuous variable. Fig. 5.2 shows the typical digital signal.

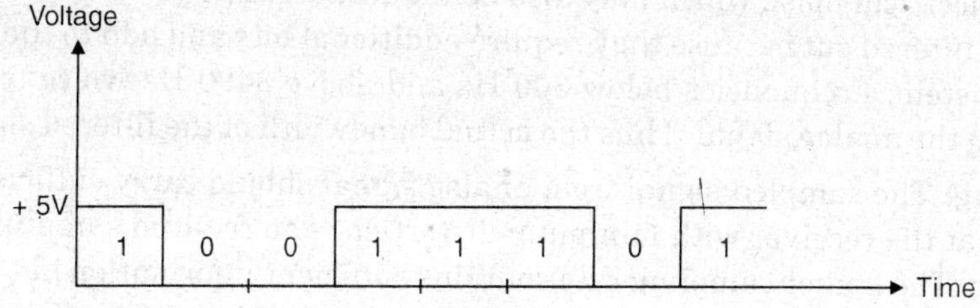

Fig. 5.2. Digital signal.

Digital signals are represented in many ways. Fig. 5.2 shows the digital signal represented by two voltages. Here 0 volts represents 0 in binary and + 5 volts represents 1 in binary. A digital signal has the following characteristics

(*a*) holds a fixed value for a specific length of time

(*b*) has sharp, abrupt changes

(*c*) A preset number of values allowed.

Each pulse (on/off) is known as binary digit (bit) the number of bits transmitted per second is the bit rates of the signal.

5.2.2. Digitization

The process of converting analog signal into digital signal is called digitization. It involves the modulation process called pulse code modulation. With digital transmission system, the quality of the system can be improved. The digital systems in comparison with the analog, provides a better switching interface, enhance easier multiplexing and produce clear signals. To convert analog signals to digital signals, a coding system called pulse code modulation (PCM) is used. The process of digitization generally involves four steps as shown in Fig. 5.3.

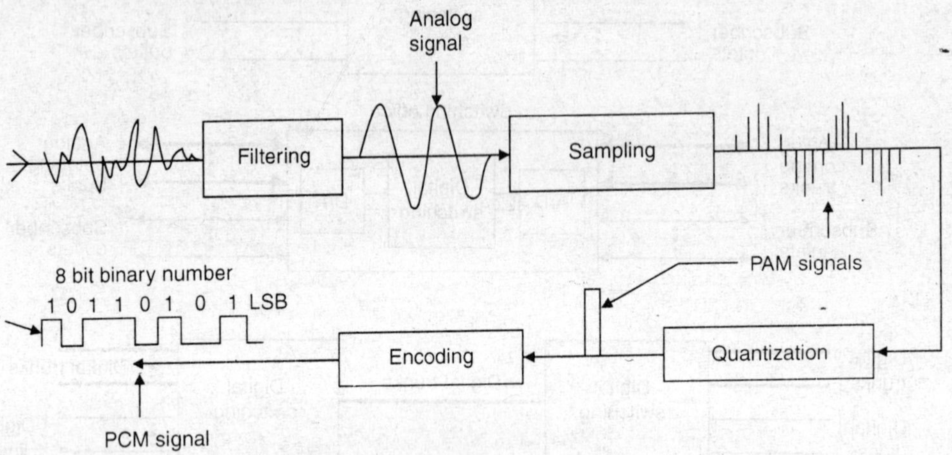

Fig. 5.3. The process of digitization.

Filtering. When signal is transmitted over wires from subscriber to local exchange, electrical noise and cross talk disturbances from neighbouring wire paths will be picked up. The signal also gets attenuated as the distance increases. Thus the speech signal transmitted is combined with noise and attenuated. By filtering process, the lower frequencies are filtered out to remove electrical noise which may also be induced from power lines. The higher upper frequencies are filtered out because they require additional bits and add to the cost of a digital transmission system. Frequencies below 300 Hz and above 3400 Hz (voice frequency range) are filtered from the analog signal. Thus the actual bandwidth of the filtered signal is 3100 Hz.

Sampling. The sampled signal from analog signal should carry sufficient information so as to receive at the receiver with minimum distortion. The reduced sampling rate contains less information. The higher sampling rate contains sufficient information but increase in rate extends the bandwidth required for transmitting the sampled signal. The Nyquist rate derived by Harry Nyquist in 1933 gave engineering compromise for the sampling rate.

By principle, the sampled signal can be recoverd exactly when

$$f_s \geq (2)(f_m) \quad \text{...(5.1)}$$

where f_s = sampling frequency

f_m = highest frequency of the band limited signal.

This minimum sampling rate is known as Nyquist rate. The speech signal is in the frequency range of 300 to 3400 Hz. For this application, a sampling rate of 8000 samples per second is almost a worldwide standard.

Fig. 5.4 shows the process of sampling. It can be realized that the sampling process is equivalent to amplitude modulation of a constant amplitude pulse train. Hence this technique is normally referred to as pulse amplitude modulation (PAM).

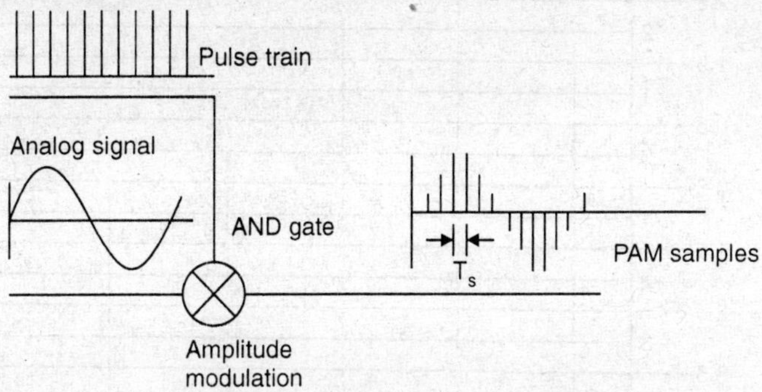

Fig. 5.4. Generation of PAM samples.

The sampling frequency $f_s = 2\,(f_m)$ ...(5.2)

$f_s \simeq 8000$ Hz ...(5.3)

As $f_s = \dfrac{1}{T_s}$ ...(5.4)

$$T_s = \frac{1}{f_s} = \frac{1}{8000} = 125\ \mu\text{ sec} \qquad ...(5.5)$$

where T_s = time period of the sampling.

In PAM, sample generator is a amplitude modulator. In simple version, it contains a AND gate. The signal to be converted is fed to one input of the AND. Pulses at the sampling frequency are applied to the other input of the AND gate to open it during the required time intervals. Thus, the output of the gate consists of pulses at the sampling rate, equal in amplitude to the signal voltage at each instant. The pulses are then passed through a pulse shaping network which gives the flat tops.

Quantization & encoding. Instead of sending a pulse train capable of continuously varying one of the parameters, the PCM technique produces a series of binary code which represents the approximate amplitude of the signal sample at that instant. This process of approximation is called signal quantization. Thus, the combined operations of sampling and quantizing generate a quantized PAM wave form.

The quantizing process is as follows. Let the signal $m(t)$ is to be quantized. Identify the signal excursion (negative peak to positive peak) which is ranges from V_L to V_H. Divide the total range into M equal intervals. In the center of each of these steps, we locate quantization levels. If the sampled amplitude is in a particular interval, the nearest value is chosen and the corresponding binary digit is approximated as equivalent to the sampled amplitude. The quality of the approximation may be improved by reducing the size of the steps, thereby increasing the number of allowable levels. Fig. 5.5 shows the process of quantization and encoding.

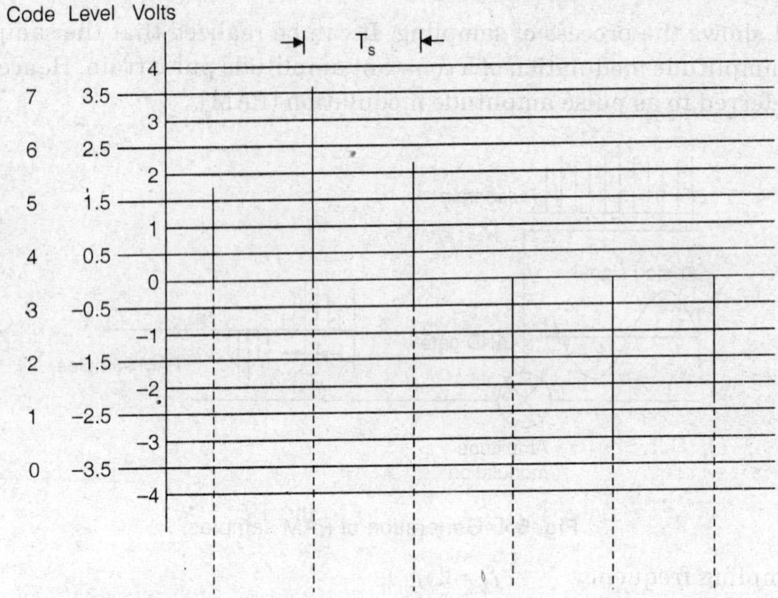

Sample value	1.8	3.6	2.2	−1.7	−3.3	−2.2
Nearest level	1.5	3.5	2.5	−1.5	−3.5	−2.5
Code	5	7	6	2	0	1
Binary representation	101	111	110	010	000	001

Fig. 5.5. Quantization with number of levels 8.

In the fourth step of the digitization process, the quantized samples are encoded into a digital bit stream (series of electrical pulses). In practical system, 256 levels can be used to obtain quality of signal. This results in the output of the encoding as 8 bit number. This binary number (or 8 bit word) is transmitted over the network as a series of electrical or optical pulses.

In the case of 256 (0 to 255) levels, the digital encoder recognizes the 255 different voltage levels of the quantized samples. Converts each into a string of eight bits (1*s* and 0*s*). The series of pulses is called a digital bit stream. The PCM process require a 64 kbps channel to encode a 4 kHz audio input signal because 8000 samples/sec × 8 bits/sample = 64000 bps.

5.3. DIGITAL TRANSMISSION

Digital transmission demand has continuously increased since its introduction in 1962 because of its high degree of accuracy. In this section, the advantages of digital transmission and the technical disadvantages of digital transmission are first discussed. There are two basic modes of digital transmission namely asynchronous and synchronous transmission. These two techniques are described in this section.

5.3.1. Advantages of Digital Transmission

The following paragraphs emphasize the advantages of digital transmission. The digital transmission use gain and phase equalization to obtain negligible inter symbol interference.

Satisfactory transmission. The principal advantage of digital transmission is its satisfactory and quality transmission even in the presence of crosstalk and noise.

Signal regeneration. In digital transmission, it is possible to use regenerative repeaters instead of analog amplifiers. During transmission, small amounts of noise, interference or distortion are added in the channel and it causes the receiver unable to identify the 0*s* and 1*s* of the binary word. Thus, the signal detection error occurs and does not represent the original data exactly.

If regenerative repeaters are inserted between transmitter and receiver spaced close enough, these nodes detect and regenerate the digital signals. Thus the regeneration process eliminates the signal degradation. The quality of the digital transmission can be further improved and increased resolution can be obtained if more bandwidth (more bits) is used. The intants at which pulses are retransmitted by a regenerative repeater are determined by local oscillator synchronized to the digit rate, which must be extracted from the received wave form.

Variations in the extracted frequency can cause a periodic variation of the times of the regenerated pulses, which is known as jitter. The variation in the times of the regenerated pulses due to changes in propagation time is known as wander.

Lower Signal to line noise ratio. In digital transmission, the noise generated in the terminal and the noise encountered in the transmission line can be separated. By the proper choice of the code (coder-decoder), the quantization noise generated by the terminal equipment can be minimised. The line noise has little effect on the message signal because of the regenerative repeater.

The signal-to-line noise ratio requirement is lower in digital than in analog transmission, better utilization of noisy media is possible. Digital transmission links provide virtually error-free performance at signal to noise ratios of 15.25 dB, depending on the type of line coding or modulation used.

Ease of multiplexing. Time division multiplexing is normally associated only with digital transmission links. The drawback of analog TDM lies in the vulnerability of narrow analog pulses to noise, distortion, crosstalk and intersymbol interference. TDM equipment costs are quite small in the case of digitized signal. Since digitization occurs only at the first level of the TDM hierarchy, high level TDM is more economical. The multiplexing operations of a digital transmission system can be easily integrated into the switching equipment.

Other advantages. Analog transmission system have difficult environments and requires special attention for processing control signals like on hook/off hook, coin deposits, clear forward etc. The digital system allow control information independently at less cost. The use of digital subscriber lines (DSL) reduces the signalling costs relative to analog subscriber lines.

The digital transmission uses larger, modern and latest technology. Hence high performance and low cost can be achieved in digital implementations. The time division multiplexer, digital signal processors, basic logic circuits, multiplexing circuits, computer systems etc. are some of the modern equipments used in digital transmission.

The integration of transmission and switching, performance monitorability, ease of encryption are the other advantages in digital transmission.

5.3.2. Disadvantages of Digital Transmission

Greater bandwidth. A major disadvantage of digital transmission is the requirement of much greater bandwidth. This is due to the fact that each sample is represented by an 8 bit codeword and each bit is transmitted as a separate discrete pulse. For example, T1 system requires approximately eight times as mush bandwidth as do 24 analog voice channels.

Need of synchronization. For efficient reception of digital signal, a timing reference, generally a 'clock' is needed. The clock specifices, when to sample the incomig signal to decide which data value was transmitted. The sample clock must be synchronized to the pulse arrival times. For networks when a number of digital transmission links and switches are interconnected, certain synchronization procedures must be established to maintain internal synchronization.

Multiplexing difficulties. Since the time of arrival of data is dependent on the distance of travel in TDM, it is not nearly as amenable to applications involving distributed sources and destinations. Also, TDM requires time slot recognition logic as each time division source must duplicate the synchrónization. Thus TDM has been primarily used in applications, where all the information sources are centrally located and a single multiplexer controls the occurrence and assignment of time slots.

Other major disadvantage is the incompatibilities with analog facilities.

5.4. MODES OF DIGITAL TRANSMISSION

The signal is normally transmitted in multiples of a fixed length unit, typically of 8 bits. The bit is represented as a specific voltage level relative to a reference level. A high signal relative to a reference indicates the transmission of a binary 1 while a low signal represents a binary 0 in positive logic.

For the receiving device to decode and interpret this bit pattern correctly, it must be able to determine the following :

(*i*) The reception of digital data involves sampling the incoming signal once per bit time to determine the binary value.

(*ii*) To sample the incoming bits properly, the receiving system should know the arrival time and duration of each bit that it receives.

(*iii*) The start and end of character or byte.

(*iii*) The start and end of each completed message.

The transmission of binary data across a link can be accomplished either in parallel or serial mode. The advantage of serial over parallel transmission is the use of common channel and thus the reduced cost. There are two basic modes of digital transmission for establishing a time base (sample clock) for effective reception of the signal. They are asynchronous and synchronous transmission. Both techniques involves fundamently different techniques. But these two approaches are common for achieving the desired synchronization.

5.4.1. Digital Signal Encoding Formats

A variety of encoding schemes are in use. Some of the commonly used digital signal encoding formats are Non return to zero-level (NRZ-L), Non return to zero inverted (NRZI), bipolar-alternative mark inversion (bipolar-AMI), Pseudoternary, manchester, Differential manchester, bipolar with 8-zero. Substitution and high density bipolar–3 zeros (HDB 3).

NRZ–L signalling is common for asynchronous transmission and Manchester or Differential Manchester encoding are generally used to accomplish synchronous transmission. Thus NRZ–L, Manchester and differential Manchester encoding schemes are shown in Fig. 5.6 and described briefly.

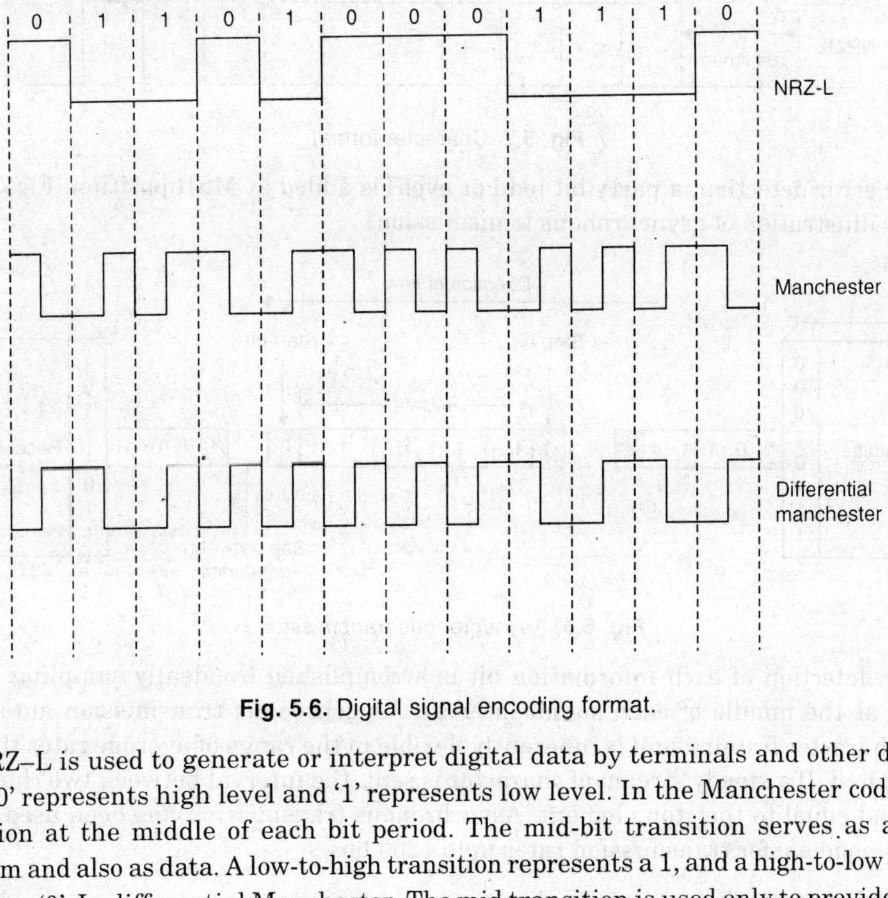

Fig. 5.6. Digital signal encoding format.

NRZ–L is used to generate or interpret digital data by terminals and other devices. In NRZ–L, '0' represents high level and '1' represents low level. In the Manchester code, there is a transition at the middle of each bit period. The mid-bit transition serves as a clocking mechanism and also as data. A low-to-high transition represents a 1, and a high-to-low transition represents a '0'. In differential Manchester. The mid transition is used only to provide clocking. The encoding of a '0' is represented by the presence of a transition at the beginning of a bit period and a 1 is represented by the absence of a transition at the beginning of a bit period.

5.4.2. Asynchronous Transmission

Asynchronous transmission involves separate transmission of groups of bits or characters. In each group, a specific predefined time interval is used for each discrete signal. The transmission times of the group are independent of each other. Hence a sample clock is reestablished for reception of each group.

To alert the receiver for the arrival of a new group, a start bit usually a '0' is added to the beginning of each byte. To indicate the completion of the group, a stop bit usually a '1' is appended to the end of the byte. Thus the size of the group is atleast 10 bits. That is 8 bit contains information, one for start bit and one for stop bit. The transition of each byte is separated by a gap of varying duration. The gap is usually identified by an idle channel or by a stream of additional stop bits. The Fig. 5.7 shows the typical character format.

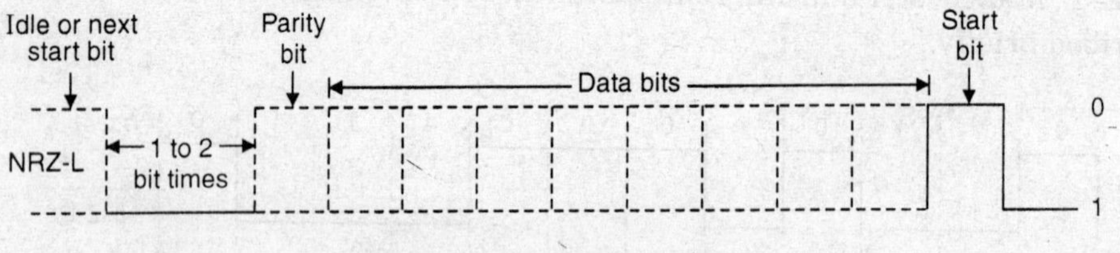

Fig. 5.7. Character format.

For error detection, a party bit (odd or even) is added at MSB position. Fig. 5.8 shows schematic illustration of asynchronous transmission.

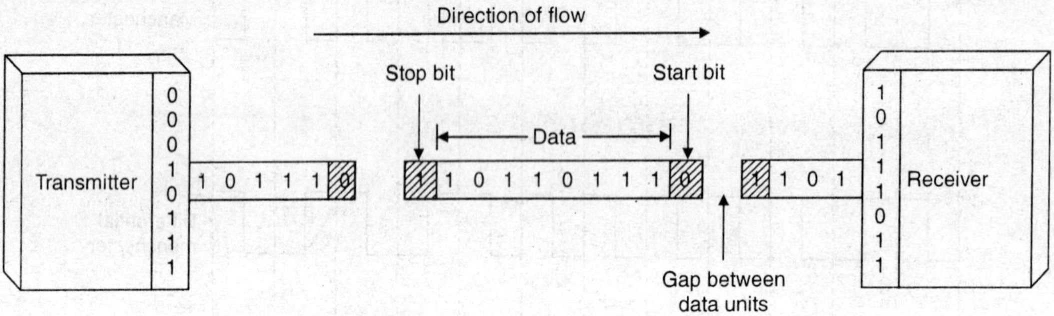

Fig. 5.8. Asynchronous transmission.

The detection of each information bit is accomplished by ideally sampling the input waveform at the middle of each signal interval. Asynchronous transmission automatically provides character framing and is inherently flexible in the range of average rates that can be accommodated. If a steady stream of character is sent, the interval between two characters is uniform and equal to the stop element. Asynchronous transmission has been used in voiced data sets (modems) for transmission rates upto 1200 bps.

Disadvantages :

1. Since the sample time for each information bit is derived from a single start bit, asynchronous system do not perform well in high noise environments. This problem can be eliminated by adding more than one start bit, but it causes the system more complex.
2. **Timing error.** The asynchronous implies a free running clock in the receiver. So any offset in the clock frequency of the receiver causes timing error. If the receiver is 5

percent slower or faster than the transmitter, the sampling of the eight information bit will be displaced by 45 percent.

3. **Framing error.** By timing error, the bit count may be out of alignment. If bit 7 is a 1 and bit 8 is a 0, bit 8 could be mistaken for a start bit. This condition is termed as a framing error, as the character plus start and stop bits some times referred to as a frame. False appearance of start bit due to noise also causes framing error.
4. The major drawback of asynchronous transmission is its poor performance in terms of error rates on noisy lines.
5. Asynchronous transmission is simple and cheap but requires an overhead of two to three bits per character. The percentage overhead could be reduced by sending larger blocks of bits between the start and stop bits.
6. The addition of stop and start bits and the insertion gaps into the bit stream made asynchronous transmission slower.

5.4.3. Synchronous Transmission

To achieve greater efficiency, synchronous transmission is used. The main advantage of synchronous transmission is speed. Hence it is more useful in high speed applications like data transfer between computer. For digital signal, synchronous transmission can be accomplished with Manchester or differetial manchester encoding. In Asynchronous transmission, digital signals are sent as separate transmission of groups whereas in synchronous transmission, the digital signals are sent continuously at constant rate. Hence the receiving terminal must establish and maintain a sample clock that is synchronized to the incoming data for an indefinite period of time.

Fig. 5.9 shows a typical synchronous frame format. The frame format starts with a preamble called flag to determine the beginning of the block of data. A postample bit is added at the end of the block of data to identify the end. Other bits are added to the block that convey control information used in data link procedures. Thus the data block plus, preamble, postamble and control information are called a frame.

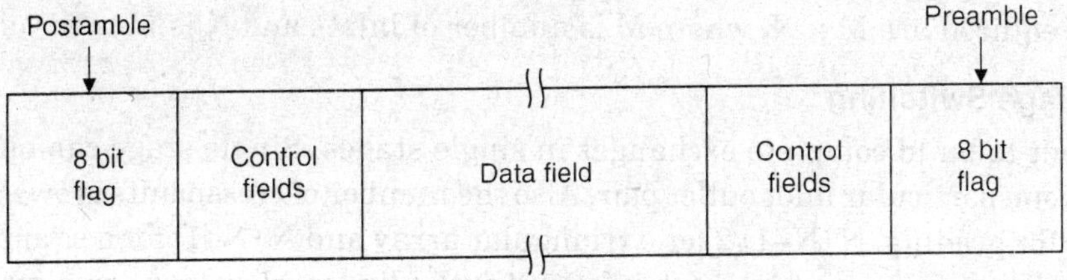

Fig. 5.9. Synchronous frame format.

In synchronous transmission, data are sent one bit after another without start/stop bits or gaps. The receiver should be capable enough to separate the bit stream into bytes for decoding purposes. Fig. 5.10 shows the schematic of synchronous transmission.

In Fig. 5.10, the data stream are shown as 8 bit block for explanation purpose. But in reality, no such portions. When the data stream reaches the receiver, it counts the bits and groups them in eight bit units. If the sender wishes to send data in separate groups, the gap between group must be filled with a special sequence of 0 and 1's that means idle.

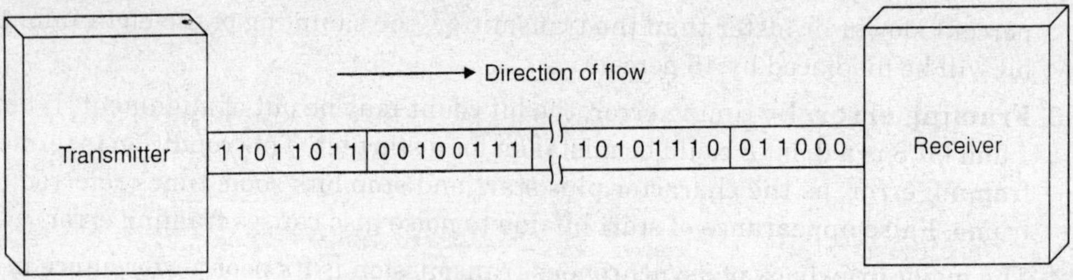

Fig. 5.10. Synchronous transmission.

As there is no start and stop bit to identify the blocks and also the grouping of bits should be performed by the receiver, it must be capable enough to keep an accurate count and error free reception.

5.5. SPACE DIVISION SWITCHING

The fundamental operation of a switch is to setup and release connection between subscribers. It involves direct connection between subscriber loops at an end office or between station loops at a PBX. The switches are hardware and/or software devices capable of creating temporary connections between two or more subscribers. In this section, the space division switching is described and in section 5.6, time division switching is explained.

In space division switching, the paths in the circuit are separated from each other spatially. It was originally designed for analog networks, but is used currently in both digital and analog switching. A crosspoint switch is referred to as a space division switch because it moves a bit stream from one circuit/bus to another. For large group of outlets, considerable savings in total crosspoints can be achieved if each inlet can access only a limited number of outlets. Such situation is called limited availability.

By overlapping the available outlet groups for various inlet groups, a technique called "grading" is established. Rectangular crosspoint array is an example of grading. For longer trunk groups, large crosspoints are expensive and not used now-a-days. The number of crosspoints required are M × N, where M is number of inlets and N is number of outlets.

5.5.1. Multistage Switching

It is inefficient to build complete exchanges in single stages. Single stage can only be used to interconnect one particular inlet outlet pair. Also the number of crosspoints grows as the square of the inputs for grading, N (N–1)/2 for a triangular array and N (N–1) for a square array. Also the large number of crosspoints on each inlet and outlet line imply a large amount of capacitive loading on the message paths. Therefore, it is usual to build exchanges in two or three stages to reduce the number of crosspoints and to provide alternative paths. The sharing of crosspoints for potential paths through the switch is accomplished by multiple stage swtiching.

Fig. 5.11 shows the three stage switching structure to accomodate 128 input and 128 output terminals with 16 first stage and last stage.

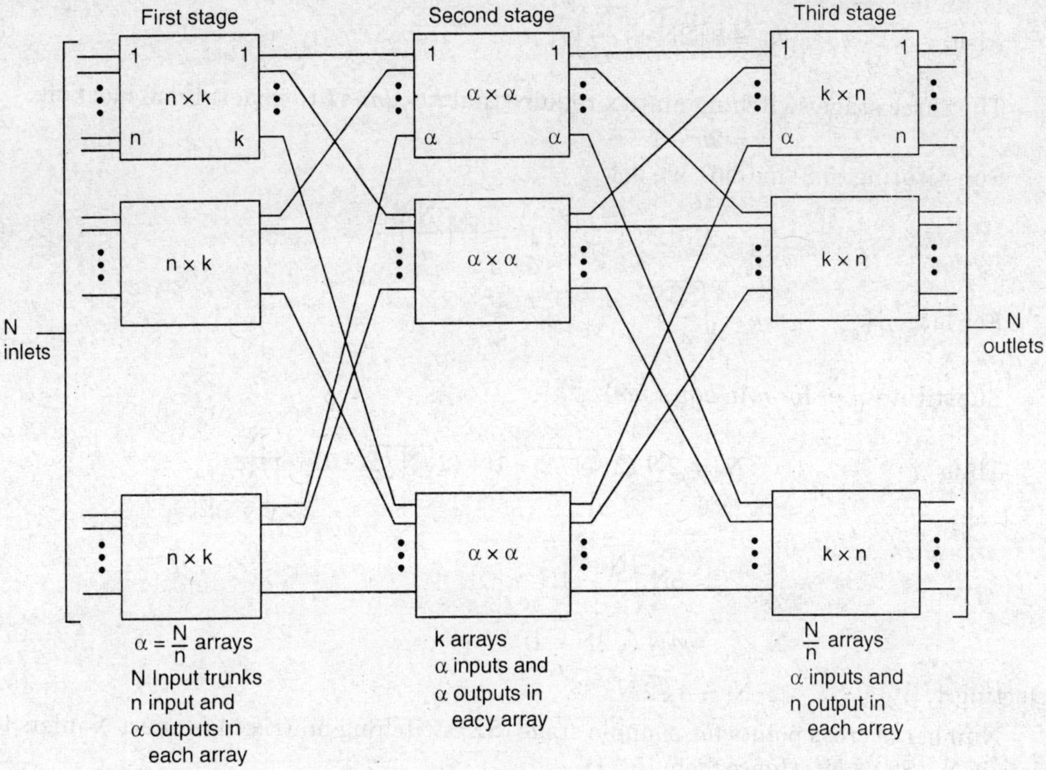

Fig. 5.11. Three stage switching structure.

The structure shown in Fig. 5.11 provides path for N inlets and N outlets. The N input lines are divided into N/n groups of n lines each. Each group of n inputs is accomodated by an n-input, k output matrix. The output matrices are identical to the input matrices except they are reversed. The intermediate stages are k in number and N/n inputs and N/n outputs. The interstage connections are often called junctors. Each of the k paths utilizes a separate center stage array. An arbitrary input can find k alternate output. Thus multistage structure provides alternate paths. Also the switching link is connected to a limited number of crosspoints. This enables the minimized capacitive loading.

The total number of crosspoints N_X for three stage is

$$N_X = 2Nk + k\left(\frac{N}{n}\right)^2 \quad ...(5.6)$$

where N = Number of inlets-outlets

n = size of each inlet-outlet group

k = number of second stage.

2Nk = number of cross points in 1st and 2nd stage

$\left(\frac{N}{n}\right)^2$ = number of cross points in each array of second stage

$k\left(\frac{N}{n}\right)^2$ = number of cross points in second stage.

Also $$N_X = k\left[2N + \left(\frac{N}{n}\right)^2\right] \quad ...(5.7)$$

The three stage switching matrix require that $k > 2n - 1$ to generate no blocking.

$$k = 2n - 1 \quad ...(5.8)$$

Substituting (5.8) in (5.6), we get

$$N_X = 2N\,(2n - 1) + (2n - 1)\left(\frac{N}{n}\right)^2 \quad ...(5.9)$$

For large N, $n = \sqrt{\frac{N}{2}}$

Substitute $\sqrt{\frac{N}{2}}$ for n in eqn. (5.9)

Thus $$N_X = 2N\,(2\sqrt{N/2} - 1) + (2\sqrt{N/2} - 1)\left(\frac{N}{\sqrt{\frac{N}{2}}}\right)^2$$

$$= 8N\sqrt{\frac{N}{2}} - 4N$$

$$N_{x(\min)} = 4N\,(\sqrt{2N} - 1) \quad ...(5.10)$$

neglecting 1, $$N_x \simeq 4\sqrt{2}\,N^{3/2} \quad ...(5.11)$$

Number of cross points for a single stage (SS) switching matrix to connect N inlets to N outlets is N_x (SS) = N^2. Hence from (5.11)

$$\lambda \equiv \frac{N_x\,(\text{min, 3 stage})}{N_x(\text{SS})} = \frac{4\sqrt{2}\,N^{3/2}}{N^2}\,.$$

$$\lambda = \frac{4\sqrt{2}}{\sqrt{N}} \quad ...(5.12)$$

where N_x(SS) = Number of cross points in single stage. In practice, this gives reasonable scaling.

Table 5.1. Number of crosspoints in a Non blocking switch for B = 0.002 and *p* = 0.7

Number of lines	Number of cross points		$\lambda = \frac{4\sqrt{2}}{\sqrt{N}}$
	Three stage switch $N_x = 4N(\sqrt{2N} - 1)$	Single stage N^2	
128	7680	16384	0.5
512	63488	262144	0.25
2048	516,096	4.2×10^6	0.125
8192	4.2×10^6	6.7×10^7	0.065
32,768	3.3×10^7	0.1×10^9	0.0313
131,072	2.6×10^8	1.7×10^{10}	0.0156

The savings in crosspoints becoming more pronounced with increasing N.

When very large number of lines must be accomodated, switching structures with more stages, even upto eight stages are used. The probability that all k links are busy is given by

$$B = [1 - (1 - P)^2]^k \quad ...(5.13)$$

where $$P = np/k \quad ...(5.14)$$

Example 5.1. *A three stage switching structure is to accomodate N = 128 input and 128 output terminals. For 16 first stage and 16 last stage, determine the number of cross points for nonblocking.*

Sol. The number of matrices at first and last stage is given by $\alpha = \frac{N}{n}$.

Hence $$n = \frac{N}{\alpha} = \frac{128}{16} = 8$$

To avoid blocking $k = 2n - 1 = 2 \times 8 - 1 = 15.$

Number of crosspoints is calculated by

$$N_x = k\left[2N + \left(\frac{N}{n}\right)^2\right] = 15\left[2 \times 128 + \left(\frac{128}{8}\right)^2\right]$$

$$N_x = 7680 \text{ cross points.}$$

Example 5.2. *If the number of crosspoints in the example is to be reduced by the factor of 3 with non blocking what is the probability that a call will be blocked ? Assume the utilization probability p = 15%.*

Sol. Number of cross points = 7680

Number of cross points reduced by factor 3 = $\frac{7680}{3}$ = 2560.

For the cross point 2560, the number of k matrices is calculated from

$$N_x = k\,(2N + (N/n)^2]$$

$$k = \frac{N}{[2N + (N/n)^2]} = \frac{2560}{256 + (128/8)^2}$$

$$k = 5$$

$$P = np/k = 8 \times 0.15/5 = 0.24$$

The probability that k links are busy is

$$B = [1 - (1 - P)^2]^k$$

$$B = [1 - (1 - 0.24)^2]^5 = 1.34\%$$

5.5.2. Blocking Probability Evaluation Techniques

All the switching systems are designed to provide a certain maximum probability of blocking for the busiest hour of the day. It is one of the aspects of the grade of service of the telephone company. There are variety of techniques to evaluate the blocking probability of a switching matrix. Depends on the accuracy, required availability, geographical area, priority, complexity and applicability of different network structures, the techniques are varying. Here, two techniques are described.

1. **Lee graphs.** It was proposed by C.Y. Lee. It is a most versatile and straight forward approaches of calculating probabilities with the use of probability graphs.

2. **Jacobaeus method.** It was presented in 1950 by C. Jacobaeus. It is more accurate than Lee graph method.

Lee graphics. C.Y. Lee's approach of determing the blocking probabilities of various switching system is based on the use of utilization percentage or loadings of individual links.

Let p be the probability that a link is busy. The probability that a link is idle is denoted by $q = 1 - p$. When any one of n parallel links can be used to complete a connection, the blocking probability B is the probability that all links are busy is given by

$$B = p^n \qquad \text{...(5.15)}$$

when n = number of links reqd. to complete connection (for 3 stage $n = 2$)

when a series of n links are all needed to complete a connection,

$$B = 1 - q^n \qquad \text{...(5.16)}$$

For a probability graph of three stage network, shown in Fig. 5.12, the probability of blocking is given by

$$B = (1 - q^2)^k \,.\, [1 - (q')^2]^k \qquad \text{...(5.17)}$$

where q' = probability that an interstage link is idle = $1 - p'$ ...(5.18)

p' = probability that any particular intersatge link is busy

k = number of centre stage arrays.

If p is known, the probability that an interstage link is busy is given by

$$p' = \frac{p}{\beta} \qquad \text{...(5.19)}$$

where $\beta = k/n$ ...(5.20)

β is the factor by which the percentage of interstage links that are busy is reduced.

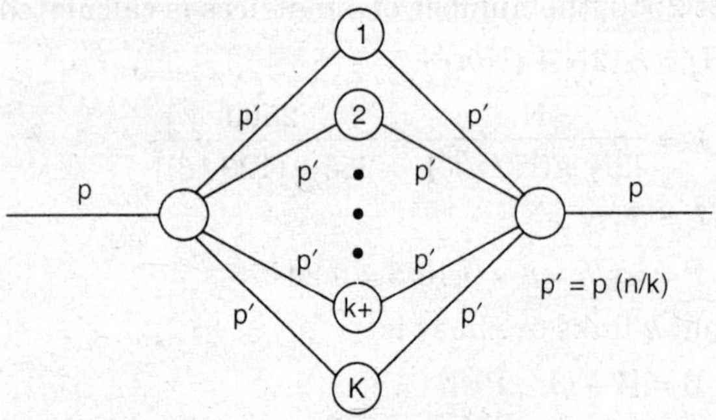

Fig. 5.12. Probability graph of three stage network.

Substituting in (5.19) in (5.18), we set $q' = 1 - \dfrac{p}{\beta}$ ...(5.21)

Substituting (5.21) in (5.17) we get complete expression for the blocking probability of a three stage switch interms p as

$$B = \left[1 - \left[1 - \frac{p}{\beta}\right]^2\right]^k \qquad \text{...(5.22)}$$

With inlets of 10% busy, the switch size of N with $n = 8$, $h = 5$, $\beta = 0.625$ requires 2560 crosspoints. The merits of this method are

(*i*) It provide accurate results

(*ii*) Its formulaes are directly relate to the network structures

(*iii*) It provides insight of the network and thus provides ideas to change the structure for high performance.

Jacobaeus. The Lee's graph approach is not much accurate. Because the probability graphs entail several simplifying assumptions. The important one which gives erroneous values of blocking is the assumption that the individual probabilities are independent. In fact the probabilities not independent and highly dependent when significant amounts of expansion are not present. According to C. Jacobaeus the blocking probability of a three stage switch is

$$B = \frac{(n\,!)^2}{k\,!\,(2n-k)\,!}\, p^k\, (2-p)^{2n-k} \qquad ...(5.23)$$

where n = number of inlets (outlets) per first (third) stage array

k = number of second stage array

p = inlet utilization.

More accurate techniques can be used for systems with high concentrations and high blocking. As the high blocking probabilities not having much practical value, those techniques are not considered here.

5.6. TIME DIVISION SWITCHING

In space division switching, crosspoints are used to establish a specific connection between two subscribers. The crosspoints of multistage space switches assigned to a particular connection is dedicated to that connection for its duration. Thus the crosspoints can not be shared.

Time division switching involves the sharing of crosspoints for shorter periods of time. This paves way for the reassign of crosspoints and its associated circuits for other needed connections. Therefore, in time division switching, greater savings in crosspoints can be achieved. Hence, by using a dynamic control mechanisms, a switching element can be assigned to many inlet-outlet pairs for few microseconds. This is the principle of time division switching.

Time division switching uses time division multiplexing to achieve switching. Two popular methods that are used in time division multiplexing are (*a*) the time slot interchange (TSI) and (*b*) the TDM bus. In ordinary time division mutliplexing, the data reaches the output in the same order as they sent. But TSI changes the ordering of slots based on the desired connections. The demultiplexer separates the stots and passes them to the proper outputs. The TDM uses a control unit. The control unit opens and closes the gates according to the switching need.

The principle of time division switching can be equally applied to analog and digital signals. For interfacing sampled analog signals but not digitized, the analog time division switches are attractive. But for larger switches, there are some limitations due to noise, distortion and crosstalk which nomally occurs in PAM signals. Thus analog switching is now used only in smaller switching systems. In this section, the analog time division switching and digital time division switching are described briefly.

5.6.1. Analog Time Division Switching

Fig. 5.13 shows a simple analog time division switching structure. The speech is carried as PAM analog samples or PCM digital samples, occurring at 125 μs intervals. When PAM samples are switched in a time division manner, the switching is known as analog time division switching. If PCM binary samples are switched, then the switching is known as digital time division switching. A single switching bus supports a multiple number of connections by interleaving PAM samples from receive line interfaces to transmit line interfaces. There are two cyclic control stores. The first control store controls gating of inputs onto the bus one sample at a time. The second control store operates in synchronism with the first and selects the appropriate output line for each input sample.

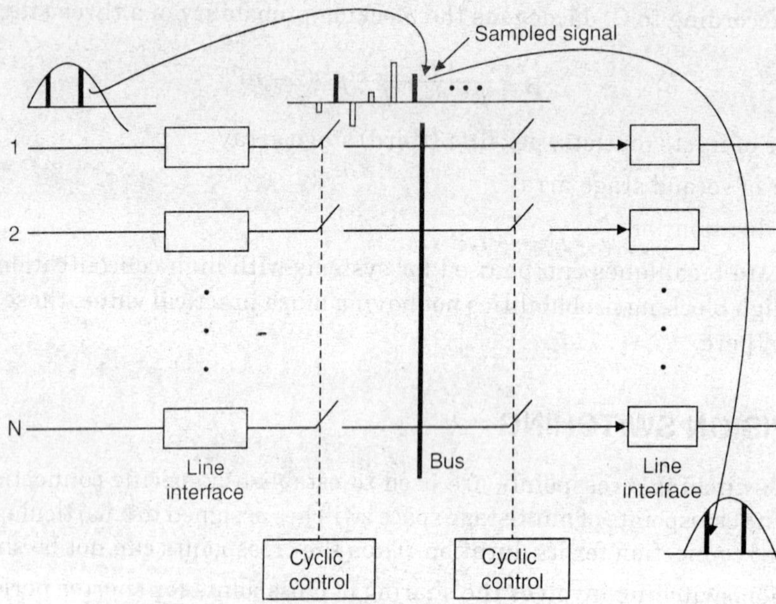

Fig. 5.13. Analog time division switching structure.

The selection of inlet/outlet is controlled by various ways. The (*a*) cyclic control and (*b*) memory based control are the important controls and described in the following paragraphs.

Cyclic control. The cyclic control is organised by using Modulo-N counter and k to 2^k decoder as shown in Fig. 5.14.

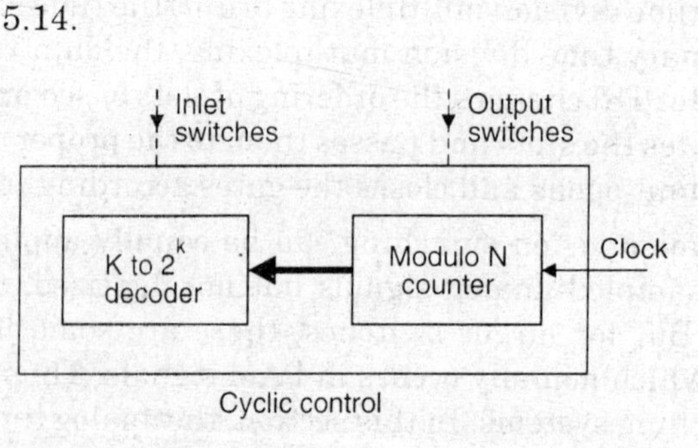

Fig. 5.14. Cyclic control.

The k and N are related by $\lceil \log_2 N \rceil = k$ ...(5.24)

where N = number of inlets/outlets

k = decoder size.

$\lceil\ \rceil$ = gives the lowest integer. It means k may be assumed lowest integer or more than that.

This kind of switching is non-blocking but lack of full availability as it is not possible to connect inlet to any outlet. The switching capacity or number of channel supported by cyclic controlled system is

$$C = \frac{125\ \mu\text{sec}}{t_s} \qquad ...(5.25)$$

The numerator 125 μ sec indicates the time taken to scan inlet and outlet and the denomintor t_s is the time in μ sec to setup connection.

Memory based control. Full availability can be achieved if any one control is made memory based. If the input side is cyclically switched and the outlets are connected based on the addresses of the outlets stored in contiguous location is referred as input controlled or input driven. If the outlets are cyclically switched, the switch is referred as output controlled or output driven. As the physical connection is established between the inlet and the outlet through the common bus for the duration of one sample transfer, the switching technique is known as time division space multiplexing. For this system,

$$C = \frac{125\ \mu\text{sec}}{t_i + t_m + t_d + t_t} \qquad ...(5.26)$$

where t_m = time to read the control memory

t_d = time to decode address and select the inlet and outlet.

t_i = time to increment the modulo N counter.

t_t = time to transfer the sample.

The capacity equations 5.25 and 5.26 are valid only for a 8 kHz sampling and non folded network (can be used for folded network with certain changes in network). The switching capacity in the memory controlled is equal to N. The use of cyclic control in input or output controlled switches restricts the number of subscribers on the system rather than the switching capacity since all the lines are scanned whether it is active or not.

No restictions on subscriber number and full availability of the switching system can be achieved by designing a switching configuration with control memory for controlling both inlets and outlets. This configuration referred to as memory controlled time division space switch is shown in Fig. 5.15.

As each word of the control memory has inlet address and an outlet address, the control memory width is $2\lceil \log_2 N \rceil$. The control memory words are readout one after another. The modulo counter is updated at the clock rate. For the path setup of kth inlet and jth outlet, the addresses are entered in control memory and path is made. Then the location is marked busy. When conversation is terminated, the addresses are replaced by null values and location is marked free. Hence

$$C = \frac{125}{t_s}\ \mu\text{ sec, where } t_s = t_i + t_m + t_d + t_t \qquad ...(5.27)$$

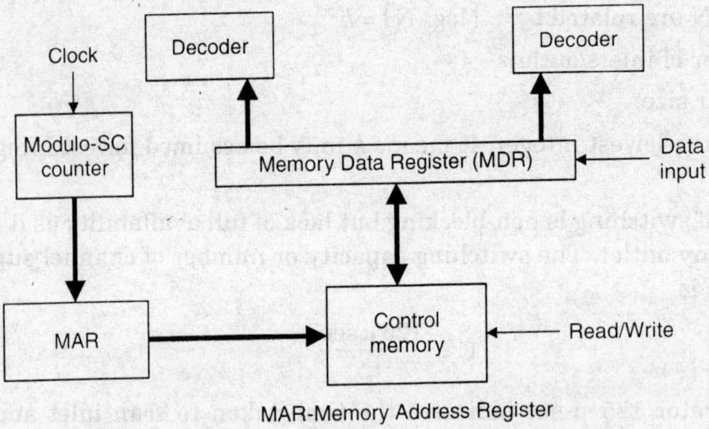

Fig. 5.15. Memory control for both inlet and outlet.

The switching matrix described above is referred to as time multiplexed switching as the switch in this configuration is replicated once for each time slot.

5.6.2. Digital Time Division Switching

The analog time division switching is useful for both analog and digital signals. The digital time division multiplexed signals usually requires switching between time slots as well as between physical lines. The switching between time slots are usually referred as time switching. Similar to analog time division switching the switching structure can be organised expect the use of memory block in place of the bus. This adds the serial to parallel and parallel to serial bit conversion circuitry's as the input to the memory block should be in parallel form. The time division switch can be controlled in any of the following three ways.

Basic operation. The basic requirement of time division switching is that the transfer of information arriving at in a time slot of one input link to other time slot of any one of output link. A complete set of pulses, arriving at each active input line is referred to as a frame. The frame rate is equal to the sample rate of each line.

A time switch operates by writing data into and reading data out of a single memory. In this process the information in selected time slots is interchanged as shown in Fig. 5.16.

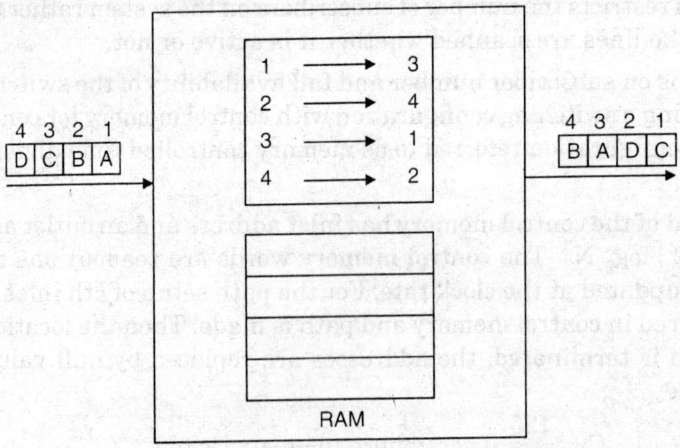

Fig. 5.16. Time slot interchange operation.

In TSI operation, inputs are sequentially controlled and outputs are selectively controlled. The RAM have several memory locations, each size is the same as of single time slot.

Fig. 5.17 shows the general arrangement of the time division time switching.

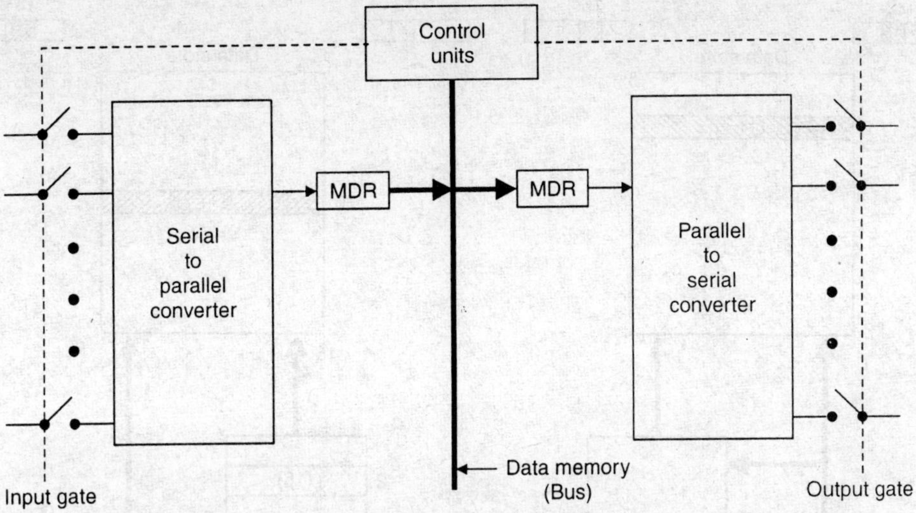

Fig. 5.17. Functional diagram of time division time switching.

The serial to parallel and parallel to serial converter are used to write the data into the memory and read the data out of memory. For convenience, two memory data system MDR are shown, but MDR is a single register. Gating mechanism is used to connect the inlet/outlet to MDR.

The input and output lines are connected to a high speed bus through input and output gates. Each input gate is closed during one of the four time slots. During the same time slot, only one output gate closed. This pair of gates allows a burst of data to be transferred from one input line to a specific output line through the bus. The contol unit opens and closes the gates according to switching need.

The time division time switch may be controlled by sequential write/random read or random write/ sequential read. Fig. 5.18 depicts both modes of operation and indicates how the memories are accessed to translate information from time slot 2 to time slot 16. Both methods use a cyclic control.

Fig. 5.18 (*a*) implies that specific memory locations are dedicated to respective channels of the incoming TDM link. Data are stored in sequential locations in memory by incrementing modulo N counter with every time slot. Thus incoming data during time slot 2 is stored in the second location within the memory. On output, information retrieved from the control store specifies which address is to be accessed for that particular time slot. Thus sixteenth word of control store contains the number 2, implying that the contents of data store address 2 is transferred to the output link during outgoing slot 16.

Random write/sequential read mode of operation is opposite to that of sequential write/ random read. Incoming data are written into the memory locations as specified by the control store, but outgoing data are retrieved sequentially under control of an outgoing time slot counter. The data received during time slot 2 is written directly into data store address 16 and it is retrieved during outgoing TDM channel number 16.

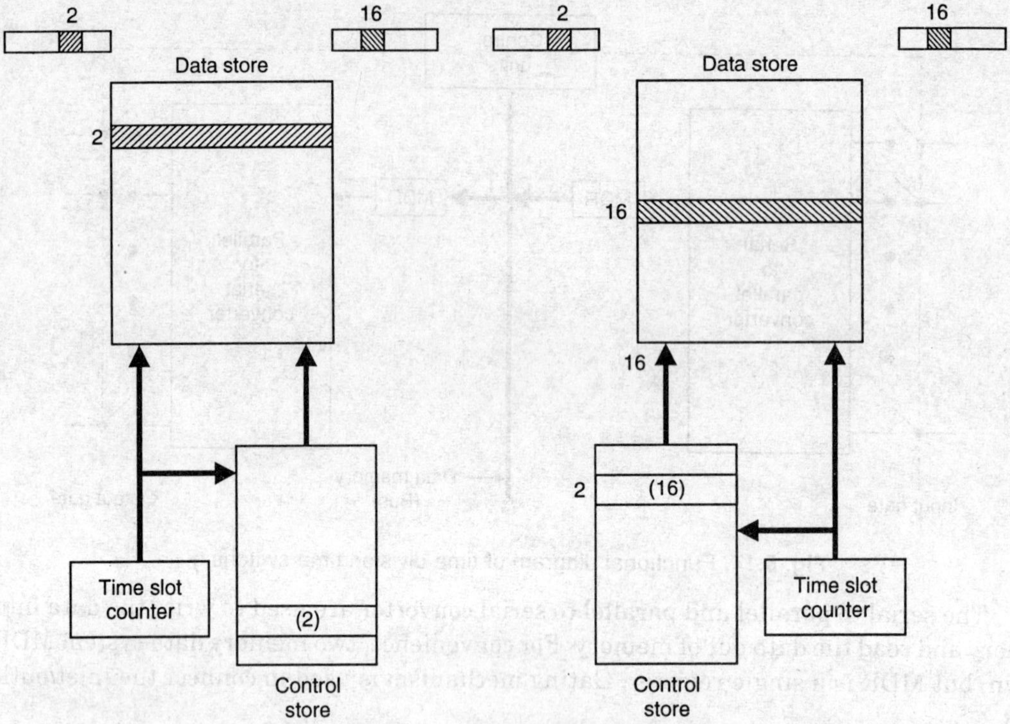

Fig. 5.18. (*a*) Sequential write/random read, (*b*) random write/sequential read.

5.7. TWO DIMENSIONAL DIGITAL SWITCHING

Combination of the time and space switches leads to a configuration that achieved both time slot interchange and sample switching across trunks. These structures also permit a large number of simultaneous connections to be supported for a given technology. Large digital switches require switching operations in both a space dimension and a time dimension. There are a large variety of network configurations that can be used to accomplish these requirements.

The incoming and outgoing PCM highways are spatially separate. So the connection of one line of local exchange obviously requires space switching to connect to the channel of outgoing highways. Thus the switching network must be able to receive PCM samples from one time slot and retransmit them in a different time-slot. This is known as time slot interchange, or simply as time switching. Thus the switching network must perform both space and time switching.

The space switching and time switching may be accomplished in many ways. A two stage combination switch may be organised with time switch as first stage and the space switch as the second stage or vice versa. The resulting configuations are referred as time space (TS) or space time (ST) switches respectively. Three stage time and space combinations of TST and STS configurations are more pupular and flexible. Very large division switches includes many combinations of time and space switches. Typical configurations are TSST, TSSSST, and TSTSTSTS. These switches support 40000 lines or more economically.

The general block diagram involving time and space switching is shown in Fig. 5.19.

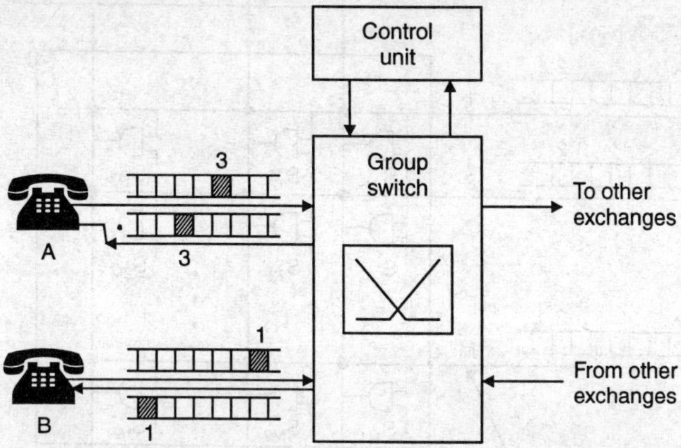

Fig. 5.19. General block diagram of combined switching.

The main task of the switching part is to interconnect an incoming time slot and an outgoing time slot. The unit responsible for this function is group switch. There are two types of building block in the digital group switch. They are time switch and space switch.

In Fig. 5.19, the subscriber makes a local call to B. The control unit has assigned time slot 3 to the call on its way into the group switch, and time slot 1 on its way out of the group switch (to B). This is maintained during the entire call. Similarly B to A also carried out. The fundamental design and structure of the two switches viz. time switch and space switch are described in the following sections.

5.7.1. Space and Time Switches

Space switch. Fig. 5.20 shows a typical space switch. It uses a space array to provide switching generally the space switch consists of a matrix of M × N switching points where M is number of inlets and N is number of outlets. A connection between an inlet and an outlet is made by the simple logic gates (AND gates). As logic gates are unidirectional, two paths through switching matrix must be established to accomodate a two way conversation. The logic gate array can serve for concentration, expansion or distribution depending on M is larger, smaller or equal to N. Fig. 5.20 shows only one voice direction. However, the corresponding components are available for the opposite direction too.

A number of M, of X slot multiplexers, provide the inputs and the outlets are connected to N, X slot demultiplexers. The gate select memory has X locations. The word containing information about which cross point is to be enabled is decoded by the translator. During each internal time slot, one cross point is activated. In the shift to the next interval time slot, the control memory is incremented by one step, and a new crosspoint pattern is formed in the matrix.

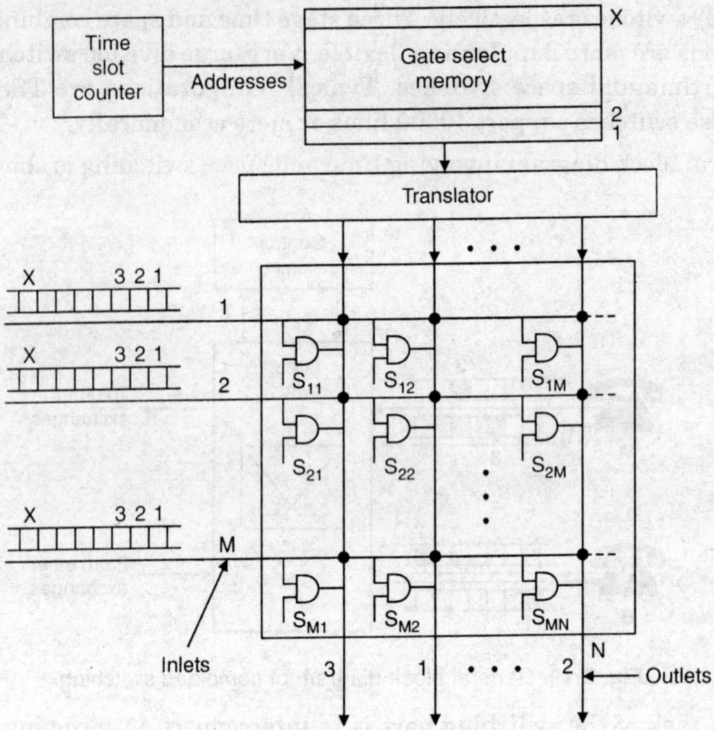

Fig. 5.20. Space switch.

Time switch. The time-slot interchange (TSI) system is referred to as time switching (T-switching). Section 5.6.2 describes the TSI switch and its associate circuits. Fig. 5.21 shows the block diagram of time switch.

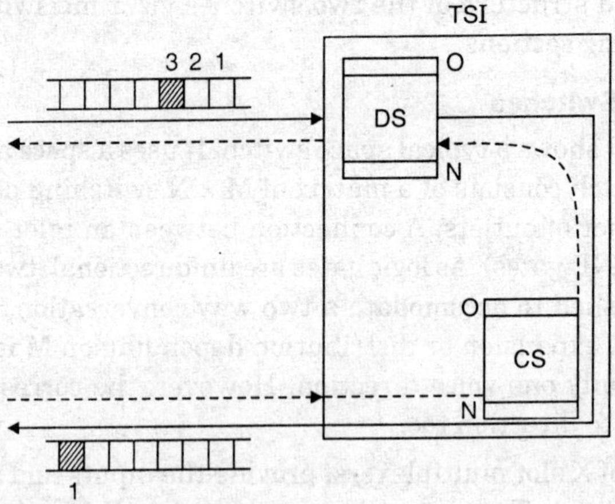

Fig. 5.21. Time switch.

Each incoming time slot is stored in sequence in a Data Store (DS). The control store (CS) determines in which order the time slots are to be read from SM. This means that a voice sample may be moved from say incoming time slot 3 to outgoing time slot 1.

5.7.2. Time-space (TS) Switching

This switch consists of only two stages. This structure contains a time stage T followed by a space stage S as shown in Fig. 5.22. Thus this structure is referred to as time-space (TS) switch. The space array have N inlets and N outlets. For each inlet line, a time slot interchanger with T slots is introduced. Each TSI is provided with a time slot memories (not shown). Similarly a gate select memory needs to be provided for the space array (not shown).

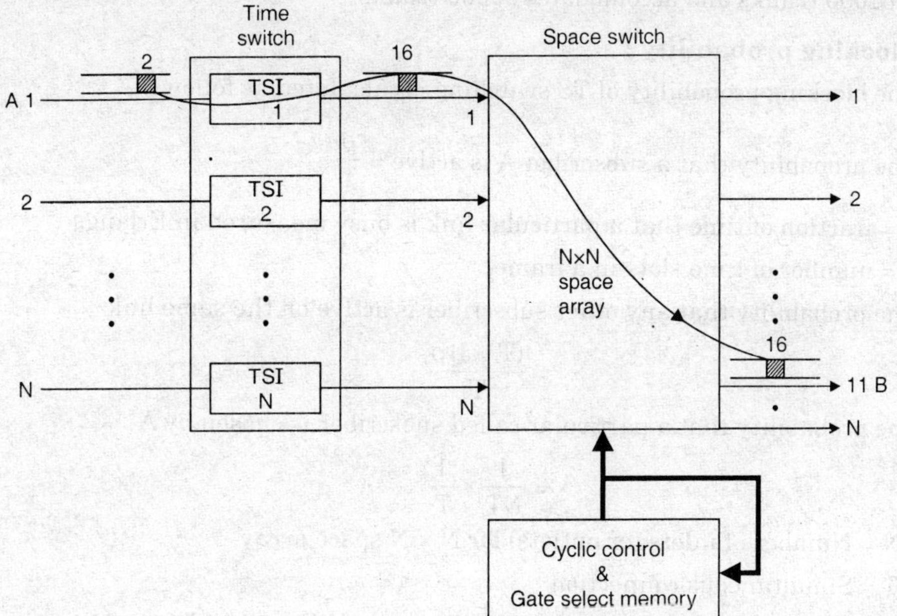

Fig. 5.22. Time space (TS) switching.

The transmission of signals carried out from sender to receiver through multiplexer input and demultiplexer output. The reverse communication also similar. Thus a hybrid arrangement is needed to isolate the transmitted signal from the received signal. The basic function of the time switch is to delay information in arriving time slots until the desired output time slot occurs.

Let the communication is to take place between subscriber A and B. Let A is assigned time slot 2 and line 1 and subscriber B is assigned time slot 16 and line 11. Then the signal moved from time slot 2 to time slot 16 by the time-slot exchanger and is transfered from line 1 to line 11 in the space array. Similarly, the signal originated by B is moved from slot 16 to slot 2 through line 11 to 1.

The cyclic control and gate select memory contains the information needed to specify the space stage configuration for each individual time slot of a frame. The time stage have to provide decays ranging from one time slot to a full frame. During each outgoing time slot, control information is accessed that specifies interstage link number to output link. During other time slots, the space switch is completely reconfigured to support other connections.

Let each time slot interchanger have T slots. If the space array is a N × N, then the simultaneous connections possible is NT. If T = 128 and N = 16, 2048 connections can be supported. This structure is not free of blocking. The control store is a parallel end around shift resister. If space array is at the inlet side and time switch is at the output side, the structure is refered as space time (ST) switching. Both TS and ST arrangements are equally effective.

TS system is used in DMS 100 digital switching system developed in canada (1979). It handles 61000 trunks and accomodates 39000 trunks.

Blocking probability :

The blocking probability of TS switching is calculated as follows.

The probability that a subscriber A is active $= \frac{\rho}{T}$ ...(5.28)

where ρ = fraction of time that a particular link is busy measured in Erlangs

T = number of time slots in a frame.

The probability that any other subscriber is active on the same link

$$= \frac{(T-1)\rho}{T} \qquad ...(5.29)$$

The probability that a particular called subscriber is chosen by A

$$= \frac{1}{NT} \times \frac{1}{T} \qquad ...(5.30)$$

where N = Number of inlets (or outlets) for N × N space array.

NT = Simultaneous connection ...(5.31)

The probability that the same time slot on a different outlet is chosen by the other subscribers on the same inlet

From 5.29 $$= \frac{(T-1)\rho}{T} \cdot \frac{(N-1)}{(NT-1)} \qquad ...(5.32)$$

From 5.30 Blocking probability $= B = \left(\frac{\rho}{T \times NT}\right)\left(\frac{(T-1)(N-1)}{T(NT-1)}\right)$

As T >> 1 and N >> 1, & NT >> 1

$$B = \frac{\rho}{NT^3} \qquad ...(5.33)$$

The TS switch can be made non-blocking by using an expanding time switch (T to T^2 slots) and a concentrating space switch (which is complex).

Implementaion complexity. In general the complexity of the switching is represented interms of number of cross points (N) and its associated cost. The number of cross points in space stage can be easily calculated which is based on the array size. The time stage uses significant amount of memory which adds the cost of the whole system. To take this into account the cost of memory bit is assumed one hundredth of the cost of cross point. Thus,

Implementation complexity $= N_X + \frac{N_B}{100}$ ...(5.34)

where N_X = Number of space stage cross points

N_B = Number of bits of memory.

The N_B not only includes the time stage memory arrays, but also the control memory (store) of the time stage and space stage. Thus,

$$N_B = N_{BX} + N_{BT} \quad ...(5.35)$$

where N_{BX} = Number of memory bits for the space stage control store

= N × (Number of control words) (number of bits per control word)

N_{BT} = Number of memory bits in the time stage equal to sum of time slot interchange and the control store bits.

= N × number of channels × number of bits per channel + N × number of control words × number of bits per control world.

Example 5.3. If N = 80, N_{BX} = 13, 440 and N_{BT} = 24,960 for a typical TS switch, calculate the implementation complexity.

$$IC = N_X + \frac{N_{BX} + N_{BT}}{100} = 80 \times 80 + \frac{13440 + 24960}{100}$$

IC = 6784 equivalent cross points.

As the number of cross points in space array is equal to 6400, the total cost is dominated by the space stage.

5.7.3. STS and TST Switching

The TS structure is of blocking nature. Let A and B are the subscribers using different time slot on the same line want to connect to two subscribers C and D using same time slot on different lines. A and B can be moved to the same time slot but during that time slot, the inlet line can be connected to C's line or D's line but not both. This is the significant limitation of the structure. Moreover, time stage switching is generally less expensive than space stage switching as digital memory is much cheaper than digital cross points (AND gates).

The multiple stages overcomes the limitations of the individual switches and cost savings can also be achieved. TST, STS, TSST, TSSSST and TSTSTSTSTSTS are the switching system configurations used in digital switching system. However, the TST structure is the most common.

STS Switching. In STS switching, the time stage is sandwiched between two space arrays. The digital switching system ITS 4/5 of USA (1976) used the STS switching configuration. It handles 3000 trunks and accomodates 1500 Erlangs of traffic. Fig. 5.23 shows the space-time-space (S-T-S) switching network for M incoming and outgoing PCM highways.

Establishing a path through an STS switch requires finding a time switch array with an available units access during the incoming time slot and an available read access during the desired outgoing time slot. The input side space stage as well as the output side space stage is free to utilise any free time switch modules. In the diagram shown in Fig. 5.23, the time slot 2 is connected to the TSM 2 where the time slot alloted is 16 and passed to the (M – 1)th line of output space array. Thus the path is provided. This structure is of non-blocking nature.

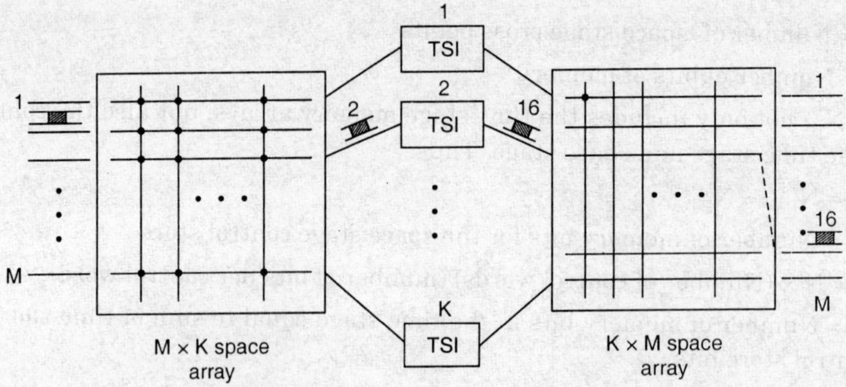

Fig. 5.23. STS switching structure.

Blocking probability. The STS switch is identical to the probability graph of three stage space switches (Fig. 5.12). Similar to that, the blocking probability of an STS switch from is

$$B = \left[1 - \left(1 - \frac{p}{\beta}\right)^2\right]^K \quad \text{...(5.36)}$$

where p = probability that a link is busy

$\beta = \frac{K}{N}$ = is the factor by which the percentage of links that are busy is reduced. ($\beta < 1$)

K = number of center stage TSM.

Implementation capacity (IC). While calculating IC, the total number of two space stage cross points, total number of two space stage control bits, number of time stage memory bits and number of time stage control bits are to be considered. Thus,

$$IC = 2KN + \frac{2KC \log_2 N + KC(8) + KC \log_2 C}{100} \quad \text{...(5.37)}$$

where K = The minimum number of centre stage TSM to provide desired grade of service, calculated from

C = number of channel.

TST Switching. In TST switching the space stage is sandwiched between two time stage switches. Of all the multistage switching, TST is a popular one. Popular digital switching systems using TST are tabulated in table 5.2.

Some important features of TST switches are :

(*i*) **Low blocking probability.** An incoming channel time slot may be connected to an outgoing channel time slot using any possible space array time slot. Thus there are many alternative paths between two subscribers. This concept reduces the blocking probability of a three stage combination switch.

(*ii*) **Stage independancy.** The space stage operates in a time-divided fashion, independently of the external TDM links. The number of space stage time slots L does not coincide with the number of external TDM time slots T.

Table 5.2. Digital switching systems using TST and its characteristics.

Type	Characteristics	
	Max. no. of trunks	**Traffic (Erlangs)**
E 10 B (France, 1970)	3600	1600
AXE 10 (Swedon 1978)	65000	30000
EWSD (Germany 1980)	60000	30000
GTD-SEAX (USA, 1982)	49000	36000
C-DOT MAX-XL (India)	40000	47000

(*iii*) **Implementation advantage.** The factors to be considered for switching design and implementation are traffic loads, modularity, testability, expandability and simple control requirements. For large switches with heavy traffic loads, the TST have good implementation advantage.

(*iv*) **More cost effective.** If the input channel loading is high, the time expansion of TST and space expansion of STS are required. Time expansion of TST can be achieved at less cost than space expansion of STS.

In comparison with STS, the TST have certain limitations. For small switches, the STS architectures are less complex to implement than TST. The control requirements of STS is simpler than TST.

The principle of operation of TST switching is shown in Fig. 5.24. In figure, two flows of time slots, one for each direction are connected together.

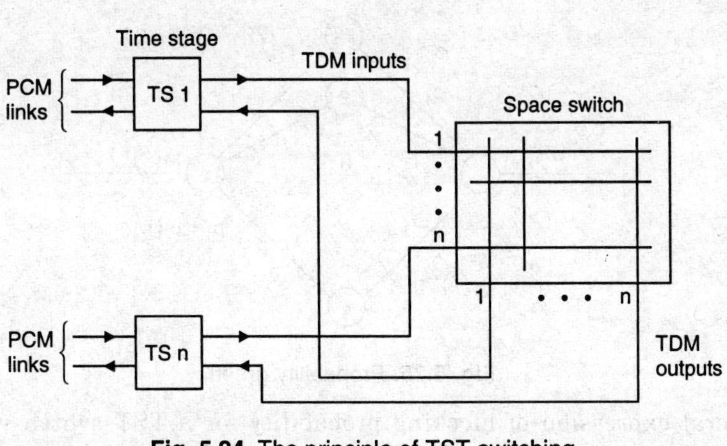

Fig. 5.24. The principle of TST switching.

The functional block diagram which explains the transfer of signals from inlet to outlet is shown in Fig. 5.25. The information arriving at the incoming link of TDM channel is delayed in the inlet times stage until an appropriate path through the space stage is available. Then the information is transferred through the space stage to the appropriate outlet time stage. Here the information is held until the desired outgoing time slot occurs. Any space stage time slot can be used to establish a connection. The space stage operates in a time divided fashion, independently of the external TDM links. There are many alternative paths between a prescribed input and output unlike a two stage network which has only one fixed path.

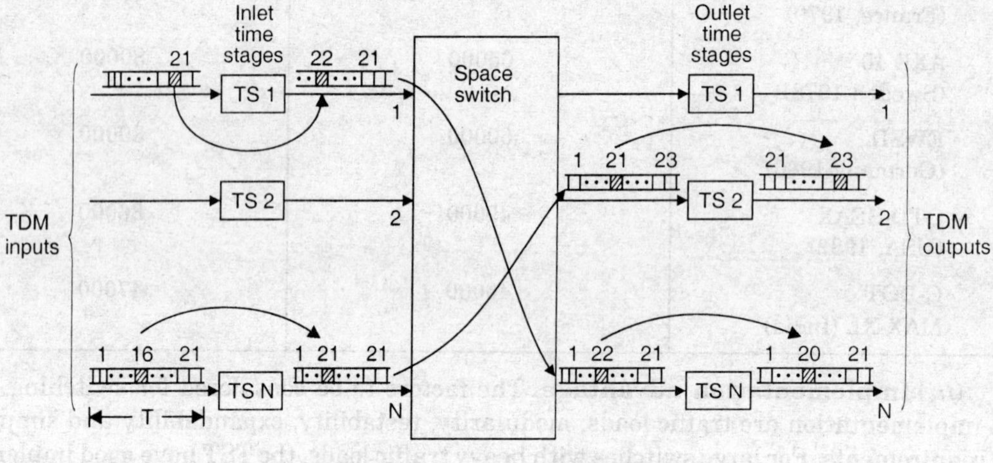

Fig. 5.25. TST switching structure.

Blocking probability. The blocking probability is minimised if the number of space stage time slots L is made to be large. By direct analogy of three stage space switches, the TST switch is strictly non-blocking if

$$L = 2T - 1 \quad \text{...(5.38)}$$

where T = number of time slot of time switch.

L = number of space slot of space switch.

The probability graph of TST switch with non-blocking stage is shown in Fig. 5.26.

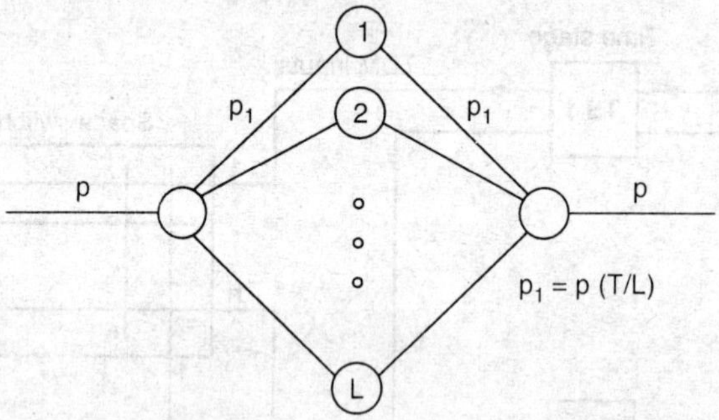

Fig. 5.26. Probability graph.

The general expression of blocking probability for a TST switch with non-blocking individual stage is

$$B = [1 - (1 - pT/L)^n]^L \quad \text{...(5.39)}$$

For 3 stage $\quad B = [1 - (1 - \rho(T/2))^2]^2$

Implementation complexity. The implementation complexity (IC) of a TST switch can be derived as

$$IC = N^2 + \frac{NL \log_2 N + 2NT(8) + 2NL \log_2 T}{100} \quad ...(5.40)$$

Example 5.4. *Determine the implementation complexity of 2048 channel TST switch with 16 TDM links and 128 channels. Let the time slot of space switch is 25.*

Sol. Given N = 16

T = 128

L = 25

$$IC = 16^2 + \frac{16 \times 25 \times \log_2 16 + 2 \times 16 \times 128 \times 8 + 2 \times 16 \times 25 \times \log_2 128}{100}$$

IC = 656 cross points.

ACRONYMS

B-AMI	—	Bipolar alternative mark in version
CM	—	Control memory
DSL	—	Digital subscriber lines
HDB-3	—	High density bipolar-3 zeros
IC	—	Implementation complexity.
KBPS	—	Kilo bits per second
MAR	—	Memory access register
MDR	—	Memory data register
NRZ-L	—	Non return to zero-level
NRZI	—	Non return to zero inverted
PAM	—	Pulse amplitude modulation
PCM	—	Pulse code modulation
SM	—	Speech memory
SPC	—	Stored program control
TDM	—	Time division multiplexing
TSI	—	Time slot interchange

RELATED WEBSITES

htttp://www.iec.org/online/tutorials/new_info//topic09.html

http://www.drewfoster.com/tutorials.html

http://www.tpvb.com/neets/tm/42-2htm

http://www.gisdevelopment.net

http://www.fiber-optics.info/article/analog-v-digital.htm

http://www.its.bldrdoc.gov/fs-1037/dir-038/_557/.htm

http://www.wi.com/pdf/technotes.

REVIEW QUESTIONS

1. Describe the evolution of digital switching.
2. Explain the process of digitization with neat diagram.
3. List the advantages of digital transmission.
4. List the disadvantages of digital transmission.
5. What are the different modes of digital transmission.
6. Explain the Asynchronous transmission with necessary diagrams.
7. List out the disadvantage of Asychronous transmission.
8. What is synchronous transmission ?
9. Explain a three stage switching (general) with neat diagram.
10. What is probability graph ?
11. Describe various blocking probability evaluation techniques.
12. Explain the principle of time division switching.
13. Distinguish analog time division switching and digital time division switching.
14. Write short notes on combinational switching.
15. With neat diagrams explain time switch and space switch.
16. Explain the TS switch with neat diagram.
17. What is internal complexity ?
18. What are the features of TST ?
19. List the practical system which uses TST, STS and TS.

6

Computer Controlled Switching Systems

6

Computer Controlled Switching Systems

6.1. INTRODUCTION

Most digital switching systems have a quasidistributed hardware architecture, since they maintain control of the switching functions through an intermediate processors. All digital switching systems employ multiprocessor subsystem for the best understanding of communication and control process. The architecture of a working digital switching system is very complex with many subsystems. All the functions cannot be dealt in a text book. Thus in the hardware configuration section, important subsystems required for communication and switching system control alone are explained.

All present day digital switching system includes minimum softwares which are necessary for implementation of call processing for all the levels of control structure. The software functionality of digital switching and other functions and the software components that are usually necessary for modern switching system are described in the switching system software organization section.

In modern digital switching systems, many call processing functions are performed by using interface controllers. Some of the call processing are call identification, call routing, path setup between subscribers, digital translation, call status, billing etc, In call processing section, various processing and state transition concepts are discussed. Switching in network environment which describes the flexibility of the switching system for new services required also discussed. Various popular digital switching systems are explained in sections 6.6, 6.7 and 6.8.

6.2. CALL PROCESSING

In this section, the basic steps involved in processing a call is discussed. Most digital system follow a similar scheme. For any switching system design, the range of signals that has to be interchanged between a terminal and system is considered. These signals described in signal exchange diagram. The sequence of operation between subscribers and system are shown in state transition diagram (s.t.d.).

6.2.1. Basic Steps to Process a Call

The sequence of processing between subscribers are described below :

1. **Idle state.** At this state, the subscriber handset is in 'on-hook' condition. The exchange is ready to detect the call request from the subscriber.

2. **Call request identification.** The exchange identifies a line requiring for a service. When the handset is lifted, current flows in the line called seize signal indicates the call request.

3. **Providing dial tone.** Once the seize signal is received, an exchange sends a dial tone to the calling subscriber to dial the numbers.

4. **Address analysis.** Once the first digit received, the exchange removes the dialtone and collect all numbers. Then the address is analysed for the validity of the number, local, STD or ISD etc. If the number is invalid, a recorded message may be sent to the calling subscriber and terminates call request.

5. **Called line identification.** The exchange determines the required outgoing line termination from the address that it has received.

6. **Status of called subscriber.** The called line may be busy or free or unavailable or even out of service. In the case of PBX, where the customer have a group of lines, the exchange tests each termination until either it finds a free one or all one found busy. For busy, number unobtainable or the handset off hook, a status signal or call progress signal is sent to the calling subscribers for line termination. Now the exchange resumes idle state.

7. **Ringing.** Once, the exchange finds the called subscriber is free, power ringing is provided to the called subscriber and audible ringing to the calling subscriber.

8. **Path setup.** When the called subscriber lifts his handset, the line is looped and ringing is removed. Once the conversation started, the exchange completes the connections between the subscribers.

9. **Supervision.** The exchange supervises the connection to detect the end of the call for charging.

10. **Clear signal.** Once the need for connection is over, either customer may replace his handset. It causes the line current seize and provides a clear signal to exchange. If the calling subscriber replaces his phone set, the clear signal sent to the exchange is called clear forward signal. If called subscriber do first, the clear signal is called clear backward signal.

6.2.2. Signal Exchange Diagram

There are two types of diagrams used to represent the sequence of events between the subscriber and exchanges. They are signal exchange diagram and state transition diagram. Both diagrams

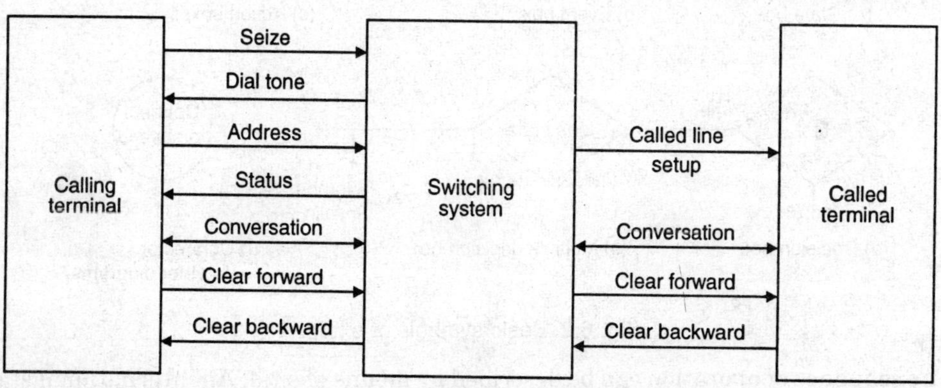

Fig. 6.1. Signal exchange diagram.

can be used to specify the behaviour of different control units in switching centre. For the local call, the steps involved in processing a call is shown in Fig. 6.1.

Normally, once the conversation is over, the exchange will be at idle state. But in general, there are two types of difficulties arises.

1. **Called subscriber held (CSH).** This condition arises when the called subscriber replaces the hand set but the caller does not. In this case, the caller does not originate a call or receiver a call.

2. **Permanent loop condition (PL).** This condition occurs when the caller replaces the phone but the called subscriber does not. Now, a loop present between called and exchange and it results in busy tone to a another call to the same called subscriber. In strowger system, this condition is called permanent glow condition.

In electromechanical system, the above conditions are removed by manual disconnection. In modern ESS systems, a time out process is used.

If the call setup between two subscribers are made through many exchanges and trunks, the originating exchange where calling subscriber is connected sends the seize and then address to the terminating exchange where the called subscriber is connected. Remaining signaling are similar to the local call, but through the originating and terminating exchanges. In electromechanical system, the signalling between exchanges are sent through same inter exchange circuits referred as channel associated signaling. In SPC controlled exchanges, inter-exchange signals are generated at orginating exchange, but processed at terminating exchange. The signals are transferred over high speed data like instead of speech connections are referred as common channel signaling.

6.2.3. State Transition Diagram

The state transition diagram (s.t.d.) specifies the response of a control unit to any sequence of events. s.t.d. is a powerful design tool. It helps the designer to consider all possibilities of occur^nce of events. Fig. 6.2 shows the basic symbols used in a state transition diagram.

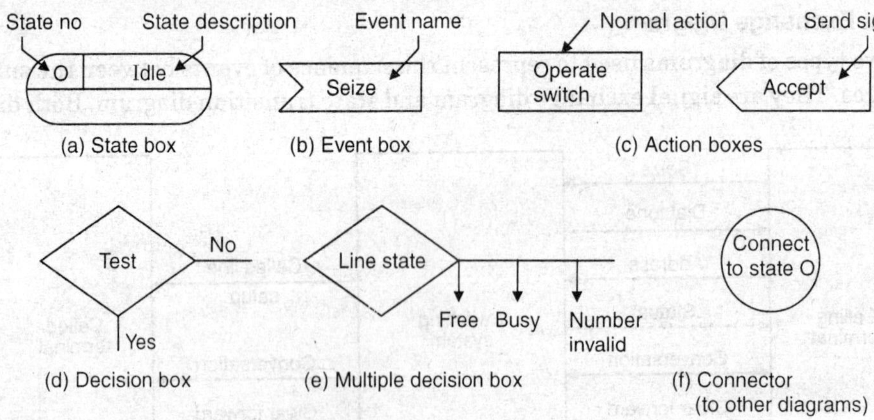

Fig. 6.2. Basic symbols of s.t.d.

The sequence of operation can be described by means of s.t.d. An international standard for such diagram has been produced by the CCITT. It is known as specification and description language (SDL).

The basic symbols are defined as follows :

State boxes. The state boxes are labelled with state number and state description. If necessary, additional information can also be included. The combination of the present state and a new event defines a task and performing this results in next state. Sometimes more than one state occurs, the choise depending on external information.

Event boxes. The intended arrow of the symbol indicate whether the event corresponds to the reciept of forward or backward signal. The forward signal and backward signal refers to the flow of signal from calling to called and called to calling subscriber through exchange respectively.

Action boxes. The rectangular box represents the action taken on the event. The portruding arrow indicates whether the signal is sent forward or backward.

Decision boxes. The diamond shaped box is used for the cases where two divisions are possible. For multiple decisions, another symbol shown in Fig. 6.2 (*e*) is used.

Connectors. This symbols are used to connect one flow chart to another diagram.

Fig. 6.3 shows the s.t.d. diagram for a typical local call. Let the calling subscriber is A and the called subscriber is B.

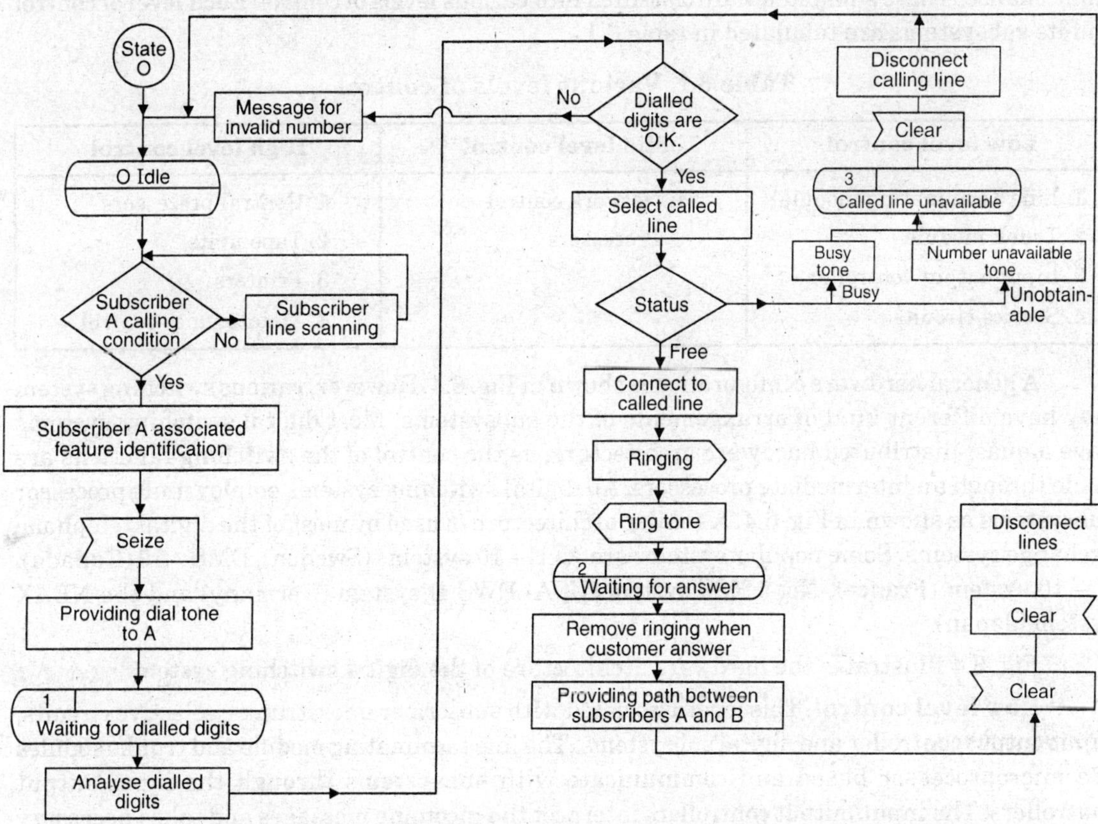

Fig. 6.3. State transition diagram for a local call.

6.3. HARDWARE CONFIGURATION

The computer controlled switching is in general referred as electronic switching system (ESS). ESS offers the greatest potential for both voice and data communications. A ESS consists of 1. computer 2. Memory or storage 3. Programming capability 4. An extremely rapid switching component. A computer based common control switching equipment implies two distinct type of units. They are 1. Control unit 2. Switching network.

The common control receives, stores and interprets dial pulses and then selects an available path through the switching hardware to complete connection. Efficient high speed common control equipment can complete many calling connections during the time of a average phone call. Thus it saves a lot of time and money. The switching network can be used to connect many lines by one common group of control devices referred as control unit. Thus the control unit is the brain of a switching system, A control unit completes its function in a small fraction of a second for a single call.

The hardware of digital switching system are broadly divided by their functions into many subsystems. The functions performed by the subsystem includes line and trunk access, line scanning, message interpretation, switching communications, pathsetup between subscribers, line supervision, line termination, billing providing advanced features and system maintenance. These subsystems are classified into various levels of control. Each level of control and its subsystems are tabulated in table 6.1.

Table 6.1. Various levels of control

Low level control	Mid level control	High level control
1. Line Terminating module 2. Trunk module 3. Input/output controller 4. Service circuits	1. Network control Processors	1. Central processors 2. Tape units 3. Printers 4. Maintenance control

A general hardware configuration is shown in Fig. 6.4. However, various switching system may have different kind of arrangements of the subsystems. Most digital switching systems have a quasi-distributed hardware architecture, as the control of the switching functions are made through an intermediate processors. All digital switching systems employ multiprocessor subsystems as shown in Fig. 6.4. A similar architecture is used by most of the digital telephone exchange systems. Some popular systems are AXE – 10 systems (Sweden), DMS – 10 (Canada), E – 10 system (France), No. 5 ESS system (USA) EWS D system (Germany) and the NEAX system (Japan).

Fig. 6.4 illustrates the hardware architecture of the digital switching system.

Low level control. This level associated with subscriber lines, trunks, selective circuits, Input/output controller and digital subsystems. The line terminating module and trunk modules are microprocessor based and communicate with subsystems through the input/output controllers. The input/output controllers interpret the incoming messages and takes necessary actions and communicate to the network control processors. All subscriber lines connected to digital switching system through the main distributing frame (MDF) are continuously scanned to detect the state of the subscriber.

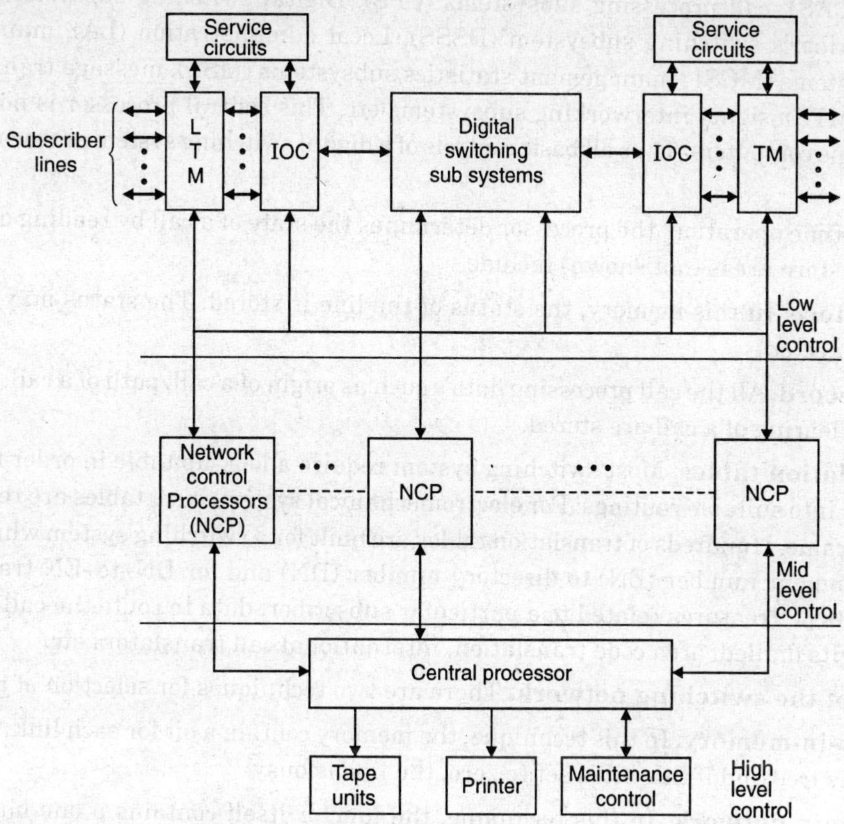

LTM : Line terminating modules, TM : Trunk modules,
IOC : Input/Output controller, NCP : Network control processors.

Fig. 6.4. Hardware architecture of digital switching system.

When the customer lifts his handset, the line scanning program detects this state and reports to the input/output controller. The IOC is the primary peripheral controller and it controls all peripherals associated with call or trunk processing.

At this level, all the requests of incoming and outgoing trunks are handled. Any advanced features to be incorporated in a digital switching system also handled at this level using IOC.

Mid level control. This level is associated with network control processors and associated circuits. The IOC is controlled by the network control processors (NCP). Many NCP's are used depends on the size of the digital switching system. A dedicated bus system is usually required for the processors to communicate with one another. Specific messaging protocols are used to communicate between processors. For messaging between the peripherals and external systems, many digital switching systems utilize standard protocols such as signaling system 7 (SS7) ; X.25 and X.75. Thus this is the most important level of control any digital switching system. Distributed processing are performed at this level.

High level control. This level associated with central processor which organizes the entire network control subprocessors. In includes many subsystems like call accounting subsystems (CAS), call processing subsystems (CPS). Digital switching subsystems (DSS). Digital subscriber's switching subsystem (DSSS), Local administration (LA), maintenance control subsystems (MCS) ; management statistics subsystems (MSS), message transmission subsystems (MTS), signal interworking subsystems etc. This central processor is normally a main frame type computers. Thus all basic controls of a digital switching system are incorporated at this level.

In real time operation, the processor determines the state of a call by reading data from memory. The store areas (not shown) include,

Line store. In this memory, the status of the line is stored. The status may be busy, free or disconnected.

Call record. All the call processing data's such as origin of a call, path of a call, duration of a call and clearing of a call are stored.

Translation tables. Most switching system require a look-up table in order to decode routing digits into suitable routings. For electromechanical system, such tables are realized by distribution frame. Hundreds of translation tables are built for a switching system which stores data for equipment number (EN) to directory number (DN) and for DN–to–EN translation. Also it consists of, freatures related to a particular subscriber, data to route the call based on the first 3 digits dialled, area code translation, international call translators etc.

Map of the switching network. There are two techniques for selection of junctors.

1. **Map-in-memory.** In this technique, the memory contain a bit for each link. If it is set to 1 the link is free and if this bit is set to zero, the line is busy.

2. **Map-in-network.** In this technique, the junctor itself contains a one bit memory element, which is read by the path setup program to check whether it is free.

The map-in-network consumes more time, but more advantages when several processors controlling the system.

6.4. SWITCHING SYSTEM SOFTWARE ORGANIZATION

In last section, three levels of controls of hardware architectures were discussed for a general digital switching system. For effective processing of a call, to perform various functions of subsystems and to interface with the other subsystems, softwares plays a vital role. The software programs enables any digital switching system input data, to give outputs in a fraction of seconds, concurrent processing of many calls in real time and performs many features other than simple pathset between subscribers for conversation.

In this section, the need for software, the software classification, basic software architecture, the involvement of software in various levels of hardware architecture, interfacing between subsystems through software and softwares presently used in various digital switching system are described.

6.4.1. Need for Software

Other than call processing, any exchange is to serve the subscriber various facilities and many administrative tasks. Fig. 6.5 shows various activities of a switching system. To carry out these activities efficiently and effectively, the use of software is unavoidable.

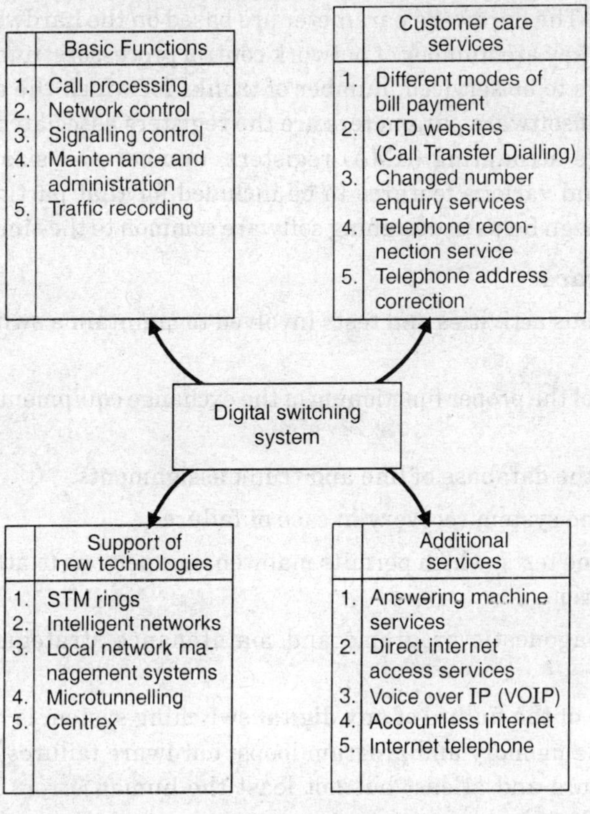

Fig. 6.5. Various activities of digital switching system.

To perform the above tasks, a large amount of software is required. However, the software for basic functions are must and remaining services are optional and requires software depends on the location of switching systems. Approximately 70% of the total software is used to perform basic functions. Only 0.1% of the total processing time is used by the 30% of the remaining service oriented software packages.

6.4.2. Software Classification and Interfacing

Classification. At various levels of hardware architecture, the softwares are used. Thus, many digital switching systems employ some system level software. Basic software systems are classfied as :

1. Maintenance software
2. Call processing software
3. Database/Administration software
4. Feature software.

Above software packages are divided into program modules. Each module dealing with specific task. Several molules are grouped together to form functional units. Various factors are associated with the development of software product. These factors include the requirements of the business, the location of telephone exchanges, customer needs, internal requirements, and parameterised design. The parameterised design includes hardware parameter and software parameters. The hardware parameter are based on the hardwares used in the central office or exchanges. They are number of network control processors, number of line controllers, number of subscribers to be serviced, number of trunks for which the exchange is engineered etc. Some examples of software parameters are the registers associated with number and size of automatic message accounting (AMA) registers, number and size of buffers for various telephony function and various features to be included for that particular exchanges. Thus, the parameterised design helps in designing software common to the similar types of exchanges.

Maintenance software

There are various activities and tests involved to maintain a switching system. Some of them are :

1. Supervision of the proper functioning of the exchange equipment, trunks and subscriber lines.

2. Monitoring the database of line and trunk assignments.

3. Efforts for the system recovery in case of failure.

4. Automatic line tests, which permits maintenance persons to attend several exchanges from one control location.

5. Effective diagonestic programs and maintenance strategies used to reduce the maintenance cost.

The root cause of the failure of any digital switching system is related to the software bugs which affects the memory and program loops, hardware failures, failure to identify the exact problem of failure and atleast but not least the human error. Thus, the first step in software build is to select the appropriate program modules which is suitable for the switching system. The points to be considered are types of lines, location of switching system, signalling systems, availability of skilled person or the level of diagonisation.

Preventive maintenance programs are activated during the normal traffic. If a fault occurs, the OS activates the maintenance program to recover the system. Effective preventive and maintenance programs and strategies helps in proper maintenance of digital switching system with reduced maintenance cost.

Call processing software. The call processing functions are controlled by a central processor. Other functions carried out by the central processor are maintenance and administration, signalling, network control. Thus, the call processing programs are usually responsible for call processing and to interface with the translation data, office data, automatic message acounting and maintenance programs. The translation data is the type of data generated by telephone companies related to subscriber. The office data is related to a particular digital switch. The call processing programs can be derived from state-transition diagrams, in specification and description language (SDL). The SDL description in text form, is machine read and stored in memory in the form of data structures and linked lists and translation tables. An interpreter programs is written to access the lists and tables and to process the call by interpreting the data within them. Fig. 6.6 shows three levels of call processing program. But it varies depends on the digital switching system.

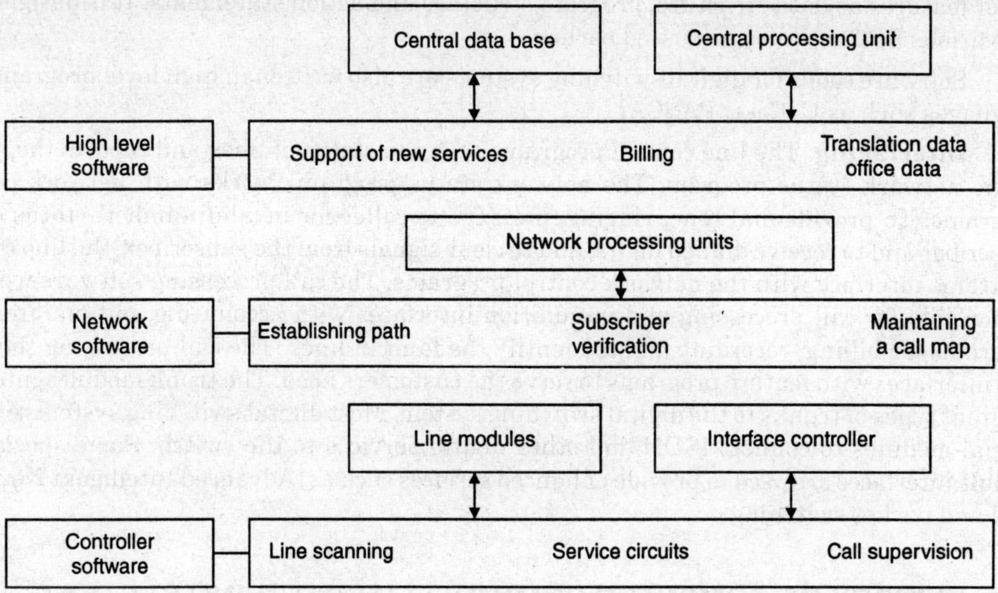

Fig. 6.6. Call processing software levels.

Data base/Administration software

For administration and data base management, large amount of software required. But these tasks are performed infrequently, it uses less than 5% of the total processing time. The administration tasks includes

1. Alarm processing
2. Traffic recording
3. Change of numbers or area codes corresponding to the change in subscriber rate and government policy.
4. Changing routing and routing codes. This decisions made on the traffic intensity of a particular exchange.
5. Generation of exchange management statistics.

Most digital switching system employ a data base system to :

1. Record office information
2. Billing information
3. Software and hardware parameters
4. System recovery parameters
5. System diagnostics.

Feature Software. In section 6.4, various feacture services were shown in Fig. 6.5. Most of the present day digital switching systems uses all packages.

Switching softwares. Softwares for digital switching systems are written in high level languages. Early electronic switching systems used assembly language programmes. In 1980, Pleanary Assembly, CCITT approved the definition of a high level language as Recommendations Z–200. This language is known as CCITT high level language (CHILL). It has three major features as data structure, program structure and action statements. It is designed for the various SPC modules discussed earlier.

Software codes for digital switching systems are also written in high level programming languages such as C, C ++, PASCAL.

Interfacing. The line control programs scan the status of lines and reports the status to the network status program. The network status programs works with network control programs. To provide dial tone, ringing, message to caller for invalid number, status of the subscriber and to receive dialled digits, and to clear signals from the subscriber, the line control programs interface with the network control programs. The call processing software which is responsible for call processing and in addition interfaces with accounting and maintenance programs for billing, recording and to identify the fault in lines. The call processing software also interfaces with feature programs to serve the customers need. The trunk modules interface different types of trunks to the digital switching system. Most digital switching systems employ special modules to connect ISDN and other digital services to the switch. Some specialized module interfaces are used to provide enhanced services such as (Advanced Intelligent Network) **AIN** and packet switching.

6.5. HISTORY OF COMPUTER CONTROLLED SWITCHING SYSTEMS

An electronic switching system went into use at the Brown engineering company at cocoa beach, florida in November 1963. It was a small installation, serving only a single company. On May 1965, first commercial electric central office was put into service at succasunna, new jercy. It served only 200 subscribers initially. It had some features of speed, provision for three party conversations and automatic transfer of incoming calls.

In 1966, Western Electric developed a computer controlled analog switch called No. 1 ESS (Electronic Switching System). Then No. 2 ESS was developed in 1967 with some advancement. It was followed by No. 1A ESS in 1976. Parallely, independent switch manufacturers also made computer controlled analog telephone switching systems. Automatic Electric made No. 1 EAX (Electronic Automated Exchange) in early 1970's and developed No. 2 EAX and No. 3 EAX in late 1970's. Western electric developed No. 2 BESS in 1974 for down sized smaller applications which is very similar to No. 1 ESS and No. 1A ESS. Then No. 3 ESS was developed for rural applications upto 4500 lines. The No. 4 ESS which is a toll switch, 4 wire tandem digital switch with 1A ESS processor was developed. All the electronic switching systems made till 1978 are of analog switch technology. However, there are still many No. 1 ESS switches still in use worldwide.

The 100% digital system were being developed after 1978. These switches are smaller and more reliable. It also allowed new technologies to be added faster by using modular techniques (adding systems to the original system without complete redesign/reinstallation). It also uses advanced computer software to include various features.

The first digital system was developed by the company called vidar in 1978. Northern Telecome (formerly Nothern, now Nortel) developed the digital matrix switch (DMS) series of digital switches. In 1979, the DMS–10 was first produced. Later DMS 100 similar to No. 5 ESS was developed. It was considerably less expensive but runs slower. Latter DMS 200, and DMS 300 were developed. In 1982, Wester Electric developed their fully developed digital switch called as No. 5 ESS. Automatic Electric developed the No. 5 EAX (better known as GTD – 5) switch.

After 1990's, many computer controlled switching systems were developed. Some of them are AXE – 10, E – 10 B, E– 12 ; Fates, NEAX, EWSD, System X, ND 10, Interface trunk terminal IT – 12 type etc. Popular digital switching systems used in various countries are tabulated.

Table 6.2. Popular ESS worldwide

Country	Type
USA	No. 1 ESS, No. 4 ESS, No. 5 ESS
	ITT–1210, ITT 1220, ITT 1230, ITI 1240
CANADA	DMS–100
GERMANY	EWSD system
JAPAN	NEAX–61
	FATES
FRANCE	E–10, E–12, E–20, E–25
U.K.	System X
SWEDEN	AXE 10
INDIA	FATES (JAPAN), NEAX (JAPAN)
	ND–10, E–10 B (FRANCE)
	CDOT–Max–XL, CDOT DSS Max
	OCB 283/CSN, CDOT SBM RAX
	CDOT 256 P RAX

In the following sections, some popular digital switching systems are described. Section 6.6 No. 1 A ESS and No. 5 ESS explained. In popular digital switching system section, DMS 100 and EWSD system are explained. The section electronic exchanges in India overviews the Indian Telecom Organizations and Switching Systems in India.

6.6. EARLY ELECTRONIC SWITCHING SYSTEM (ESS)

In this section, No. 1A ESS introduced in 1976 and No. 5 ESS introduced in 1982 are described. The No. 1A ESS was the updated version of No. 1 ESS which is the first electrronic switching system developed by wester Electric and introduced to market in 1966. The switching systems developed till 1979 are analog switches and the switch No. 5 ESS developed in 1982 includes all the features of prior systems and still pupular. Thus 1A ESS and 5 ESS are described in this section.

6.6.1. No. 1 A ESS

The No. 1A ESS was developed in 1976. It replaced No. 1 ESS with advanced technology. The primary difference between No. 1 ESS and No. 1A ESS were speed and capacity. No, 1A ESS is based on the smaller magnetically latched remreed whereas No. 1 ESS uses ferreed relay. These new network are plug compatible with the old networks so that they can be intermixed with the old design.

The original No. 1 ESS design serves maximum of 65000 lines and a maximum calling rate of 25000 calls in the busy hour. It uses maximum of 16 line switch networks. The No. 1A ESS designed with 28 line switch networks and trunk switch networks are 2048 by 2048 size. This serves a maximum of 1,30,000 lines and process 110,000 calls per busy hour.

Hardware Architecture. No. 1 A ESS consists of two subsystems. They are 1. Processor subsystems and 2. Peripheral subsystems. Each subsystems are described below :

Processor subsystem. This subsystem contains duplicated processors with call and program memories. The 1A ESS uses an improved processor which doubles the capacity of the exchange in comparison with No. 1 ESS. The 1A processor is designed as a plug in replacement for existing processors and is the control unit of the No. 1A ESS. The 1A processor is based on highspeed integrated circuit technology and achieves a speed increase by the factor of 4 to 8. Fig. 6.7 shows the processor subsystem of No. 1A ESS.

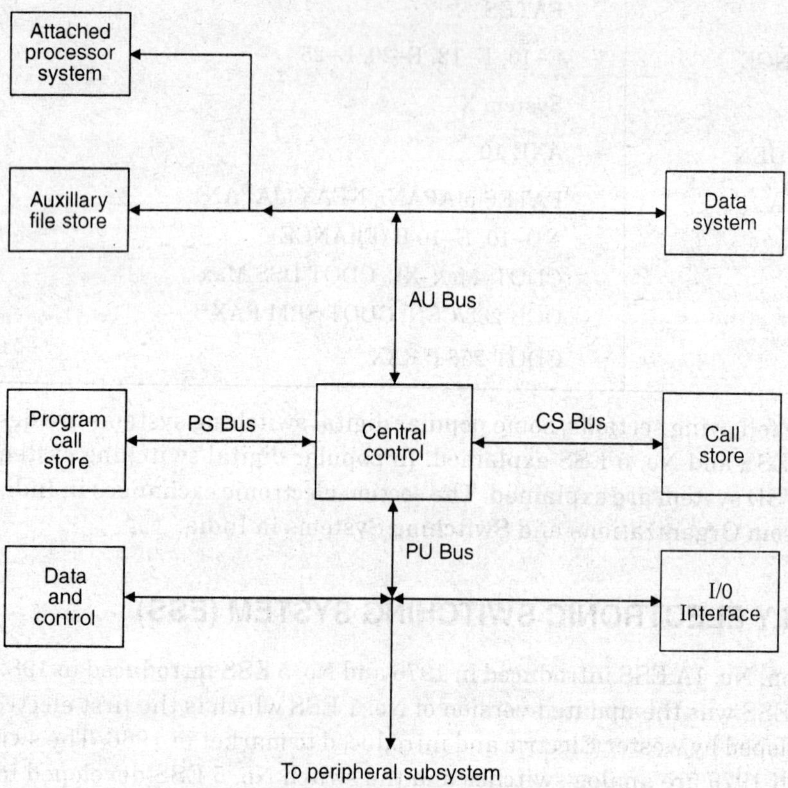

Fig. 6.7. No. 1A ESS processor subsystem.

The 1A processor has an additional set of buses. The 1A processor is most specifically used for the following :

(*i*) to control the 1A ESS switch

(*ii*) to support future switching systems

(*iii*) to accomodate bulk memory systems

(*iv*) to provide real time and continuous control through highly automatic maintenance.

In addition to address bus and data bus, 1A processor uses auxillary unit (AU) bus, Call store (CS) bus, Program store (PS) bus and peripheral unit (PU) bus.

The central control interfaces with the 1A, and performs the processing funfctions of the 1A. It also executes all maintenance routines. The program store uses the high speed semi-conductor memories that stores programs, instructions and system configuration systems. The call store is used for storage of translation data and frequently changed call processing data such as status of trunks and switching network, records of network termination and maintains data related to programmed diagnostic tests. Call store also includes an emergency system recovery programs.

In auxillary file store, copies of program are held on separate discs used for program backup. In auxillary data system, magnetic tape system is used to store and retrieve data such as system reinitialization, memory dumps etc. The attached processor system uses 3B 20D computer, of which one or more are used as slave processor used for multitasking. The I/O interface is used to connect 1A to the terminals for input control messages and to receive status messages.

Peripheral subsystem. The peripheral subsystem containing the switch networks, junctors, senders and receivers. It also includes panel system and features like alarm facility, status and control. The peripheral subsystem is shown in Fig. 6.8.

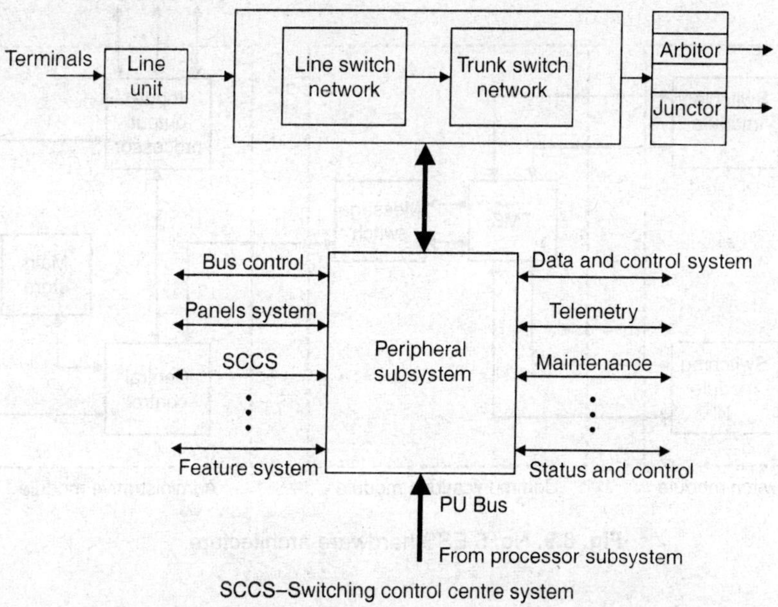

Fig. 6.8. No. 1A ESS peripheral sub systems.

The switch networks are built from the small sized remreed relays. In No. 1A ESS a maximum of 28 time switch network are possible. Trunk switch networks connects 2048 inlets to 2048 outlets. The junctors are designed to provide necessary power and tone feeding conditions and necessary signalling conditions for links and trunks.

Software. The development of new software tools (such as high level programming languages, use structured programmes) has increased the productivity, speed of operation and allows to carry more features.

6.6.2. No. 5 ESS

In 1982, Wester electric developed their fully developed digital switch called No. 5 ESS and put into service. The 5 ESS is a digital SPC switching system which utilizes distributed control, a TST switching network and modular hardware and software design. No. 5 ESS is fully digital switch and did everything of No. 1A ESS and more. Each TSI's of No. 5 ESS have their own processor. This makes the 5 ESS one of the fastest switches.

No. 5 ESS supports POT and centrex services as well as the advanced services such as ISDN, STP, SCP and AIN. Though No. 5 ESS primarily used as a local central office, it can be used as an operator services switch or as low to medium traffic volume tandem.

Hardware architecture. Its architecture can be classified as quasi-distributed since it maintains control of the system via various modules. The major components are

1. Administrative module
2. Communications module
3. Switching module.

The 5 ESS switching system uses a modular software structure and is UNIX based switch. Fig. 6.9 shows the 5 ESS system Architecture.

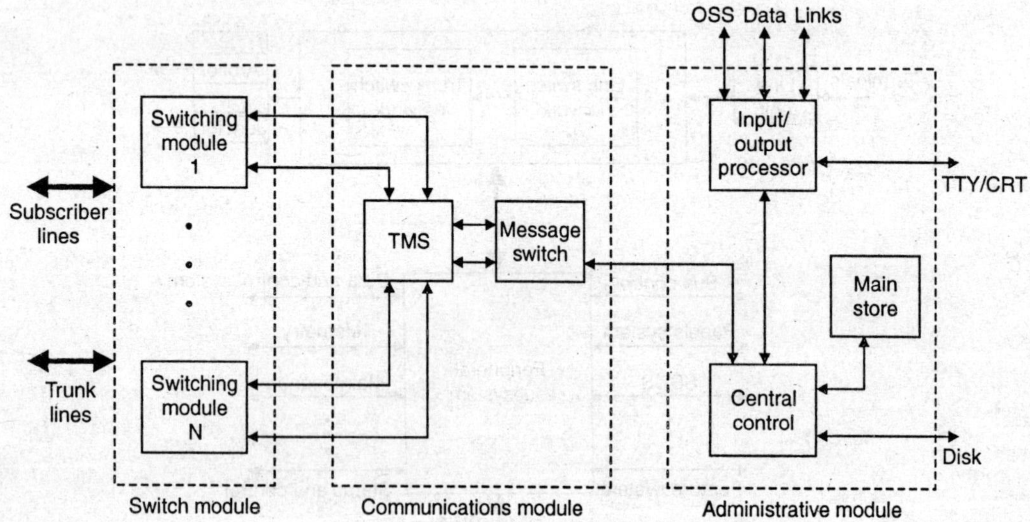

Fig. 6.9. No. 5 ESS hardware architecture.

Administrative module (AM) : The function of the AM is to assist in call processing functions, system maintenance, software recovery, error detection and system initialization. The AM is based on a duplexed AT & T main frame computer. It contains three major parts namely, central control, Input/output processor, mainstore and automatic message accounting (AMA) units.

Central control. It uses two 3B 20S processors. Of the two, one will be in active mode and other will be in stand by mode. It interfaces with message switch of communication module and Input/Output processor. By interfacing, it performs to control video displays, printer's tape units and monitoring master control center.

Main store. It stores programs and data. The costomer's details such as telephone number, related features, billing option (like electronic clearance) etc are stored here.

Input/Output processors. It provides TTY and data link interfaces to the 3B20S processors, 5 ESS network, Master control center (MCC), and various operational support systems (OSS).

Automatic Message Accounting (AMA) units : It uses data links to transport calling information to central revenue accounting office and AMA type.

Other functions of AM include (*i*) routing of calls to a particular switching module and traking their availability (*ii*) Controls and allocates time slots for the time multiplexed switch (TMS) and (*iii*) Supports hard disk access and maintenance system software.

Communication Module (CM) : CM provides communication between the AM and the switching module (SM). There are two basic components of CM. They are

(*i*) **Message switch.** It provides control message transfer between the 3B20S processor and interface modules. It contains clock for synchronizing the network. (referred as network control and timing links (NCT)). Through NCT's, the message switch performs packet switching functions between CM and its SM.

(*ii*) **Time multiplexed switch (TMS).** It performs space division switching between switching module's. It provides permanent time slot paths between each SM and the message switch for control messages between the processor and SM's.

Switching modules (SM). It provides call processing and being the first stage of switching. The common components of the switch module are shown in Fig. 6.10. Port control for lines and trunk allocations, setting up and releasing calls, scanning are performed by SM.

Interface Units. The switching module's are equipped with four types of interface units.

(*i*) **Line units (LU).** It contains a solid state two stage analog concentrator that provides access to 64 output channels. It is used for terminating analog lines. The concentrator can be fully equipped to provide 8 : 1 concentrators or can be fully equipped to provide 6 : 1 or 4 : 1 concentration.

(*ii*) **Trunk units (TU).** It is used for terminating analog trunks. Each TU requires 64 time slots. Depending on the applications of lines and trunks, SM's can be configured differently.

(*iii*) **Data line Trunk Unit (DLTU) and Data Control Line Unit (DCLU).** It is used for terminating digital trunks and remote switch module's (RSM). Each fully equipped DLTU and DCLU requires. 256 time slots.

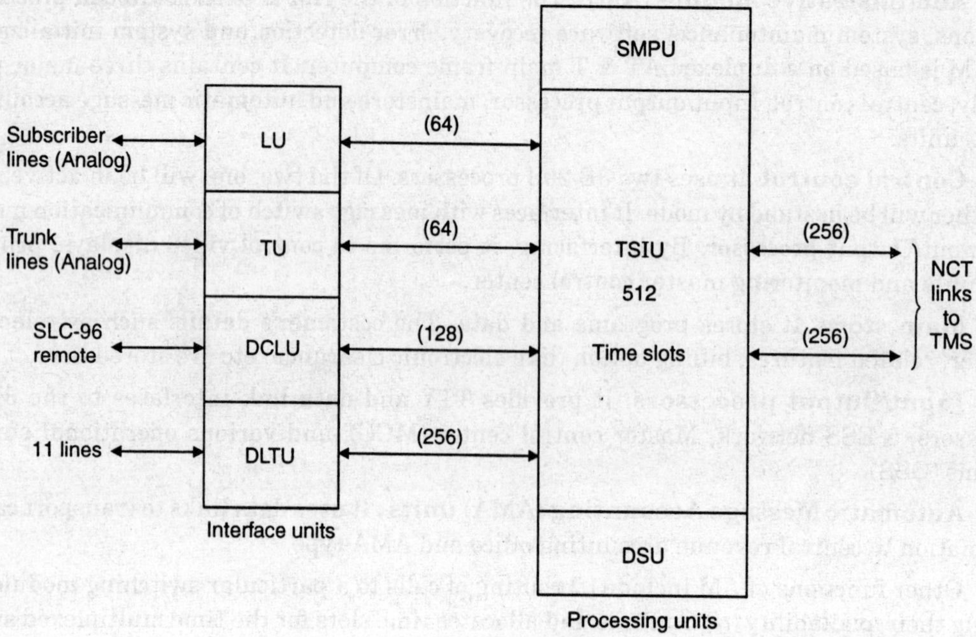

Fig. 6.10. Switching module.

Processing units. It contains three parts namely switch module processor unit (SMPU), Time slot interchange Unit (TSIU) and digital Service Unit (DSU).

(*i*) **SMPU.** It contain microprocessors which perform many of the call processing functions for trunks and links terminated on the SM.

(*ii*) **TSIU.** The function of the TSIU is to provide time division switching within each SM. The NCT linkage between the SM and CM is time slotted by the TSIU via the TMS. It is of 512 time slot capacity. It switches time slots from interface units to one of the NCT links (for intermodule calls). It switches time slots from one interface unit to another within the SM (for intramodule calls).

(*iii*) **Digital Service Unit (DSU).** Local DSU provides high usage service circuits such as tone decoders and generators for lines and trunks terminated on the SM. Global DSU provides low usage service circuits such as 3-port conference circuits and the transmission test facility, for all lines and trunks in the office.

Software Architecture. The 5 ESS is a UNIX based switch. The operating system provides process management, interprocess communication, timing services and task scheduling. The operating system supports the AM processors and SM processors. Most of the software that performs administrative functions at system level, some frequently required features related software, TMS software are resides in AM. The SM contains all programs necessary for the control of switching periphery. SM also keeps the software related to the status information of lines, trunks and terminals associated with other subsystems.

6.7. POPULAR DIGITAL SWITCHING SYSTEMS

In this section, two digital switching systems which are popular in India and around the world are described.

6.7.1. DMS-100 System

The digital multiplexing system (DMS) versions are popular in Canada. Since its introduction, there are six version of DMS entered into market. A brief descriptions of the DMS switches are given below :

Different versions of DMS systems

DMS-10. It is a digital switch and cost effective. It is in service at suburban and rural areas. It allows access to local and long distance service. It can handle upto 12000 subscribers. It is the smallest of the DMS family.

DMS-100. It is a class 5 local office. It is designed to deliver services over subscribers lines and trunks. It provides various services.

DMS-200. The DMS-200 switch has toll capabilities. It is used for toll center applications. It provides TOPS (telephone operator position system) which is the world's premier operator service, from Northern Telecom. The DMS 100/200 system combines the toll capabilities and applications of DMS-200 and DMS-100 S public networking.

DMS-250. This is the long distance tandem switch that connects long distance calls. It is used by the interexchange carriers. It is powerful.

DMS-300. This is the international exchange, which gates calls internationally. It provides the most advanced range of international services. This system can interface with any system. This is also known as the **International Gateway system.**

The most common version of DMS system is DMS-100. This system is described below.

DMS-100. It is a class 5 local office with the ability to handle 1000 to 100,000 lines. It was first put into service in 1979. This system supports many features. Some of the features and benefits are :

1. It supports hardware/software, located at the telephone company's premises.

2. It services all sizes of organizations, from small businesses using only a few lines, to the most complex network systems.

3. This system monitors and controls its own operations automatically. It diagnoses faults and problems and does its own repairs.

4. **Automatic Route Selection.** This system automatically routes long distance calls over the most economic route available.

5. **Station Message Detail Recording.** It provides a detailed record of long distance changes, including the originating number, time and duration, authorization code, etc.

6. **Direct Inward System Access (DISA).** This system enables company personel to use cost saving company facilities for long distance calling, even from outside the company.

7. **Call Dial Rerouting (CDR).** This facility provides a network management, controls and restrict long distance calling.

8. This system is fully modular. It meets present needs and new features.

Its architecture can be classified as quasi-distributed since it maintains control of the system via a central processor. Its switching fabric is time multiplexed and is classified as a TST network. The hardware architecture and software architecture of the DMS-100 system is described below.

Hardware Architecture. The hardware architecture of the DMS switch consists of the following main components :

1. Processor Modules
2. Switching Module
3. Maintenance Administration Position (MAP)
4. Peripheral Module.

Fig. 6.11 shows the general hardware architecture of the DMS-100 system.

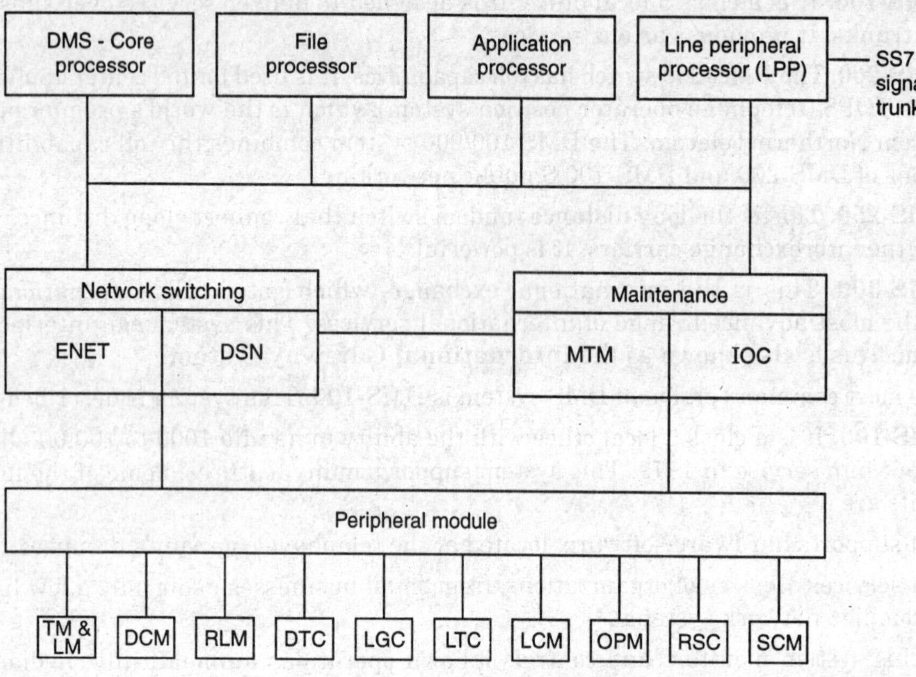

Fig. 6.11. Hardware architecture of DMS-100.

Processor Modules. The main components of processor modules are DMS-Core Processor, file processor, Application Processor and LPP. The DMS-core processor supports the call processing, performs control function, down loading DMS super-node software and supports other peripheral controllers. The core processors is a set of central processing units. The file processor provides access to large data files for applications such as SCP and AMA. The application processors are intended for special applications. LPP supports CCS7 and advanced data applications. It is used to support SSP, STP, SCP, ISDN and other services.

Switching Module. The DMS-100 system is equipped with two types of switching matrix. They are dual shelf network (DSN) or enhanced network (ENET). With this arrangement, it takes up less than 16% of the space for the same number of step by step system 20% of cross bar system. DMS offers remote switching with a bunch of remote modules. The use of remote modules extends the advantage like plant floor savings, cost reduction, less maintenance etc. The use of DSN and ENET supports narrow band and wide band services.

Maintenance and Administration Position (MAP) : MAP provides the operation management, administration and maintenance. MAP is a integrated multifunction machine interface that switch maintenance, line and trunk network management and service order changes can be carried out. The billing system supports automatic message accounting (AMA). It records AMA data on tape and disk.

Peripheral Module (PM) : Peripheral modules (PM's) are used as interface between digital carrier spans (DS-1), analog trunks, and subscriber lines. The PM's are used for scanning lines for changes of circuit state, doing timing functions used for call processing, creating dial tones, sending, receiving signalling and controlling information to and from the other modules and checking the network. Before 1984, only four types of PM's gave trunk interfaces (TM, LM, DCM and RLM). Latter many modules are added. All those modules are discussed below :

Trunk Module (TM) and Line Module (LM). The TM changes incoming speech into digital format. TM has the ability to handle 30 analog trunks. The PCM information is combined with trunks supervising and control signals and then transmitted to 2.56 Mb/s over speech links to the network. Each trunk can carry 36 ccs. The TM also uses service circuits such as MF receivers, announcement trunks and test circuits.

LM gives an interface for a maximum of 640 analog lines and condenses the voice and signalling into 2, 3, or 4 DS-30, 32 channel speech links. 4 speech links have the ability to handle 3,700 Average Busy Season Busy Hour (ABSBH) CCS per LM.

Remote Line Module (RLM). It is a LM operating in a remote location from the DMS host. The RLM's can be located up to 150 miles from the host office, depending on the transmission facilities.

Digital carrier Module (DCM). Gives digital interface between the DMS switch and the DS-1 digital carrier. The DS-1 signal consists of 24 voice channels. The DCM takes out and puts in signalling and control information on the DS-1 bit stream which then makes them DS-30, 32 channel speech links. The DCM can interface 5 DS-1 lines, 5 × 24 = 120 voice channels, into 4, 32 channel speech links. The DCM can carry a maximum of 36 CCS of traffic on each trunk.

Digital Trunk Controller (DTC). It provides digital trunk inter-connection between DMS-100 digital switching system and other central offices. It has the ability to interface 20, DS-1 lines. Then the DS-1 lines are linked to the network by a maximum of 16 DS-30 speech links, each trunk is able to handle 36 CCS.

Line Group Controller (LGC). It does medium level processing tasks with the ability to use host and remote subscriber line interfaces. The LGC can interface upto 20 DS-30 speech links from the LCMs or up to 20 DS-1 links with the ability to serve RSCs, RLCMs or OPMs.

Line Trunk Controller (LTC). It combines the function of the LGC and DTC. The LTC has the ability to handle the LCM, RSC, RLCM, OPM and digital trunk interfaces. The LTC has the ability to give interfaces to a maximum of 20 outside ports from DS-30 A speech links or DS-1 links to 16 network side DS-30 speech links.

Line Concentrating Module (LCM). LCM when used with the LGC or LTC is just an expanded version of the line module. An LTC provides interface between the subscriber lines and the line group controller. Both of these interfaces provide line concetration. It can serve upto 640 subscriber lines interfaced with two to six DS-30A speech links. Using 6 speech links 5,390 CCS can be handled per LCM.

Outside Plant Module (OPM). It is an outside plant remote unit. The OPM can handle 640 lines over six DS-1 links.

Remote Switching Centre (RSC). It interfaces subscriber lines at a remote location to a DMS-100 host. It consists of (*a*) LCM similar to LCM of host DMS-100 (*b*) Remote cluster controller (RCC) when gives DS-1/LCM interface, local switching inside the remote and local intelligence (*c*) Remote trunking which handles the use of RSC originating or terminating traffic for digital trunking off the RSC. It can give trunking to a CDO co-located with the RSC or within the service range of the RSC, PABXs or Direct inward dialling (DID) trunks. (*d*) Remote-off-remote which lets the RLCMs and OPMs connect to the RCC through DS-1 interfaces. It lets RLCM and OPM subscribers to use the same lines to the host as the RSC subscribers and (*e*) Emergency Stand-Alone (ESA) which allows you to call internal to the RSC, if communication with the DMS-100 is lost. It will give station to station and station to trunk calls for DOTS, IBN and electronic business sets.

RSC can handle 16,200 CCS with the use of 16 DS-1 links. RSC had the ability to handle interface for 5,760 lines and is used a replacements for dial offices or PBXs. The Remote line concentrating module (RLCM) is just LCM used in a remote location from the DMS-100 host. The RLCM can handle 640 lines. This is sometimes used as a replacement of CDOs or PBXs.

Subscriber Carrier Module (SCM). It interconnects digital loop carriers to the switching fabric. It gives a direct interface for remote concentrators. Different versions of SCM are SCM-100 R, SCM-1000 U, SCM 100 S and SCM-100. The SCM 100 R can interface upto five Northern Telecom DMS-1 rural remote terminals (RTS). A DMS-1 rural remote terminal can interface upto 256 lines. The SCM-100 U can interface up to three DMS-1 urban RTs. A DMS-1 Urban can interface up to 576 POTS or special service lines. SCM100s can interface upto four mode I (non-concentrated) SLC-96 systems or up to six Mode II (concentrated) systems.

Software Architecture. The DMS 100 is programmed in BNR proprietary PASCAL, in PROTEL and in Assembler (rarely these days). It is upgraded by software loads that take place every six months. The software for the central processor is written in PROTEL, a highlevel PASCAL based language. Peripheral processors use a XMS-PASCAL software language.

The DMS supernode software is arranged in four layers as shown in Fig. 6.12.

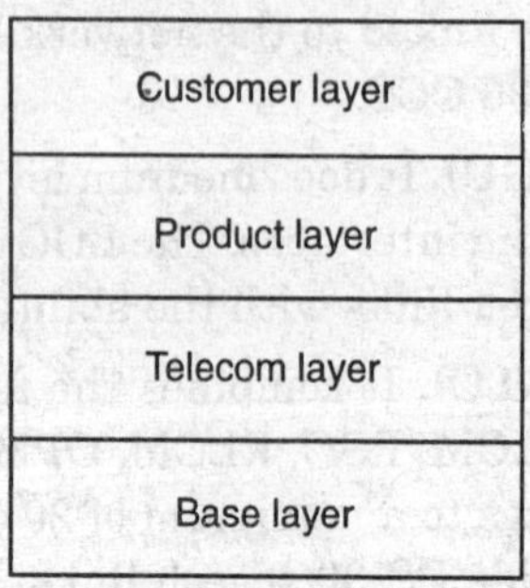

Fig. 6.12. Software architecture.

The base layer contains DMS super node operating system. The telecom layer supports all basic telecommunication functions. The product layer supports different products that work under this architecture. Some of the products are DMS-100, DMS-250 etc. The customer layer supports customer software and trial services specific to customer.

6.7.2. EWSD System

The evolutionary changes and further demands in telecommunication such as new services for data images and personnel communication demands the switching system for increased processing power, additional line connectivity and open connections to a new generation of multiservice access platform. The siemens EWSD switching system is well ahead of the challenge with a uniquely flexible architecture that anticipates change and adapts easily. EWSD system is an integral part of powerful network solutions. The EWSD switch is designed for incremental expansion in processing power connectivity and services.

The EWSD system meets all of today's service needs in even the largest metropolitan centers. It minimises the investment required to meet each new level of service demand. It makes the convergence of voice and data a reality. Hence, because of its inherent versatility, the EWSD system has become the universal switch, capable of responding to the full range of telecommunication standards and all variety of global service demands. Over 160 million lines of EWSD switch capacity are now in service in more than 100 counties. It was first put into service in 1980.

Features of EWSD system :

1. EWSD system is the leader in the availability of industry standard national ISDN. With national ISDN, telephone companies and carriers can

(*i*) Offer integrated voice and data to meet the needs of the most businesses and residential professionals.

(*ii*) provide dial-up video conferencing

(*iii*) offer distance learning services by universities.

(*iv*) offer medical imaging to health care providers

(*v*) also guarantee their customers full portability of expensive on-premises terminals and equipment.

2. The AIN 0.1 and AIN 0.2 capabilities on EWSD ststems, enables the network service providers to create their own customer services. The switch offers full SSP access to SCP, allowing services to be deployed across the entire network without changes to individual switching centers.

3. The EWSD switch provides a wide variety, of business and residential services encompassing ISDN, centre and AIN capability. These services are complemented by comprehensive operation, Administration, maintenance and provisioning (OAM and P) capabilities such as user friendly craft machine interface, diagnostics, billing, fault handling, traffic and maintenance measurements.

4. EWSD systems can combine digital and analog lines, allowing full interoperability between digital and analog terminals.

5. With EWSD technology, business customers can have all the conveniences and functionality of EKTS systems without the need to purchase and maintain customer premises switches.

6. EWSD technology allows carriers to deliver ACD services directly from the network. Customers can avoid the cost of on site ACD switching equipment.

7. EWSD system provides a high capacity interface which will prove essential in connecting ISDN, personal communications, video and other advanced services.

8. EWSD also provides Global system for Mobile communication (GSM) based PCS.

9. EWSD systems connect to any of the new generation of multi-service remote digital carrier terminals.

Hardware Architecture. The EWSD switching system is based on a modular hardware platform completely integrated under generic software. Processing is distributed throughout the modular components, and the components can be assembled into a single central office, or they can be distributed to move call processing close subscriber communities. The basic subsystems of the EWSD switch are

- Switching network
- Line trunk group
- CCS network
- Digital line unit
- Channel bank
- Packet handler
- OAM and P
- Message buffer, central clock
- Generator and coordination processor.

Fig. 6.13 shows the Siemens Stromberg-Carlson EWSD switch hardware architecture.

The switching network (SN) is the switching fabric of the EWSD. It is time multiplexed and is classified as a TST network. It connects lines trunks and signalling connection between subsystems. This LTG interface multiplexes and controls traffic between line and trunk interfaces and the switching network.

The architectural design provides maximum flexibility. Line types such as single or two party analog, coin, TR-08, basic rate ISDN and XDSL can be varied simply by changing the line card. Connectivity can be increased or relocated by deploying additional digital line units (OLU) or Remote Control Units. DLU interconnects analog subscribers, ISDN basic access and PBX lines to the LTG and also supports SLC access. The channel bank interface connects analog trunks to the LTG.

The services encompassing ISDN, centrex and Advanced Intelligent network capabilities are complemented by comprehensive Operation, Administration and maintenance (OAM and P) capabilities such as user friendly craft machine interface, dignostics, billling fault handling, traffic and maintenance measurements. This subsystem also supports the human mechine interface systems. The CCNC interface supports SS7 signalling between the EWSD switch and SSPs, STPs and SCPs. MB, processes signals between the network and the coordination processor. The CCG provides clock signals for the EWSD switch.

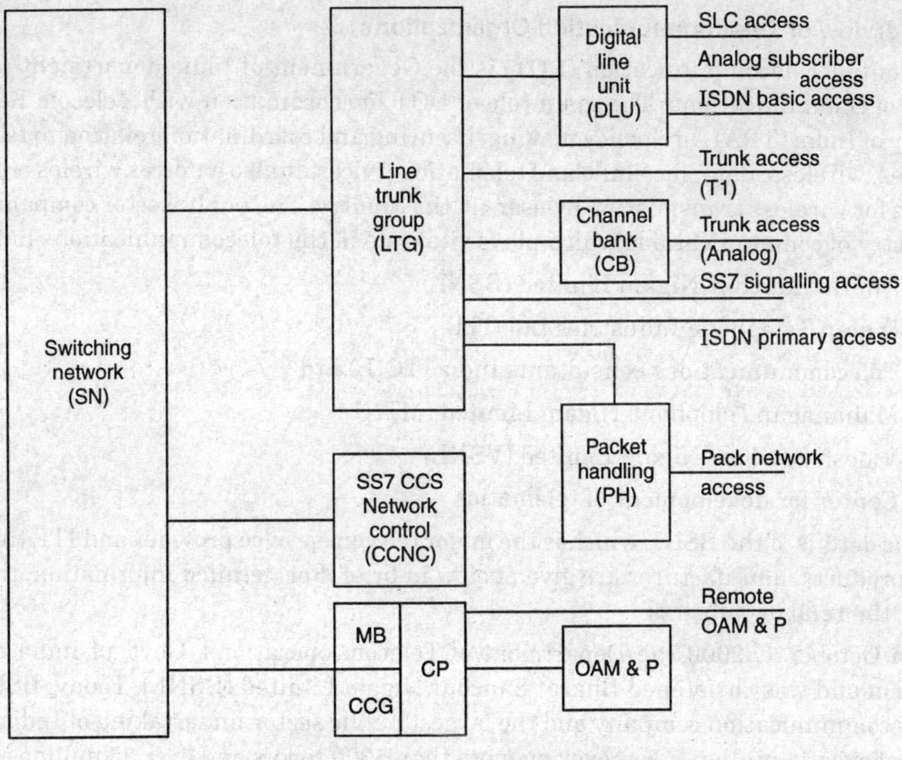

Fig. 6.13. Hardware architecture of EWSD.

Software Architecture. The software is developed utilizing well defined and loosely coupled functional subsystems. Software additions and changes to one function do not have direct impact on other functions except by design. The structure of the EWSD software is highly modular. The EWSD software can broadly be classified into OAM and P software, exchange software, support software and customer premises software. The exchange software supports call processing, the coordination processor, and all other peripheral processors. All peripheral units such as the MB, DLU, LTG and CCNC are loaded with specific software for their respective function alities. The switching network is also loaded with its specific software.

Because of the quality of manufacturing and exhaustive pre-testing of software releases, the EWSD switch has proven a reliability leader in many countries. EWSD system proved low downtime rates, extemely high reliability and high protection for service providers switching investment.

6.8. ELECTRONIC EXCHANGES IN INDIA

In this section, the overview of the Indian Telecommunications organizations, switching systems in India and some of the switching systems in brief are described.

6.8.1. Overview of Telecommunication Organizations

Department of Telecommunication (DOT) is the Government of India department under the ministry of communications. The main role of DOT in coordination with Telecom Regulatory Authority of India (TRAI) are policy making, licencing and coordination relating to telegraphs, telephones, wireless, data, facsimile and telematic services. It also enforces wireless regulatory measures for wireless transmission by users in the country. The public sector companies under the ministry of communications which plays vital role in the telecommunications in India are

1. Bharat Sanchar Nigam Limited (BSNL)
2. Indian Telephone Industries Ltd (ITI)
3. Telecommunications consultants India (TCIL) Ltd
4. Mahanagar Telephone Nigam Limited (MTNL)
5. Videsh Sanchar Nigam Limited (VSNL)
6. Center for development of telematics

The details of the BSNL which is the major telecom service provider and ITI, the leading telecom products manufacturer are given below in brief. For detailed information, the reader can refer the related websites.

On October 1, 2000 the Department of Telecom operations, Govt. of India become a corporation and was christened Bharat Sanchar Nigam Limited (BSNL). Today, BSNL is the No. 1 telecommunication company and the largest public sector undertaking of India. It has a network of over 45 million lines covering more than 5000 towns and over 35 million telephone connections. The main functions of BSNL includes planning, engineering, installation, maintenance, management and operation of voice and non-voice telecommunication services all over the coutry.

ITI established in 1948 is a Telecom company manufacturing the entire range of telecom equipments which includes telephones, large digital switches, transmission systems like microwave, Fiber optic systems and satellite communication systems. Its highly satisfied customers in India include BSNL, MTNL, defence services, parliamentary, police and internal security organisations, railways etc. Many African and south Asian nations are its overseas customers. Related to digital switches, ITI in collaboration with ALCATEL, France manufactures large digital switches and with C-DOT India, manufactures small, medium and Large digital switches.

TCIL is a premier telecommunication consultancy and engineering company under the ministry of communications. TCIL with its number of joint venture companies manufactures computer hardware, copper and optical fibre cables, developing software packages and providing consultancy and engineering services to other computer, information technology, telecom and software companies.

6.8.2. Switching Systems in India

ITI has contributed to 73% of the installed base of public switching lines and two thirds of the installed base of large switches in India. ITI provides similar service support for these products outside India also, which will be cost effective. The indigenously develop switching systems used in India are :

- CDOT 256P RAX
- CDOT SBM.
- CDOT SBM-XL
- CDOT MAX-L
- CDOT MAX-XL
- CDOT TAX-XL
- CDOT AN-RAX
- CDOT RLC
- CDOT CNMS

The digital switching systems in collaboration with other countries are :

- OCB-283 M/S ITI, M/S Alcatel
- 5 ESS M/S lucent
- EWSD M/S HTL, M/S Siemens
- AXE-10 M/S Ericsson
- FETEX-150 M/S Fujitsu
- NEAX-61E M/S NEC

The Lastest switching technologies presently working in BSNL network are listed in table 6.3

Table 6.3

Switching Technology	Latest Version
OCB-283	R 24
5 ESS	13.1
EWSD	13
AXE-10	IN56
FETEX-150	IND4
NEAX-61E	–
E 10 B	2L
CDOT 256	4–2–1
CDOT SBM/MAX	2–2–1–4
CDOT AN-RAX	Release 11
CDOT RLC	3.2

The switching techniques currently under testing are

1. CDOT 5BM-VE
2. CDOŢ MAX-XL Version 2–3–1–6
3. OCB-283 Version R25
4. ALCATEL 1000 E 10 MM

Telecommunication Engineering Centre (TEC) is an 'S & T' (standards and testing) institution of Department of Telecommunications, Ministry of Communications and information technology. Govt. of India. The primary responsibilities of TEC are :

(*i*) Setting standards and specifications for Telecom equipment and services.

(*ii*) Conducting field trials for new equipment and services.

(*iii*) Studying new technology and services and give technical advice to DOT for their introduction in the network.

(*iv*) Technical and Advisory support for DOT.

The switching division of TEC is responsible for all activities related to the switching products either working in the BSNL/MTNL network. The TEC switching division supports.

1. Preparation of specifications of state of the art digital switching systems.
2. Validation of switching systems to be inducted in the network.
3. Interface requirements of switches of private operators networks.
4. Testing of hardware and software upgrades of various switching systems.
5. Field support to switching systems in the BSNL./MTNL network.
6. Technical support to DOT.

For various digital switchings used in India, the readers are advised to refer the related websites.

ACRONYMS

AIN	—	Advanced intelligent network
AM	—	Administrative module
AMA	—	Automatic message accounting
CAS	—	Call accounting subsystems
CHILL	—	CCITT High Level language
CLNC	—	SS7 CCS network control
CPS	—	Call processing subsystems
CSH	—	Called subscriber held
CTD	—	Cell transfer delay
DLTU	—	Data line trunk unit
DSS	—	Digital switching subsystems
DSSS	—	Digital subscribers switching subsystems
DSU	—	Digital service unit
EN	—	Equipment number
IOC	—	Input/Output controllers
LTM	—	Line terminating modules
LU	—	Line unit
MCC	—	Master control center
MCS	—	Maintenance control subsystems
MDF	—	Main distribution frame
MSS	—	Management statistics subsystem
MTS	—	Message transmission subsystems
NCP	—	Network control processors
OSS	—	Operational support systems
DCLU	—	Data control line unit
PL	—	Permanent loop condition
SDL	—	Specification and Description language

SM	—	Switching module
SMPU	—	Switch module processor unit
SPC	—	Stored program control
S.t.d.	—	State transition diagram
TM	—	Trunk modules
TMS	—	Time multiplexed switch
TSI	—	Time slot intercharge
TSIU	—	Time slot intercharge unit
TU	—	Trunk unit

RELATED WEBSITES

http://library.telecommagazine.com/data
http://www.eventhelix.com
http://www.tmcnet_com/it/0504/
http://whitepapers.zdne/.co.vk
http://www.acad.humberc.on.ca/~infotech
http://www.morehouse.org/hin/ess
http://www.darkfall.demon.co.vk
http://www.dmine.com/phworld/network/tandem.htm
http://www.siemens.com.vn/marke/places
http://www.dotindia.com/

REVIEW QUESTIONS

1. List the basic steps to process a call.
2. What is state transition diagram ?
3. Draw the basic symbols of s.t.d.
4. Draw and explain the s.t.d. flow chart for a typical local call.
5. What is meant by hardware configuration ?
6. Explain with neat diagram, the various levels of general hardware configuration.
7. What is the need for software in switching ?
8. How the software systems are classified ?
9. Explain the hardware and software architecture of No. 1 ESS.
10. In what way the No. 5 ESS is superior than No. 1 ESS ?
11. Explain the various modules of No. 5 ESS hardware architecture with neat diagrams.
12. List the popular digital switching systems.
13. List the popular versions of DMS family.
14. Explain DMS-100 switching system with necessary hardware and software architecture.
15. What are the features of EWSD system ?
16. List the digital switching systems which are popular in India.
17. Draw and explain the hardware architecture of EWSD system.
18. List the organizations involved in telecommunication related activities.

7

Signalling Techniques

7

Signalling Techniques

7.1. INTRODUCTION

A subscriber can be able to talk with or send data to someone in any part of the world almost instantly and an exchange is able to set path and clear it after the conversation instantly by an effective signalling system. A signalling system link the variety of switching system, transmission systems and subscriber equipments in a telecommunication network to enable the network to function as a whole.

The signalling are classified according to the internal signalling of an exchange, signalling between exchanges and signalling between an exchange and subscriber. Thus a signalling system must be obviously be compatible with the switching systems which itself partitioned into subsystems in a network.

Traditional exchanges sent signals over the same circuit in the network. The introduction of SPC in exchanges enhances the services and introduced new services to the subscriber. These services require more signals to be transmitted and hence needs a separate data channel. The former method of signalling is referred as channel associated signalling and the latter is common channel signalling.

In the following sections, different forms of signalling, types of signalling, various types of inchannel signalling, common channel signalling, networks and the world wide popular signalling system 7 (SS7) are described.

7.2. SIGNALLING CLASSIFICATIONS

Communication networks generally connect equipments such as telephones and fax machines via several line sections, switches and transmission media for exchange of speech, text and data. To achieve this, control information has to be transferred between exchanges for call control. Call control is the process of establishing and releasing a call. This is referred as signalling. In general, signalling is defines as follows. "Signalling is the process of generating and exchanging information among components of a telecommunication system to establish, monitor or release connections and to control related network and system operations".

7.2.1. Forms of Signalling

The information that must be transmitted between subscribers, and between switching centres falls broadly under three classes.

(*i*) **Supervisory signals or line signals.** These are the signals necessary to initiate a call setup and to supervise it, once it has been established. It is also referred as subscriber loop signalling. Various signalling between an exchange and a subscriber is explained in section 6.2.2. Line signals can be transmitted by the use of a single control channel in each direction line.

(*ii*) **Routing signals or register signals.** Information transfer related to call setup is usually referred to as register signals. The basic information is the dialled code which indicates to the subsequent switching centres the required routing. In addition to the basic information, signals such as route information, terminal information, register control signals, acknowledgement signals, status of called terminal etc are also involved.

(*iii*) **Management signals or interregister signalling.** These signals are used to convey information or control between exchanges. This signalling also referred as inter exchange signalling. This signalling involves remote switching of private circuits, routing plans, modification of routing plans, traffic over load, priority of the call, class of service etc. The signalling may be performed by link-by-link basis, which passes signals exchange to exchange or end-to-end signalling which is between originating and terminating exchange referred as line signalling.

7.2.2. Classification

Traditional signalling uses the same channel to carry voice/data and control signals to carry out the path setup for speech or data transfer. This signalling is referred as inchannel signalling or per trunk signalling (PTS). An alternative to inchannel signalling is called as common channel signalling (CCS). CCS, uses a separate common channel for passing control signals. It couples the signals for a large number of calls together and send them on a separate signalling channel. Hence basically, signalling technique is classified into

1. In channel signalling/per trunk signalling (PTS)
2. Common channel signalling (CCS)

The inchannel signalling is classified further into four categories as shown in Fig. 7.1.

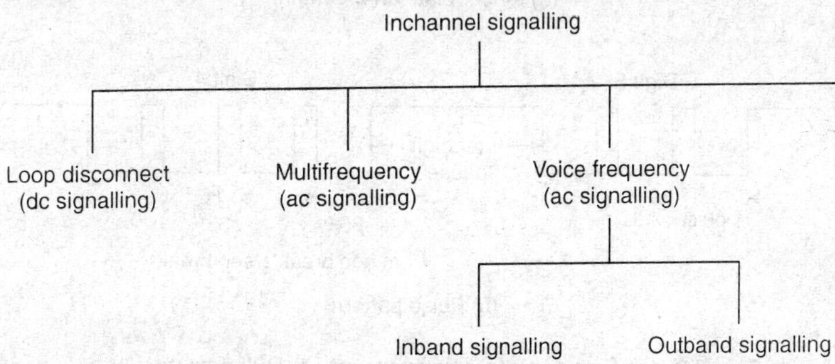

Fig. 7.1

The common channel signalling (CCS) is classified according to the transmission of signlas between exchangers. They are :

1. associated signalling
2. non-associated signalling

3. quasi associated signalling

Both inchannel and common channel signalling are described in the following sections.

7.3. INCHANNEL SIGNALLING

For data transfer and path setup for conversation, various forms if signalling were developed in automatic switching system. Various classifications are mentioned in last section. In this section, the inchannel signalling which uses same path for control signals and data/speech transfer are discussed.

7.3.1. Loop Disconnect - dc Signalling

The earliest and still the most common telephone set is the rotary dial telephone. The mechanism to transmit the identity of the called subscriber to exchange is pulse dialling. The basic idea is to interrupt the d.c. path of the subscriber's loop for a specified number of short periods to indicate the number dialled. This is called loop-disconnect (or rotary) signalling. Fig. 7.2. shows the finger plate arrangement of rotary dial telephone and a typical pulse pattern.

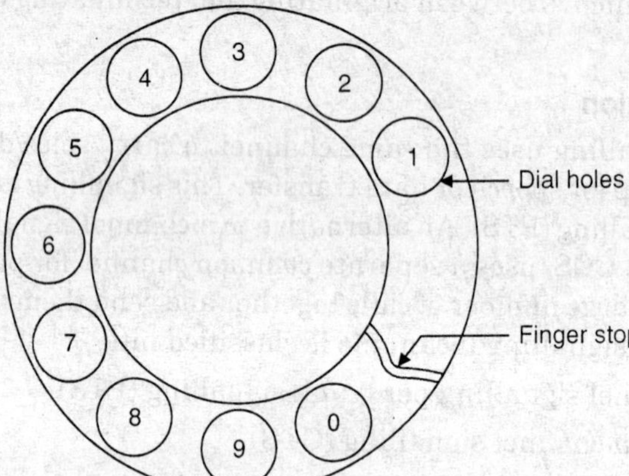

(a) Finger plate arrangement

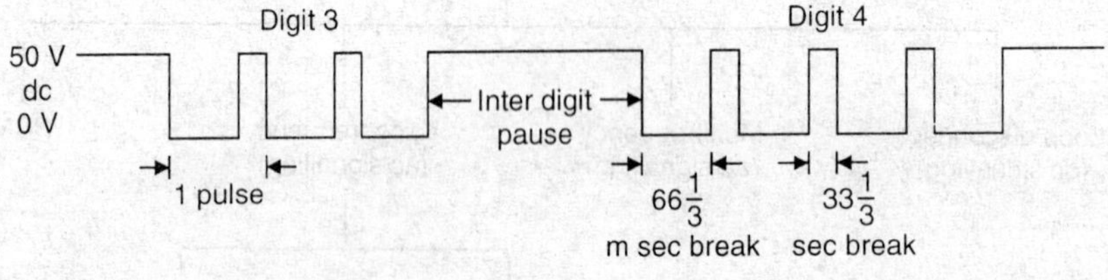

(b) Pulse pattern

Fig. 7.2. (*a*) Finger plate arrangement (*b*) Pulse pattern.

When the dial is rotated, the corresponding digit is pulled round to the finger guard and the dial is released. A governor inside the dial causes it to rotate back automatically at a fixed speed, causing a series of pulses to be sentdown the subscribers line. After the dial has returned to its rest position the next digit is dialled. The dial operates at about 10 impulses per second with a break of about 66 1/3 m sec and a make of about 33 1/3 m sec. The mechanical design of the dial ensures the minimum period between two consecutive digits called as interdigit pause.

The clear signal is produced when the subscriber replaces the handset, by breaking the d.c. path. Any break of more than 100 m sec is assumed as a clear signal. This signalling is relatively cheap but slow. For long distance lines due to the characteristics of lines, performance variance of equipments, the pulse shape may be defreded. This cause more errors in dc signalling.

7.3.2. Multi frequency (mf) – AC Signalling

The slow dialling and hence slow call processing of the rotary dial telephone reduces the productivity of the switching system. The touch tone phone created by AT & T's Western Electric Division sends out frequency based tones instead of electrical impulses. The touch tone dialling scheme is shown in Fig. 3.7. This is called a dual tone multi frequency (DTMF).

When a number is pressed, two separate tones are generated and sent to the local exchange simultaneously. As each number has distinct set of dial tones, the switching system recognizes the number dialled. By rotary dial, the pulses are sent at a rate of 10 per second (*i.e.* 10 pulses per sec), whereas tone dialling generated at 23 pulses per sec (PPS). Thus tone dialling are must faster and hence quicker caller processing and increased productivity. The extra keys available in touch tone key pad enables additional/new features. The detection of the digit is carried out by using frequency filters. The frequencies are within the voice band and care must be taken to reduce the risk of speech imitations.

The path setup for a rotary call is around 43 seconds, whereas with tone dialling, the call can be setup in approximately 6 to 10 seconds. An *mf* key phone has a little inter-digit pause because the numbers can be keyed very quickly.

7.3.3. Voice Frequency (vf) – AC Signalling

The base band of the telephone channel is 0-4000 Hz. Normally, the speech band occupies the bandwidth of 300-3400 Hz. If the signalling frequencies are chosen within the range of base band of telephone channel, then signalling is referred as voice frequency signalling. The choice of frequency for the control signal depends on various parameters. Based on the selection of frequency the voice frequency signalling are classified into two classes. They are

1. In band signalling
2. Out band signalling

This signalling is also referred as frequency division multiplex (fdm) system.

In band signalling. If the control signal frequencies are within the speech band (300–3400 Hz), the signalling is called inband signalling. Typicaly in the range of 2280 or 2600 Hz. As it uses the same frequency band as the voice (300–3400 Hz), it must be protected against imitations of speech. For example, if a tone is around 2600 Hz.and lasts more than 50 m sec, the switching equipment assume it as a line disconnect signal. Thus, the control signal frequencies must be selected carefully to avoid this limitation. The equipment must be able to distinguish speech and signal. Various parameters are available to distinguish signal from speech. Some of them are

(*i*) Selection of frequency for signal.
(*ii*) Signal recognition time
(*iii*) Voice characteristics.

Selection of frequency. Every voice generates a different pattern of amplitude and frequency changes. Female voice pattern typically generates more changes in amplitude than frequency and male voice pattern generates more changes in frequency than amplitude. If we consider the English language, the energy level of speech is predominant at lower frqeuencies and it falls gradually over the rest of the band. At higher frequencies, there is an increase in crosstalk. Thus a compromise is made and the frequency selection is in the range 2280–2600 Hz. The most prevalent example is 2600 Hz tone used as an on-hook signal for inter office trunks. Although, normal speech rarely produces a pure 2600 Hz, the accompanying frequencies are used to distinguish it from the control signal.

Signal recognition time. If the durations of signals are made longer than the period for which the speech frequency is likely to persist in speech, the chance of signal imitation is reduced significantly.

Voice frequency receiver. Fig. 7.3. shows a guard circuit by which the probability of the signal receiver responding to a speech frequency can be reduced to a low level. If the recognition delay time of the circuit is above 50 m sec, the speech imitation can be reduced.

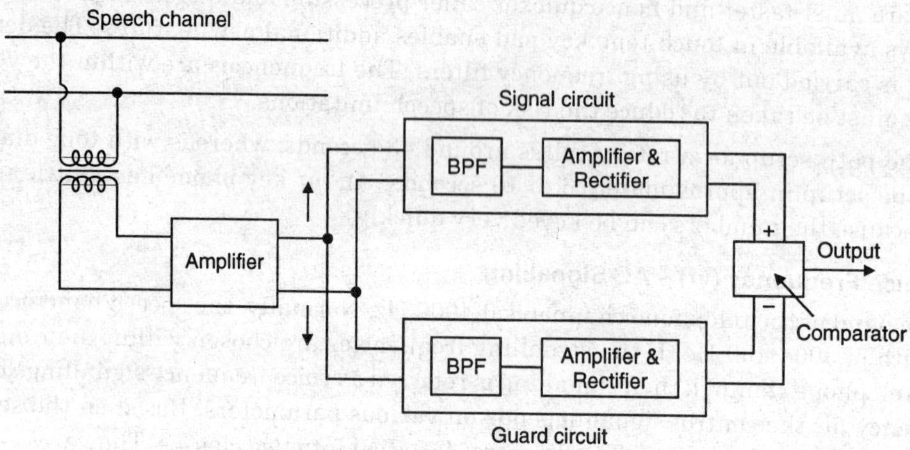

Fig. 7.3. Voice frequency receiver.

The speech/signal received from speech channel is amplified and then passed along two paths. One path contains a signal circuit and other path contains guard circuit. Both circuits outputs are connected to the comparator. Depends on the specifications set by the designer, it gives the output. The signal circuit and guard circuit contains a band pass filter (BPF), amplifier and rectifier. This BPF of signal circuit accepts the signal frequency and rejects the remaining. The BPF of guard circuit is designed to accept all other frequencies and reject signal frequency. The amplified and rectified outputs of both circuits are compared at comparator. If the output from the signal circuit exceeds that from the guard circuit, the receiver operates. If the output from guard circuit exceeds the signal circuit the receiver gives output signal.

Advantages of Inband signalling:

1. Inband signalling can be used on any transmission medium.
2. The control signals can be sent to every part where a speech signal can reach.
3. Owing to the flexibility of operation, it is the most widely used signalling system for long distance telephone networks.
4. It's operations are simpler.

Disadvantages of Inband signalling :

1. More possibility of speech signals imitating control signals. This problem can be reduced using suitable guard circuit.

2. The inband signal may 'spill-over' from one link to the another and causes error in that signalling system. This limitation occurs when several transmission links one connected end-to-end. The spill over problem can be eliminated by operating a line split to disconnect link whenever a signal is detected. The line split is designed generally to operate with in 35 ms.

Outband signalling

This signalling has frequenies above the voice band but below the upper limit of 4 kHz. The CCITT recommended frequency for outband signalling is 3825 Hz, but 3700 Hz and 3850 Hz are also used. The general layout of an outband signalling is shown in Fig. 7.4.

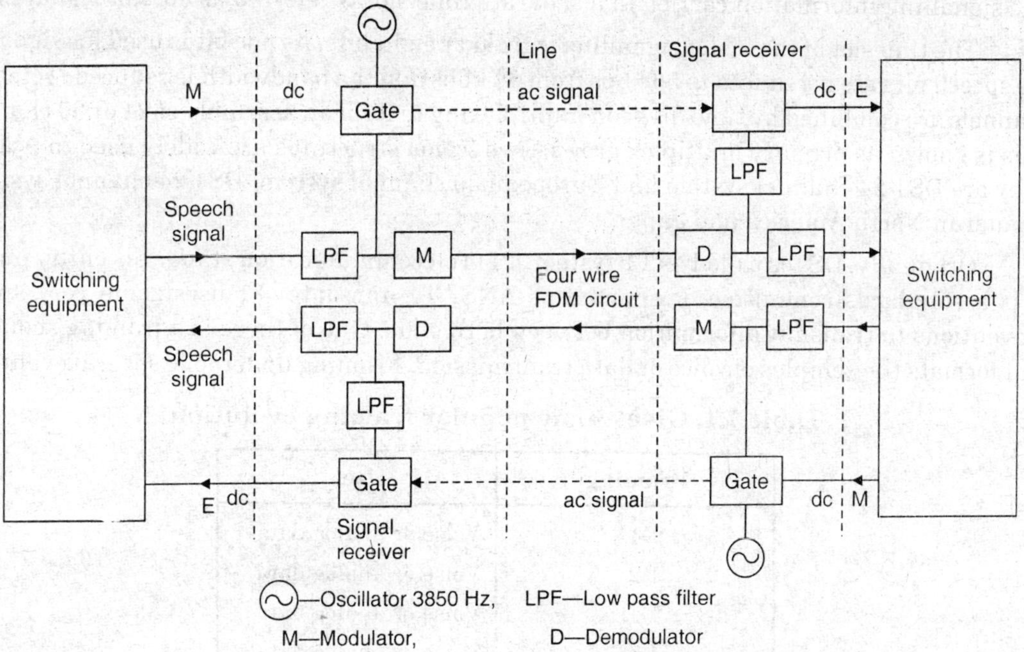

Fig. 7.4. General layout of an outband signalling.

The system shown above uses 4 wire E and M trunk (two wire voice path, an E-lead, and an M-lead with earth ground returns). Other types of E and M trunk is defined with 8 wires 4 wire voice path, an E-lead, with an associated return lead (SGD), and an M lead with an associated return (SB). In any type of E and M interface, supervision signalling is always conveyed on the E and M leads and not on the the voice pair. The E lead always carries signal from the signalling apparatus to the switching equipment and the M lead carries signals from the switching equipment to the signalling aparatus.

Advantages:

1. The requirement of line splits are not necessary to avoid sigal imitation.
2. Signals and speech can be transmitted simultaneously without disturbing the conversation.
3. Simple and consequently cheap.

Disadvantages:

1. Very narrow bandwidth is available for signalling.
2. Filtering circuits are needed to handle the signalling bands.
3. More dependent on the transmission system.

7.3.4. PCM Signalling

In PCM systems, signalling and speech are sampled, coded and transmitted within the frame of PCM channels. Thus, with PCM, a convenient way of transmission is possible. The signalling information and speech information carried in the same time slot is referred as inslot signalling. The signalling information carried in a separate time slot is referred as outslot signalling.

The timeslot of the inslot signalling is fixed at eight bits. As one bit is used for signalling, the speech bit rate is reduced to 56 kbps from 64 kbps and the bandwidth is reduced. Telephone channels are combined by time division multiplexing to form an assembly of 24 or 30 channels. This is known as primary multiplex group. Two frame structures are widely used in practice. They are DS1 24 channel system and European 30 channel system. DS1 24 channel system is popular in North America and Japan.

Originally, DS1 is called as T1 system. T1 (Telecommunication standards entity number 1) is a standards committee designated as ANSI T1-nnn-date. T1 uses some very specific conventions to transmit information between both ends. One of these is a framing sequences that formats the semples of voice or data transmission. Framing undergone several evolutions.

Table 7.1. Gives some popular framing evolution

Framing	Use
D1	Voise or analog data
D2	Voise or analog data
D1D	Voise or analog data
D3	Voise or analog data
D4	Voise and digital data

Starting with a DS1 signal as fundamental block, the table 7.2 lists the digital TDM signals of North America and Japan. DS0 is a 64 kbps signal that makes up the basis for the DS1.

Table 7.2. Digital TDM signals of North America and Japan

Designator	Number of Voice circuits	Bit Rate (Mbps)	Equivalent DS 1
DS0	1	0.064	—
DS1	24	1.544	1
DS1C	48	3.152	2
DS2	96	6.312	4
DS3	672	44.736	28
DS4	4032	274.176	168

DS1 24 channel system. DS0 is a 64-kbps signal that makes up the basis for the DS1. Twenty four DS0s combined to produce DS1. Fig. 7.5 shows the architecture of a T1 link and DS-1 frame format.

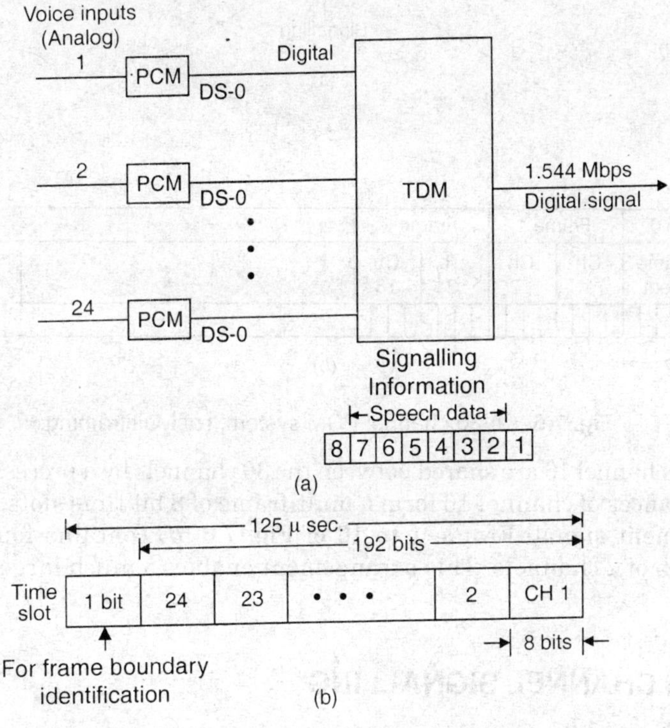

Fig. 7.5. (*a*) Architecture of T1 link, (*b*) DS-1 frame format.

Incoming analog signals were time division multiplexed and digitized for transmission. Each individual TDM channel are assigned 8 bits per time slot. Each frame is made of 24 × 8 bits = 192 bits plus one additional bit added to each frame to identity the fame boundaries. Thus each frame contains 193 bits. The frame interval is 125 μ sec. Hence the basic T1 line rate is 1.544 Mbps. This line rate has been established as the fundamental standard for digital transmission.

The earlier version of the 24 channel system consists of 7 speech bits plus 1 signalling bit. The recent version of 24 channel system uses a 12 frames together called as multi frame.

In multi frame, in every 6th frame, the eight bit of each channel is used for signalling instead of speech. The 193rd bit of the PCM frame is used on alterate frames in multi frame arrangement. Except the bits mentioned above, remaining all bits are used for speech. This arrangement reduces the quantization distortion. For a sampling rate of 8 kHz, this arrangement provides a signalling channel capacity of 1.33 kbits/sec. This signalling channel is further subdivided into two separate channels of capacity 667 bits/sec.

30 channel PCM system. It is based on 30 speech channels transmitted within a frame of 32 time slots (0 to 31). The total bit rate is 32 × 8 × 8 kbits/sec = 2048 kbps. Channel 0 is used for providing the framing signal and channel 16 is used for transmitting the signalling information relating to speech channels 1 to 15 and 17 to 31. Fig. 7.6 shows the 30 channel PCM system.

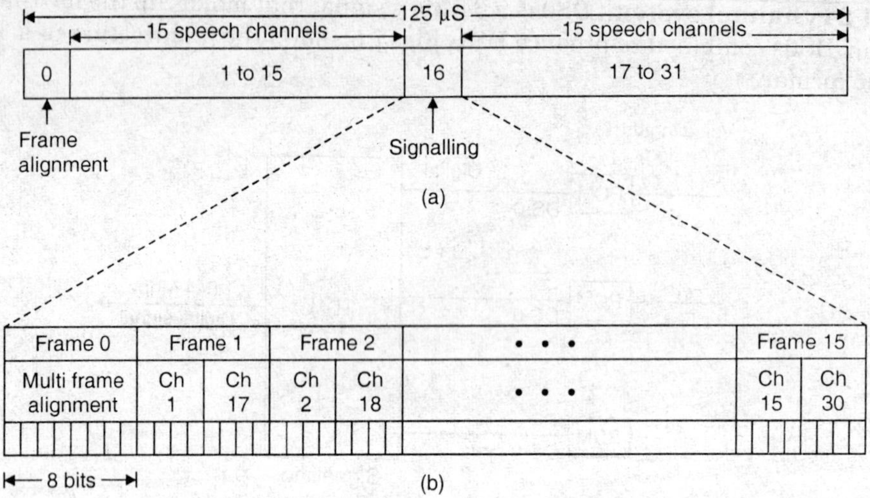

Fig. 7.6. (*a*) 30 channel PCM system, (*b*) Multiframing.

The 8 bits of channel 16 are shared between the 30 channels by a process of multiframing. 16 successive appearances of channel 16 form a multiframe of 8 bit time slots. Frame 0 contains a multiframe alignment signal. Frame 1 to 16 of Fig. 7.6 (*b*) contains four bits of signalling infomation for each of 2 channels. This arrangement enables a much larger number of signals to be exchanged.

7.4. COMMON CHANNEL SIGNALLING

Introduction of SPC digital switching systems with high speed processors in the telecom network has necessiated modernisation of signalling. Also, in order to meet the transfer of varieties of informations for call management and network management and to satisfy the subscribers requirement on various features, the uninterrupted, high speed signalling has become inevitable. The rapid dvelopment of digital systems paved way for the new signalling system called common channel signalling.

Instead of using the same link for signalling information and message as in inchannel signalling, the common channel signalling (CCS) uses a dedicated line for the signalling information between the stored program control elements of switching systems. Fig. 7.7 shows the basic schematic of CCS.

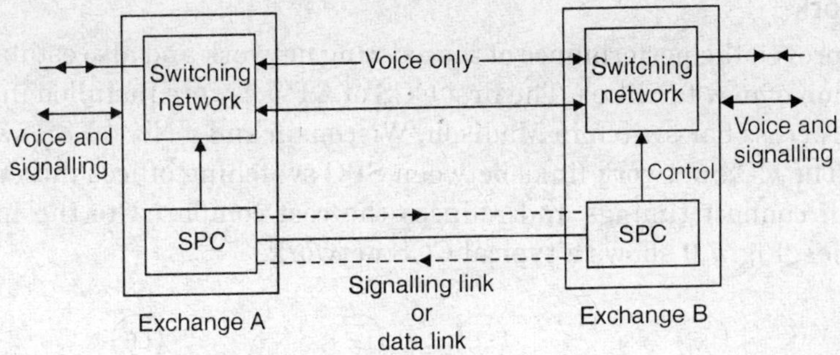

Fig. 7.7. Basic schematic of CCS.

The data link sends messages that identify specific trunks and events. Two signalling channels, one for each direction are used in a dedicated manner to carry signalling information. Hence, they are capable of carrying information for a group of circuits. At the bit rate of 2.6 kbps, CCS can carry signals for 1500–2000 speech circuits. CCS network is basically a store and forward network. In CCS network, the signalling information travels in a link-by-link basis along the route. The information arrived at a node is sorted, processed and forwarded to the next node in the route. The CCS technique is also called the transparent mode for signalling.

7.4.1. CCS Signalling Message Formats

CCS signalling information is transferred as signalling units, which is of varying length with one or more fixed length. A signalling unit is divided into a number of fields. The fields may be address unit, centralised service message unit, acknowledgement unit, synchronisation or idle unit, management message unit etc. The signalling information includes data related to routing, addressing digits, inter exchange information, routing status to the originating exchange maintenance request/details etc. Fig. 7.8 shows the typical CCS signalling message format. Data transfer is made top to bottom and left to right serially on the channel.

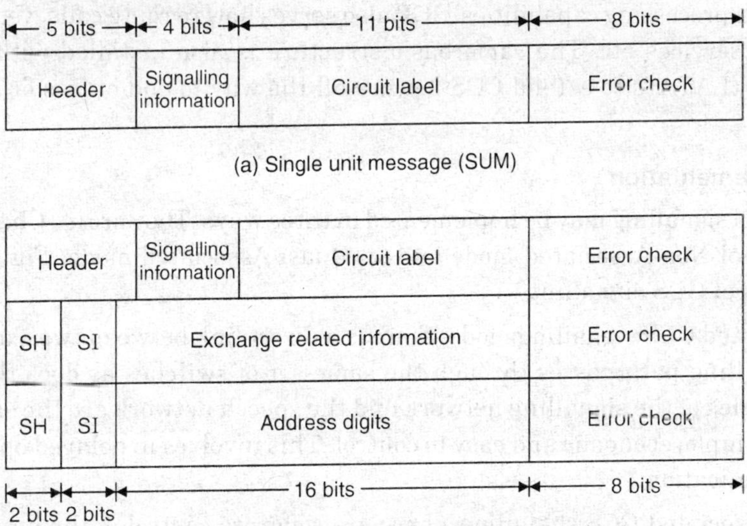

(a) Single unit message (SUM)

(b) Multi unit message (MUM)

Fig. 7.8. Basic CCS signalling message formats.

7.4.2. CCS Network

CCS network improved the performance of the existing network and also establishes a platform for the introduction of new facilities. The first CCS of AT & T were installed in the toll network between a No. 4 A cross bar switch in Madison, Wisconsin and a No. 4 ESS switch in Chicago, Illinois in 1976. The CCS network links between SPC switching offices increases the speed of long distance call connect timings and reduces the cost compared to the inchannel SF/MF signalling facilities. Fig. 7.9 shows a typical CCS network.

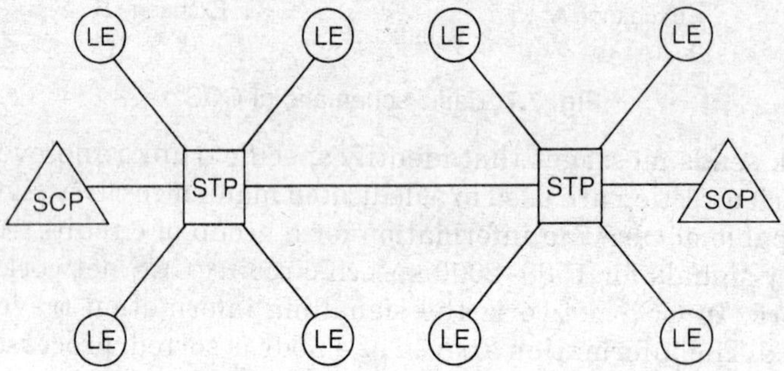

Fig. 7.9. CCS basic structure.

The CCS network consists of two types of nodes. They are signal transfer points (STP) and Signal control points (SCP). These nodes are interconnected by signalling links. The original communication protocols used between CCS entities was CCITT signalling system No. 6 (CCS 6). The modified protocol introduced in 1980 by CCITT is signalling system No. 7.

The STP's are the packet switching nodes of the CCS network. They receive and route incoming signalling messages towards the proper destination. They also perform specialized routing functions. Signal control points (SCP) are data bases that provide information necessary for advanced call processing capabilities. SCP also serves how to route calls, verify credit cards, process special services etc. The same basic structure is also installed within local Access Transport Area (LATA)'s to extend CCS feature all the way to end offices or local exchanges (LE).

7.4.3. CCS Implementation

Common channel signalling may be implemented in three ways. They are (*a*) Channel Associated mode (*b*) Channel Non-Associated mode and (*c*) Quasi-Associated mode. Fig. 7.10. shows all the three modes of CCS signalling.

In associated CCS signalling mode, there is a direct link between two exchanges. In this mode, the signalling path passes through the same set of switches as does the speech path. Network topologies of the signalling network and the speech network are the same. This mode of operation is simple, economic and easy to control. This involves in delayed operation for long distance communication.

In non-associated CCS signalling, there are separate control of the networks from the switching machines themselves. In multiexchange network, signal message passing through several intermediate nodes is referred as non-associated signalling. The network topologies for the signalling and the speech networks are different. Between exchanges, many STP's are placed. This approach is flexible as far as the routing is concerned. It demands more comprehensive scheme for message addressing than is needed for channel associated signalling.

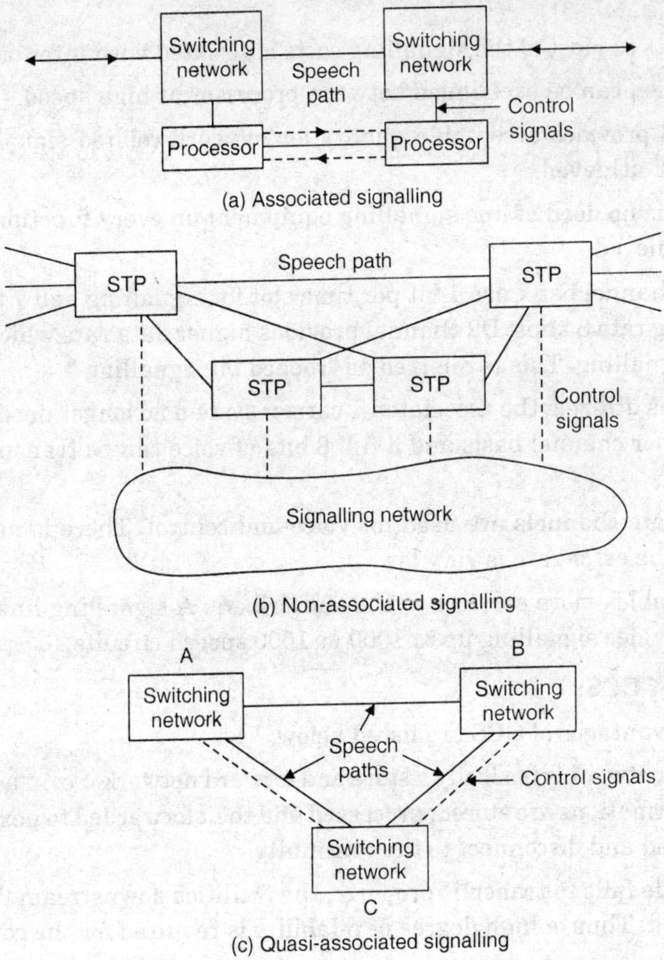

Fig. 7.10. Three modes of CCS signalling.

In practice, CCS messages are routed through one intermediate node for short distance communication. This is known as quasi-associated signalling. It establishes simplified predetermined paths between exchanges. The signalling paths are not associated but are fixed for given speech connections.

7.4.4. Advantages and Disadvantages of CCS

Advantages. The major advantages of CCS are listed below:

1. For each associated trunk group, only one set of signalling facility is required. The channel used for CCS need not be associated with any particular group. Fig. 7.7 shows a CCS network that is disassociated from the message network structure.

2. The introduction of SPC switching machines and CCS provides efficient routing procedure.

3. CCS allows for signalling at any time in the entire duration of a call, not only at the beginning.

4. CCS removes most of the signalling costs associated with inter office trunks.

5. Information can be exchanged between processor at high speed.

6. The CCS provides acceptable quality for network related signalling tones such as DTMF, MF and SF achieved.

7. As there is no need of line signalling equipment on every function, considerable cost savings can be achieved.

8. The D1 channel bank use 1 bit per time slot for signalling and 7 bits for voice which provides signalling rate 8 kbps. D2 channel provides higher data rate which use 1 bit in every sixth frame for signalling. This is referred as "robbed bit signalling."

When CCS is utilized, the associated T carrier system no longer need to carry signalling information on a per channel basis and a full 8 bits of voice can be transmitted in everytime slot of every frame.

9. As separate channels are used for voice and control. There is no chance of mutual interference and the error rate is very low.

10. CCS enables more services to the subscribers. A signalling link operating at 64 k bit/s normally provides signalling up to 1000 or 1500 speech circuits.

Disadvantages of CCS:

Some disadvantages of CCS are listed below:

1. The CCS network is basically a store and forward network. So, in a established circuit, the signalling informations are stored, processed and then forwarded to next node. This causes additional overhead and disconnect to the continuity.

2. If one node fails to transmit properly, the facilities downstream from the disconnect will not be released. Thus, a high degree of relability is required for the common channel.

3. Proper interfacing facility is necessary, as most of the present day telephone networks are equipment with inchannel signalling systems.

4. As the signalling information is not actually sent over speech paths in CCS, the integrity of speech path is not assured. As a remedy, routing testing of idle paths and the continuity test of an established path become necessary in CCS.

5. Different trunks in a group may terminate at different switch, say local exchange, other foreign exchange circuits etc. With CCS, all trunks are first terminated to the local central office and then forward to the different destination.

In channel and common channel signalling comparison is tabulated in Table 7.3

Inchannel signalling	Common channel signalling
Automatic propagation of signalling information enables the simultaneous process and release of associated facilities.	Signalling information must be relayed from one node to the next in a store and forward fashion.
Integrity of speech and signalling are maintained as they are on the same path.	Special equipments should be provided for the integrity as they are travelling on separate path.
Separate signalling equipment is required for each trunk and hence is expensive.	Only one set of signalling equipment is required for a whole group of trunk circuits and therefore CCS is economical.
Transfer of information such as address digits is from common control network originating office — Voice channel — receiving office — common control network.	Transfer of information is directly between control elements (processors of SPC systems).
Trunks are held up during signalling.	Trunks are not required for signalling.

7.5. SIGNALLING SYSTEM 7

A callers ability to talk with some one in any part of the world almost instantly (path setup) and clearing the path after the conversation instantly is depends on the networks capability. The concept of CCS has been extended by ITU-T in mid 1960's to improve the network capability. The first signalling system developed is called as signalling system number 6 (SS6). SS6 used a fixed length signal unit (28 bit signal units). The first development of the SS6 in north America was used in the U.S. on a 2400 bps data link. Later these links were upgraded to 4,800 bps. SS6 networks are very slow.

Signalling system 7 (SS7) was first defined in 1980, with revisions in 1984 and 1988. SS7 has been designed to be an open ended CCS standard that can be used over a variety of digital circuit switched networks. SS7 is a high speed communication network. Except in North America, SS7 is referred as **common channel inter office, signalling system number 7 (CCS 7 for short)**. One of the most significant reasons, the carriers employs this system is to save time and money on the network.

The CCS7 (*i.e.* SS7 or C7) is a global standard for telecommunication defined by ITU-T. The standard defines the procedures and protocols by which network elements in the PSTN transfers information over a digital signalling network to effect wireless (cellular) and wireless call setup, routing and control.

The CCS-7 signalling was introduced in India with the commissioning of new technology switches in major cities by the end of 1995–96. These new technology switches are manufactured by multi national companies such as LM Ericsson of Sweden, Siemens of Germany, AT & T of the U.S.A., Alcatel of France, and Fijitsu of Japan.

In the following sections the SS7 network architecture, SS7 link types, layers of SS7 protocol and signal unit structure are described.

7.5.1. Purpose of SS7 Network

1. An interesting feature of SS7 is that, it is a prerequisite for introduction of the following new services :

(*a*) The SS7 provides the internal control and network intelligence essential to an ISDN.

(*b*) Intelligent Network (IN) which uses network elements such as SSP, SCP and intelligent peripheral (IP), for achieving improved subscriber substation.

(*c*) Personal Communication Systems (PCS) which uses SS7 to provide personalised voice, data, image and video communication services that can be accessed regardless of location, network and time utilizing advanced microcell technology. It enables personalised billing, personalised numbering (one number for one person at any place at any time) and time and location independent charging.

2. The primary purpose of SS7 was to access remote database to lookup. This lookup process has several benefits because the carriers do not have to maintain a full database at each switching node, but they know how to get the remote data base end find the information quickly.

3. SS7 achieves, increased revenue generation, additional call charging capabilities, variety of calling-card services.

4. SS7, in integration with SPC systems enables quick and efficient call setup and teardown across the network in less than one second. Also, this integration provides for better supervision, monitoring and billing systems integration.

5. SS7 enables full use of the channel for the talk path because the signalling is done out of band on its own separate channel.

6. SS7 achieves enhanced call features such as call forwarding, calling party name/ number display and three way calling. Also with SS7, efficient and secured world wide communication is possible. It also provides, toll free and toll wireline services.

7. It handles the rerouting of network by using automatic protection switching services such as those found in SONET.

8. SS7 uses packet switching concept. Hence SS7 network is capable of preventing the misrouted calls, duplication of call requests and lost packets (requests for service).

7.5.2. Features of SS7

Introduction of this form of signalling is essential for the development of telecom networks in any country because of the following features:

1. Internationally standardised by the ITU.

2. SS7 is suitable for any transmission medium *i.e.*, can be operated over both terrestial and satellite links.

3. Even though SS7 is optimised to work with digital SPC exchanges utilising 64 kbps digital channels, it is suitable for operation over analog channels.

4. SS7 is suitable for various communication services such as telephony, text, data, images and video.

5. Transport mechanism is application independent.

6. High performance and flexibility along with a future oriented concept which meets new requirements.

7. High reliability for message transfer.

8. Processor friendly structure of the messages (signal units in multiples of 8 bits).

9. SS7 networks are much faster, efficient in the call setup and teardown process.

10. Flexible structure which accomodates all the technical advancement in telecommunication systems.

7.5.3. SS7 Network Architecture

The SS7 network is built out of the essential components interconnected by signalling links. SS7 messages are exchanged between network elements over 56 or 64 kbps bidirectional channels called signalling links. In this section, the essential components used in the signalling network, signalling link types and the basic SS7 architecture are described.

Essential components of SS7 network. There are three essential components used in SS7 network. They are SSP, STP and SCP. Fig. 7.11 (*a*) shows the symbol that are used to depict these three key elements of any SS7 network. STP and SCP's are customarily deployed in pairs. While elements of a pair are not generally co-located, they work redundantly to perform the same logical function. When drawing complex network diagram, these may be depicted as a single element for simplicity as shown in Fig. 7.11 (*b*).

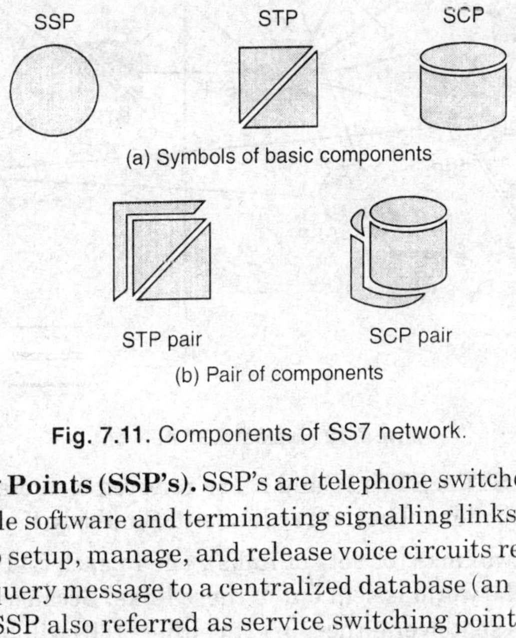

Fig. 7.11. Components of SS7 network.

Signal Switching Points (SSP's). SSP's are telephone switches (end offices or tandems) equipped with SS7 capable software and terminating signalling links. An SSP sends signalling message to other SSP's to setup, manage, and release voice circuits required to complete a call. An SSP may also send a query message to a centralized database (an SCP) to determine how to route a call. Sometimes SSP also referred as service switching point.

Signal Transfer Point (STP's). STP's are the packet switches of the SS7 network. An STP routes each incoming message to an outgoing signalling link based on routing information contained in the SS7 message. Because it acts as a network hub, an STP provides improved utilization of the SS7 network by eliminating the need for direct links between signalling points.

An STP may perform global title translation, a procedure by which the distination signalling point is determined from digit present in the signalling message. An STP can also acts as a "fire wall" to screen SS7 messages exchanged with other networks.

Signal Control Points (SCP's). SCP's are databases that provide information necessary for advanced call processing capabilities. An SCP sends a response to the originating SSP containing the routing number(s) associated with the dialled number. An alternate routing number may be used by the SSP if the prime number is busy or the call is unanswered within a specified time.

SCP's and STP's are generally deployed in mated pair configurations in separate physical locations to ensure network wide service in the event of isolated failure.

Basic signalling architecture. The essential components STP, SSP and SCP are combined together to form the architecture of SS7 network. Fig. 7.12 shows a simple example of how the basic elements of SS7 network are deployed to form two interconnected networks.

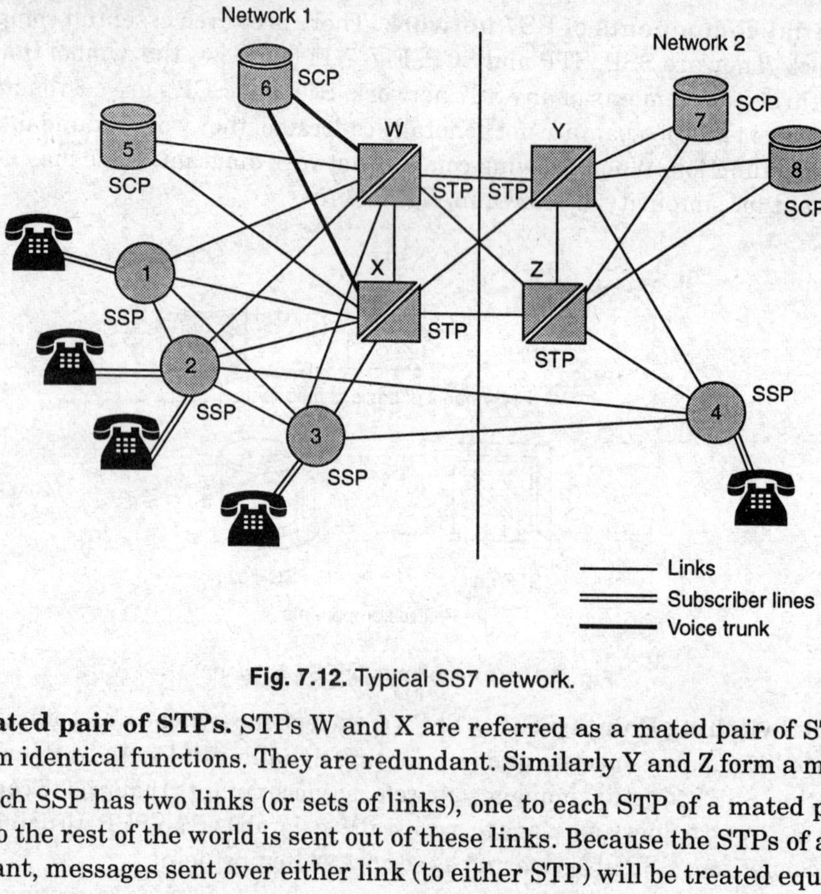

Fig. 7.12. Typical SS7 network.

1. **Mated pair of STPs.** STPs W and X are referred as a mated pair of STPs because both perform identical functions. They are redundant. Similarly Y and Z form a matched pair.

2. Each SSP has two links (or sets of links), one to each STP of a mated pair. All SS7 signalling to the rest of the world is sent out of these links. Because the STPs of a mated pair are redundant, messages sent over either link (to either STP) will be treated equivalently.

3. The STPs of a matched pair are joined by a link (or set of links).

4. **Quad.** Two mated pairs of STPs interconnected by four links (or set of links) are referred as a quad.

5. **Matched pairs of SCPs.** SCPs are usually (though not always) deployed in pairs. As with STPs, the SCPs of a pair are intended to function identically. Pairs of SCPs are referred as matched pairs of SCPs. Note that they are not directly joined by a pair of links.

SS7 signalling link types. Signalling links are logically organized by link type (A through F) according to their use in the SS7 signalling network. SS7 signalling links are characterized according to their use in the signalling network. Virtualy all links are identical in that they are 56 kbps or 64 kbps bidirectional data links that support the same lower layers of the protocol. Fig. 7.13. shows the signalling link architecture.

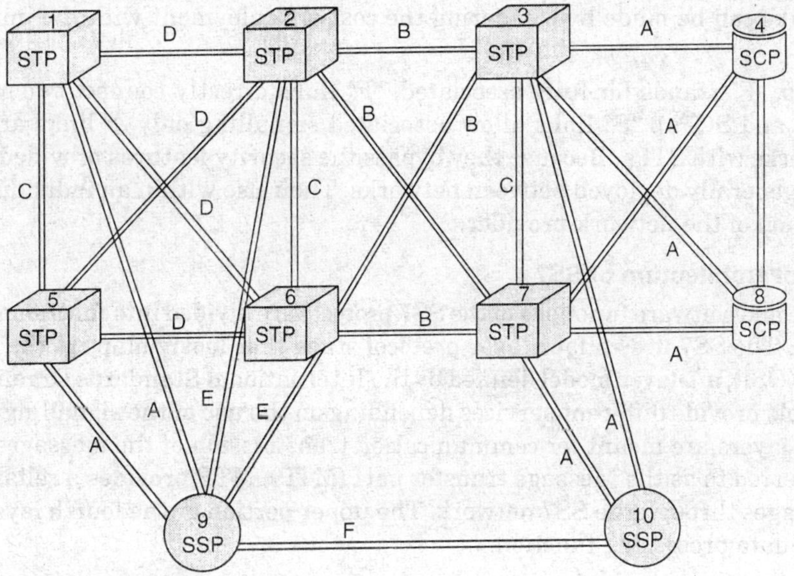

Fig. 7.13. Signalling link architecture.

The different types of links are defined as follows:

A links. 'A' links interconnect an STP and either an SSP or SCP which are collectively referred to as signalling end points. 'A' stands for access, because only messages orginating from or destined to the signalling end point are transmitted on an 'A' link. Examples of 'A' links are 1–9, 3–10, 3–4, 3–8, 4–7, 5–9, 7–8 and 7–10 in Fig. 7.13.

Signalling that SSP or SCP switches to send to any other node is sent on either of its A links to its home STP, which in turn, processes or routes the messages. Examples are 4–7–10 (or) 8–7–10 in Fig. 7.13. Similarly, message intended for an SSP or SCP will be routed to one of its home STP which will forward them to the addressed node over its 'A' links.

B links. 'B' stands for bridge. 'B' link connects one STP to another STP. Bridge describes the quad of links interconnecting peer pairs of STPs. 'B' links, or D links interconnecting two mated pairs of STPs are referred as B/D links.

C links. 'C' stands for cross. A cross (c) links connects. STPs performing identical functions into a mated pair. A 'C' link is used only when an STP has no other route available to a destination signalling point due to link failure. Thus the 'C' links are used to enhance the reliability of the signalling network.

D links. The link 'D' denotes diagonal and describes the quad of links interconnecting mated pairs of STPs at different hierarchical levels. Because, there is no clear hierarchy associated with a connection between networks, interconnecting links are referred as either B, D or B/D links.

E links. 'E' link stands for extended. This link connects an SSP to an alternate STP. 'E' link provide an alternate signalling path if an SSP's home STP can not be reached via an A link. Hence 'E' links provide backup connectivity to the SS7 network. In Fig. 7.13, 9–2 and 9–6 are alternative E links. A network includes "A", "B", "C", "D" and E links. But E links may or may not be deployed at the discretion of the network provider. The decision of whether or not to deploy "E" links can be made by comparing the cost of deployment with the improvement in reliability.

F links. "F" stands for fully associated. "F" links directly connect two signalling end points (SSP's and SCPs). "F" links allow associated signalling only. F links are not usually used in networks with STPs. Because they bypass the security features provided by an STP, F links are not generally deployed between networks. Their use within an individual network is at the discretion of the network provider.

7.5.4. Protocol architecture of SS7

The hardware and software functions of the SS7 protocol are divided into functional abstractions called 'levels'. The SS7 uses a four layer protocol stack that loosely maps to the Open System Interconnect (OSI), a 7-layer model defined by the International Standards Organization (ISO). These protocols provide different services depending on the use of the signalling network. The bottom three layers are meant for communication transmission of the messages. These three levels are referred to as the Message transfer part (MTP). MTP provides a reliable service for routing messages through the SS7 network. The upper portion or the fourth layer of the stack performs the data processing function.

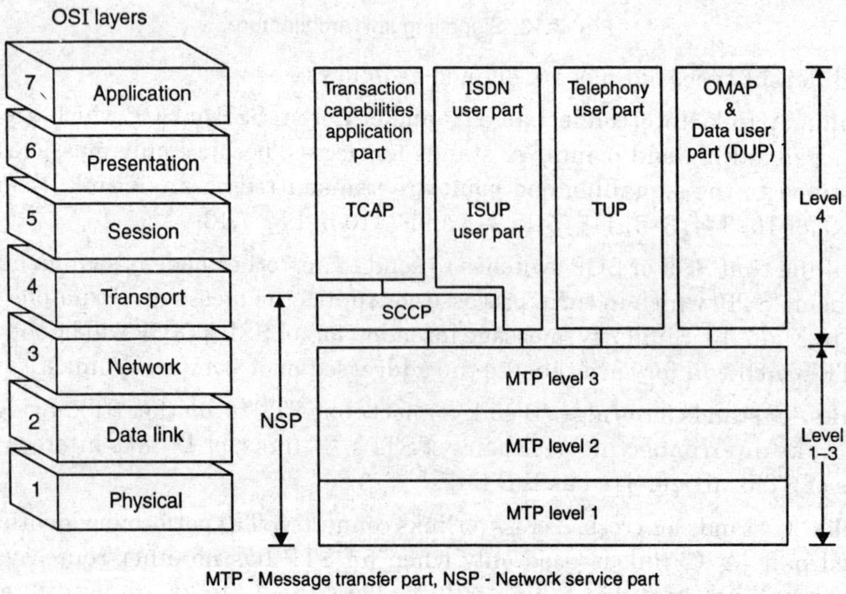

Fig. 7.14. OSI and SS7 protocol architecture.

The SS7 architectural principles are similar to the OSI architecture proposed for data networks (discussed in chapter 11). Many of the functions in each of these layers are common in these two architectures. The SS7 network is an interconnected set of network elements that is used to exchange messages in support of telecommunication functions. The SS7 protocol is designed to both facilitate these functions and to maintain the network over which they are provided. Fig. 7.14 shows the protocol architecture of SS7 and OSI reference model.

Message transfer part. The MTP provides a reliable service for routing messages through the SS7 network. The MTP is divided into three levels as described below.

MTP level 1. The lowest level MTP 1 is equivalent to the OSI physical layer. MTP level 1 defines the physical, electrical and functional characteristics of the digital signalling link. It uses time slot of a 2 Mbit/s PCM system or time slot 24 of a 1.5 Mbit/s system. Physical interfaces defined include E–1 (2048 kb/s, 3264 kb/s channels), DS-1 (1544 kb/s, 2464 kb/s channels), V-35 (64 kb/s), DS-0 (64 kb/s), and DS-0A (56 kb/s).

MTP level 2. MTP level 2 provides link layer functionality. The main purpose of this layer is to turn a potentially unreliable physical link into a reliable data link. It (*a*) ensures accurate end-to-end transmission of a message across a signalling link (*b*) implements flow control, message sequence validation and error checking. When an error occurs on a signalling link the message is retransmitted. It is equivalent to OSI data link layer.

MTP level 3. This level is equivalent in function to the OSI network layer. It extends the functionality provided by MTP level to provide network layer functionality. MTP level 3 provides message routing between signalling points in the SS7 network. MTP level 3 reroutes traffic away from failed links and signalling points and controls traffic when congestion occurs. The functions of MTP level 3 includes node addressing, routing alternate routing and congestion control.

User part. The user part (UP) known as level 4 in the layered structure of the signalling system. It is application dependent and includes the message, message coding and protocols necessary to support telephony and ISDN. This consists of the processes for handling the service being supported by the signalling system. The message transfer part is capable of supporting many different user parts. The defined user parts are telephone user part (TUP), ISDN User Part (ISCP), Signalling Connection Control Part (SCCP) and Transaction Capabilities Application Part (TCAP). TUP and ISUP are interactive and allows efficient signalling in a digital environment. The SCCP and TCAP interfaces UP and MTP.

1. **ISDN User Part (ISUP).** ISUP is used for both ISDN and Non-ISDN calls. The ISUP defines the protocol used to setup, manage, and release trunk circuits that carry voice and data between terminating line exchanges. It manages the trunk network on which they rely. Calls originate and terminate at the same switch do not use ISUP signalling.

2. **Telephone Use Part (TUP).** The TUP is involved in response to actions by a subscriber at a telephone. The control signalling associated with TUP deal with establishment, maintenance, and termination of telephone calls. TUP handless analog circuits only. In many countries ISUP has replaced TUP for call management.

3. **Signalling Connection Control Part (SCCP).** SCCP was added to SS7 specifications in 1984. The SCCP and MTP together are referred to as Network Service Part (NSP). SCCP provides connectionless and connection-oriented network services. SCCP provides two major functions that are lacking in the MTP.

(*a*) It has the capability to address applications within a signalling point. MTP does not deal with software applications within a node. The SCCP allows subsystems such as service call processing, calling-card processing, advanced intelligent network (AIN) and custom local area signalling services (CLASS) to be addressed explicitly.

(*b*) The second function provided by the SCCP is the ability to perform incremental routing using a capability called global title translation (GTT). GTT frees originating signalling pionts from the burden of having to know every potential destination to which they might have to route a message. GTT selects the correct destination to which the message should be routed. GTT effectively centralizes the problem and places it in a node (the STP) that has been designed to perform this function.

Intermediate GTT (in which it uses tables to find another STP further along the route to the destination) minimizes the need for STP to maintain extensive information about nodes that are far away from them. GTT is also used at the STP to share local among mated SCPs in both normal and failure scenarios.

4. **Transaction Capabilities Application Part (TCAP).** TCAP supports the exchange of non-circuit related data between applications across the SS7 network using the SCCP connectionless service. It is used for data base services such as calling card, service calls, AIN, switch to switch services including repeat dialling and call return. Queries and responses sent between SSPs and SCPs are carried in TCAP messages. For example SSP sends a TCAP query ˋo determine the routing number associated with a dialled service number to check the personal identification number of a calling card user. Because TCAP messages must be delivered to individual applications within the nodes they address, they use the SCCP for transport.

5. **Operations, Maintenance and Administration part (OMAP).** OMAP module deals with messages relating to the network management, operations and maintenance. This is the area for future definitions. Presently OMAP services may be used to verify network routing databases and to diagonise link problems. OMAP defines messages and protocol designed to assist administrators of the SS7 network. OMAP includes messages that use both the MTP and SCCP for routing.

7.5.5. SS7 Signalling Units

There are three types of signalling units (SU) defined in SS7. They are message signal unit (MSU), Link status signal unit (LSSU) and Fill in signal unit (FISU). SUs of each type follow a format unique to that type. A high level view of those formats is shown in Fig. 7.15. All three SU types have a set of common fields that are used by MTP level 2. The SU is based on the high level data link control (HDLC) protocol (described in chapter 11).

The MSU transfers information supplied by a user part (level 4) via the signalling network level (level 3). The LSSU is used for link initialization and flow control. The FISU is sent to maintain alignment when there is no signal traffic.

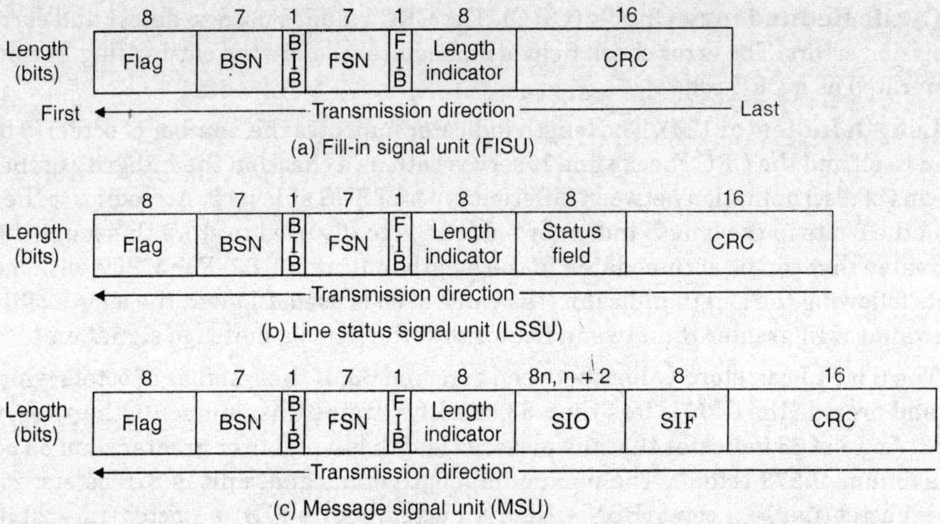

Fig. 7.15. Signalling units.

The common fields used in all the signalling units are described below :

Flag. The flag indicates the beginning of a new signal unit and implies the end of the previous signal unit (if any). The binary value of the flag is 0111 1110. The same sequence may occur in messages and wrongly interpreted as flags. By the technique known as bit stuffing and unstuffing, false flag can be prevented. By this technique before transmitting a signal unit, MTP 2 addes a zero bit after any sequence of five one bits. Upon receiving, MTP 2 removes any zero bit following a sequence of five one bits to restore the contents of the messages.

Backward Sequence Number (BSN). The BSN is used to acknowledge the receipt of signal units by the remote signalling point. The BSN contains the sequence number of the signal unit being acknowledged.

Backward Indicator Bit (BIB). A negative acknowledgement is indicated by inverting the BIB bit, which remains unchanged for all subsequent positive acknowledgement.

Forward Sequence Number (FSN) and Forward Indicator Bit (FIB). The FSN contains the sequence number of the signal unit. The FSN identifies the SU uniquely using modulo 128 count. The FIB is used in error recovery like the BIB. When a signal unit is ready for transmission, the signalling point increments the FSN by 1. The cyclic redundancy check (CRC) checksum value is calculated and appended to the forward message.

Upon receiving the message, the remote signalling point checks the CRC and copies the value of the FSN into the BSN of the next available message scheduled for transmission back to the initiating signalling point. If the CRC is correct, the backward message is transmitted. If the CRC is incorrect, the remote signalling point indicates negative acknowledgement by toggling BIB prior to sending the backward message. When the originating signalling point receives a negative acknowledgement, it retransmits all forward messages, beginning with the corrupted message, with the FIB toggled.

Cyclic Redundancy Check (CRC). The CRC value is used to detect and correct data transmission errors. The error check field is immediately before the closing flag. It contains 16 bits generated as a CRC code.

Length Indicator (LI). The length indicator indicates the number of octets (8 bit bytes) between itself and the CRC check sum. It serves both as a check on the integrity of the SU and as a means of discrimination between different types of SUS at level 2. Accroding to the protocol only 6 of the 8 bits in the length indicator field are actually used to store this length. Thus the largest value that can be accumodated in the length indicator is 63. For MSUs with more than 63 octets following the length indicator, the value of 63 is used. LI gives the length of the signal unit. A value of LI greater than two indicates that the SU is a message signal unit.

The 6 bit LI can store values between zero and 63. If the number of octets which follow the LI and precode the CRC is less than 63, the LI contains this number. Otherwise, the LI is set to 63. An LI of 63 indicates that the message length is equal to or greater than 63 octets (up to a maximum of 273 octets). The maximum length of a signal unit is 279 octets: 273 octets (data) + 1 octet (flag) + 1 octet (BSN + BIB) + 1 octet (FSN + FIB) + 1 octet (LI + 2 bits spare) + 2 octets (CRC).

The remaining two important fields service information octet (SIO) and signalling information field (SIF) used in MSU are described below :

Service information octet (SIO). It indicates the user part according to the message (*e.g.*, telephone, data or ISDN). The SIO field in an MSU contains the 4 bit **subservice field** followed by the 4 bit **service indicator.** FISU and LSSUs do not contain an SIO. The subservice field contains the network indicator (*i.e.*, national or international) and the message priority. The service indicator specifies the MTP user, thereby allowing the decoding of the information contained in the SIF. Table 7.4 shows the service indicator and MTP user.

Table 7.4. Service indicator values

Service indicator	MTP User
0	Signalling Network Management Message (SNM)
1	Maintenance Regular Message (MTN)
2	Maintenance Special Message (MTNS)
3	SCCP
4	TUP
5	ISUP
6	Data user part (call and circuit related messages)
7	Data user part (facility registration/cancellation messages)

Signalling Information Field (SIF). SIF may consist of upto 272 octets and contains the information to be transmitted. The SIF in an MSU contains the routing label and signalling information. LSSUs and FISUs contain neither a routing label nor SIO as they are sent between two directly connected signalling points.

MTP level 3 routes messages based on the routing label in SIF of MSU. The routing label is comprised of the destination point code (DPC), originating point code (OPC) and signalling link selection (SLS) field. Point codes are numeric addresses which uniquely identify each signalling point in the SS7 network. When the DPC in a message indicates the receiving signalling point, the message is distributed to the appropriate user part indicated by the service indicator in SIO.

Messages destined for other signalling points are transferred provided that the receiving signalling point has message transfer capabilities (like an STP). The selection of outgoing link is based on the information in the DPC and SLS.

7.5.6. Basic Call Setup with ISUP

Fig. 7.16 shows the example where a subscriber on switch A places a call to a subscriber on switch B.

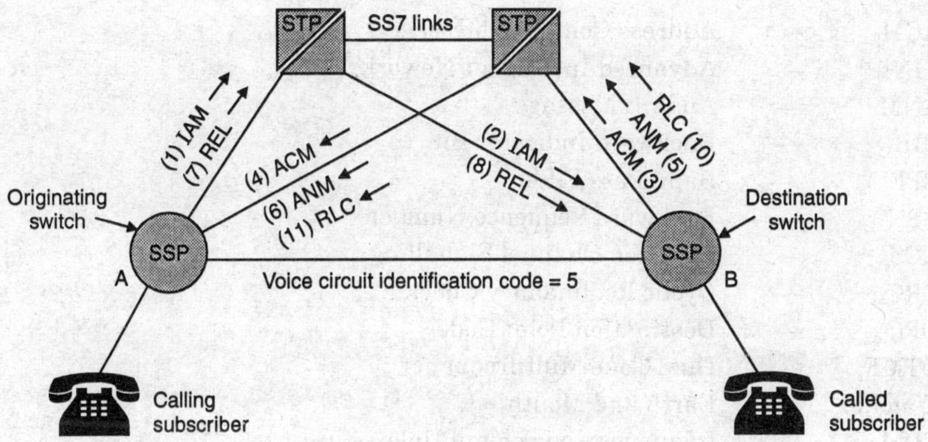

Fig. 7.16. Call setup with ISUP.

1. When calling subscriber calls a subscriber, the originating SSP transmits an ISUP initial address message (IAM) to reserve an idle trunk from switch A to B. The IAM includes the originating point code, destination point code, calling and called numbers.

2. The IAM is routed via the home STP of the originating switch to the destination switch. Same signalling links are used for the full duration of a call unless a link failure forces to change link.

3. On reception of IAM by called subscriber (if idle), the destination switch transmits an ISUP address complete message (ACM) to its home STP.

4. The destination switch rings the called subscriber and send a ringing tone over the trunk to the originating switch.

5. When the called panty picks the phone, the destination switch terminates the ringing tone and transmitts an ISUP answer message (ANM) to the originating switch via its home STP.

6. The STP routes the ANM to the originating switch and initiates billing.

7. If the calling subscriber hangs up first, SSP (A) sends an ISUP release message (REL) to release the trunk circuit between the switches.

8. The STP routes REL to the destination switch.

9. If the called subscriber hangs up first, the destination switch sends an REL to the originating switch (not shown).

10. Upon receiving REL, the destination switch disconnects the trunk from the called party's line. Then, it transmits an ISUP release complete message (RLC) to the originating switch to acknowledge the release of the trunk circuit.

11. When originating switch receives RLC, it terminates billing cycle and sets the trunk state to idle.

ACRONYMS

ACM	—	Address Complete Message
AIN	—	Advanced Intelligent Nework
ANM	—	Answer Message
BIB	—	Backward Indicator Bit
BPF	—	Band Pass Filter
BSN	—	Backward Sequence Number
CCS	—	Common Channel Signalling
CRC	—	Cyclic Redundancy Check
DPC	—	Destination Point Code
DTMF	—	Dual Tone Multifrequency
E and M	—	Earth and Mouth
FDM	—	frequency division multiplex
FIB	—	Forward Indicator Bit
FISU	—	Fill in Signal Unit
FSN	—	Forward Sequence Number
GTT	—	Global Title Translation
IAM	—	Initial Address Message
ISUP	—	ISDN User Part
LI	—	Length Indicator
LSSU	—	Link Status Signal Unit
Mf	—	Multi frequency
MSU	—	Message Signal Unit
MTP	—	Message Transfer Point
NSP	—	Network Service Point
OMAP	—	Operations, Maintenance and Administration Part
OPC	—	Originating Point Code
OSI	—	Open System Interconnect
PCM	—	Pulse Code Modulation
PPS	—	Pulses Per Second
REL	—	Release Message

RLC	—	Release Complete Message
SCCP	—	Signalling Connection Control Part
SCP	—	Signal Control Points
SH	—	Signalling Header
SI	—	Signalling Information
SIF	—	Signalling Information Field
SIO	—	Service Information Octet
SLS	—	Signalling Link Selection
SS7	—	Signalling System 7
SSP	—	Signalling Switching Points
STP	—	Signal Transfer Point
SU	—	Signalling Units
TCAP	—	Trasaction Capabilities Applications Part
TUP	—	Telephone User Part
Vf	—	Voice frequency

RELATED WEBSITES

http://www.aliancedatacom.com/technologies/signaling systems
http://encyclopediathefreedictionary.com
http://resource.intel.coraltelecom
http://www.ericsson.com/support/telecom

REVIEW QUESTIONS

1. What are the different forms of signalling ?
2. Distinguish Inchannel signalling and CCS ?
3. What is Inband and Outband Signalling ?
4. List the advantages and disadvantages of Inband Signalling ?
5. Draw the CCS Signalling message formats.
6. What is STP, SSP and SCP ?
7. Draw a SST network architecture.
8. What are the signalling link types ?
9. Explain with necessary diagrams, each type of the voice frequency (vf) signalling.
10. Explain architecture of T1 link and DS-1 frame format in relation to the PCM signalling.
11. What are the three ways of implementing CCS ? Explain each types of signalling with neat diagrams.
12. List the advantages and disadvantages of CCS.
13. List the purpose and features of SS7.
14. Explain all the signalling link types of SS7 link architecture.
15. Explain with neat diagram the protocol architecture of SS7.
16. Name three types of signallig units used in SS7. With neat diagrams explain each fields associated with the signalling units.

8

Traffic Engineering

8

Traffic Engineering

8.1. INTRODUCTION

The telecommunication system has to serve the voice traffic and data traffic. The traffic is defined as the occupancy of the server. The basic purpose of the traffic engineering is to determine the conditions under which adequate service is provided to subscribers while making economical use of the resources providing the service. The functions performed by the telecommunication network depends on the applications it handles. Some major functions are switching, routing, flow control, security, failure monitoring, traffic monitoring, accountability internetworking and network management.

To perform the above functions, a telephone network is composed of variety of common equipment such as digit receivers, call processors, interstage switching links and interoffice links etc. Thus traffic engineering provides the basis for analysis and design of telecommunication networks or model. It provides means to determine the quantum of common equipment required to provide a particular level of service for a given traffic pattern and volume. The developed model is capable to provide best accessability and greater utilization of their lines and trunks. Also the design is to provide cost effectiveness of various sizes and configuration of networks.

The traffic engineering also determines the ability of a telecom network to carry a given traffic at a particular loss probability. Traffic theory and queuing theory are used to estimate the probability of the occurrence of call blocking. Earlier traffic analysis based purely on anytical approach that involved advanced mathematical concepts and complicated operations research techniques. Present day approaches combine the advent of powerful and affordable software tools that aim to implement traffic engineering concepts and automate network engineering tasks.

In this chapter, the traffic design requirements, various probability distributions, loss systems, delay systems and combination of delay and loss systems are described.

8.2. TRAFFIC DESIGN REQUIREMENTS

In the study of teletraffic engineering, to model a system and to analyse the change in traffic after designing, the static charateristics of an exchange should be studied. The incoming traffic undergoes variations in many ways. Due to peak hours, business hours, seasons, weekends, festival, location of exchange, tourism area etc., the traffic is unpredictable and random in nature. So, the traffic pattern/characteristics of an exchange should be analysed for the system design. The grade of service and the blocking probability are also important parameters for the traffic study.

8.2.1. Traffic Statistics

The following statistical information provides answer for the requirement of trunk circuits for a given volume of offered traffic and grade of service to interconnect the end offices. The statistical descriptions of a traffic is important for the analysis and design of any switching network.

1. **Calling rate.** This is the average number of requests for connection that are made per unit time. If the instant in time that a call request arises is a random variable, the calling rate may be stated as the probability that a call request will occur in a certain short interval of time.

If 'n' is the average number of calls to and from a terminal during a period T seconds, the calling rate is defined as

$$\lambda = \frac{n}{T} \text{ calls/hour} \quad \text{...(8.1)}$$

In telecommunication system, voice traffic and data traffic are the two types of traffic. The calling rate (λ) is also referred as average arrival rate. The average calling rate is measured in calls per hour.

2. **Holding time.** The average holding time or service time 'h' is the average duration of occupancy of a traffic path by a call. For voice traffic, it is the average holding time per call in hours or 100 seconds and for data traffic, average transmission per message in seconds. The reciprocal of the average holding time referred to as service rate (μ) in calls per hour is given as

$$\mu = \frac{1}{h} \text{ calls/hour} \quad \text{...(8.2)}$$

Sometimes, the statistical distribution of holding time is needed. The distribution leads to a convenient analytic equation. The most commonly used distribution is the negative exponential distribution. The probability of a call lasting atleast t seconds is given by

$$P(t) = \exp(-t/h) \quad \text{...(8.3)}$$

For a mean holding time of $h = 100$ seconds, the negative exponential distribution function is shown in Fig. 8.1.

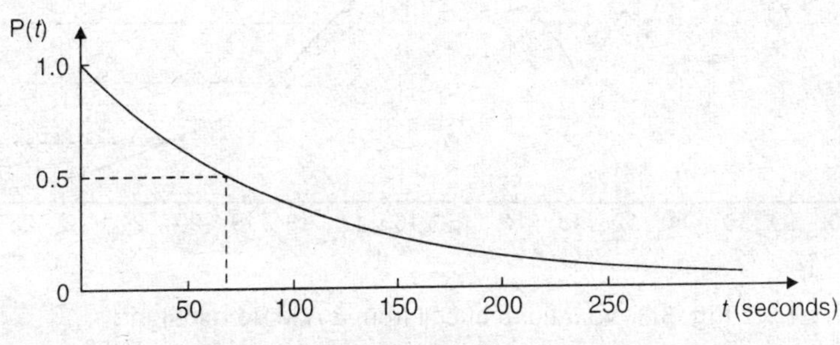

Fig. 8.1. Negative distribution function for h = 100 sec.

Figure shows that, 50% probability call lasts longer than 70 sec.

3. **Distribution of destinations.** Number of calls received at a exchange may be destined to its own exchange or remote exchange or a foreign exchange. The destination distribution is described as the probability of a call request being for particular destination. As the hierarchical structure of telecommunication network includes many intermediate exchanges, the knowledge of this parameter helps in determining the number of trunks needed between individual centres.

4. **User behavior.** The statistical properties of the switching system are a function of the behavior of users who encounter call blocking. The system behaves differently for different users. The user may abanden the request if his first attempt to make a call is failed. The user may makes repeated attempts to setup a call. Otherwise the user may wait some times to make next attempt to setup a call. These behavior varies person to person and also depends on the situation.

5. **Average occupancy.** If the average number of calls to and from a terminal during a period T seconds is 'n' and the average holding time is 'h' seconds, the average occupancy of the terminal is given by

$$A = \frac{nh}{T} = \lambda h = \frac{\lambda}{\mu} \text{ Erlangs} \qquad \text{...(8.4)}$$

Thus, avarage occupancy is the ratio of average arrival rate to the average service rate. It is measured in Erlangs. Average occupancy is also referred as traffic flow or traffic intensity or carried traffic.

8.2.2. Traffic Pattern

An understanding of the nature of telephone traffic and its distribution with respect to time (traffic load) which is normally 24 hours is essential. It helps in determining the amount of lines required to serve the subscriber needs. According to the needs of telephone subscribers, the telephone traffic varies greatly. The variations are not uniform and varies season to season, month to month, day to day and hour to hour. But the degree of hourly variations is greater than that of any other period. Fig. 8.2 shows the typical variations of calls from 8.00 A.M. to midnight.

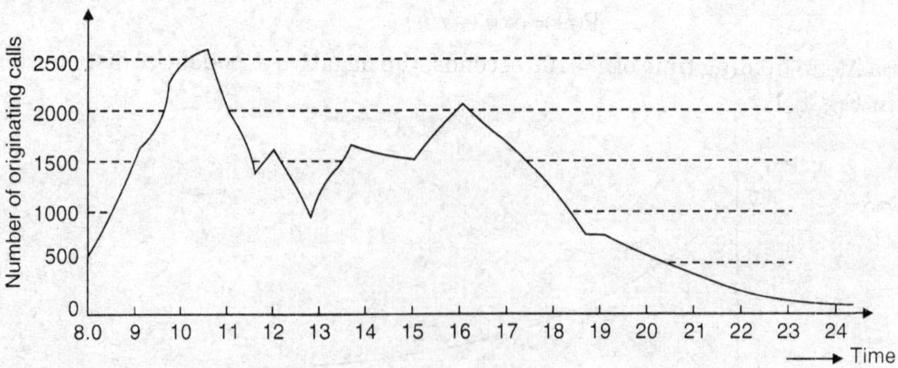

Fig. 8.2. Variations of call from 8 A.M. to midnight.

If the behavior of the traffic shown above is systematic for period of time or season, good judgement about the design of switching system or lines or trunks or any common shared equipments can be made. Thus, the combination of historical records, experience, location of exchanges (business area or residential area), vacations, govt. policies, holidays etc., decides the design of telecommunication network. Various parameters related to traffic pattern are discussed below :

Busy hour. Traditionally, a telecommunication facility is engineered on the intensity of traffic during the busy hour in the busy session. The busy hour vary from exchange to exchange, month to month and day to day and even season to season. The busy hour can be defined in a variety of ways. In general, the busy hour is defined as the 60 minutes interval in a day, in which the traffic is the highest. Taking into account the fluctuations in traffic, CCITT in its recommendations E.600 defined the busy hour as follows.

1. **Busy hour.** Continuous 60 minutes duration for which the traffic volume or the number of call attempts is greatest.

2. **Peak busy hour.** It is the busy hour each day varies from day to day, over a number of days.

3. **Time consistent busy hour.** The 1 hour period starting at the same time each day for which the average traffic volume or the number of call attempts is greatest over the days under consideration.

In order to simplify the traffic measurement, the busy hour always commences on the hour, half hour, or quarter hour and is the busiest of such hours. The busy hour can also be expressed as a percentage (usually between 10 and 15%) of the traffic occuring in a 24 hour period.

Call completion rate (CCR). Based on the status of the called subscriber or the design of switching system the call attempted may be successful or not. The call completion rate is defined as the ratio of the number of successful calls to the number of call attempts. A CCR value of 0.75 is considered excellent and 0.70 is usually expected.

Busy hour call attempts. It is an important parameter in deciding the processing capacity of an exchange. It is defined as the number of call attempts in a busy hour.

Busy hour calling rate. It is a useful parameter in designing a local office to handle the peak hour traffic. It is defined as the average number of calls originated by a subscriber during the busy hour.

Day-to-day hour traffic ratio. It is defined as the ratio of busy hour calling rate to the average calling rate for that day. It is normally 6 or 7 for rural areas and over 20 for city exchanges.

8.2.3. Units of Telephone Traffic

Traffic intensity is measured in two ways. They are (*a*) Erlangs and (*b*) Cent call seconds (CCS).

Erlangs. The international unit of traffic is the Erlangs. It is named after the Danish Mathematician, Agner Krarup Erlang, who laid the foundation to traffic theory in the work he did for the copenhagen telephone company starting 1908. A server is said to have 1 erlang of traffic if it is occupied for the entire period of observation. More simply, one erlang represents one circuit occupied for one hour.

The maximum capacity of a single server (or channel) is 1 erlang (server is always busy). Thus the maximum capacity in erlangs of a group of servers is merely equal to the number of servers.

Thus, the traffic intensity which is the ratio of the period for which the server is occupied to the total period of observation is measured in erlangs.

Example 8.1. *If a group of 20 trunk carries 10 erlangs and the average call duration is 3 minutes, calculate (a) average number of calls in progress (b) total number of calls originating per hour.*

Given data : *No. of trunks* $= 20$

traffic intensity $= 10$ *erlangs*

holding time $h = 3$ *minutes*

observation period $T = 60$ *minutes (generally).*

Sol. (*a*) Traffic intensity per trunk $= \dfrac{10 \text{ erlangs}}{20} = 0.5$ erlangs/trunk.

Average no. of calls per trunk for 1 erlang for 60 minutes $= 20$

For 0.5 erlang, average no. of calls in progress $= 10.$

(*b*) Traffic intensity

$$A = \frac{nh}{T} = 10 \text{ erlangs}$$

total number of calls origineting per hours

$$n = \frac{10 \times 60}{3} = 200 \text{ calls.}$$

For the present day networks which support voice, data and many other services, erlang is better measure to represent traffic intensity.

Cent call seconds (CCS). It is also referred as hundred call seconds. CCS as a measure of traffic intensity is valid only in telephone circuits. CCS represents a call time product. This is used as a measure of the amount of traffic expressed in units of 100 seconds. Sometimes call seconds (CS) and call minutes (CM) are also used as a measure of traffic intensity. The relation between erlang and CCS is given by

$$1E = 36 \text{ CCS} = 3600 \text{ CS} = 60 \text{ CM} \quad ...(8.5)$$

Example 8.2. *Consider a group of 1200 subscribers which generate 600 calls during the busy hour. The average holding time is 2.2 minutes. What is the offered traffic in erlangs, CCS and CM.*

Given data : $n = 600$

$h = 2.2$ *minutes*

$T = 60$ *minutes.*

Sol. Traffic intensity in erlangs $A = \dfrac{nh}{T}$ erlangs $= \dfrac{600 \times 2.2}{60} = 22$ erlangs.

Traffic intensity in CCS $A = \dfrac{600 \times 2.2 \times 60 \text{ seconds}}{100 \text{ seconds}} = 792$ CCS.

Traffic intensity in CM $= 792 \times 100$ CS (cent = 100) $= 79200$ CS

$$= \frac{79200}{60} \text{ CM} = 1320 \text{ CM.}$$

8.2.4. Grade of Service (GOS)

For non-blocking service of an exchange, it is necessary to provide as many lines as there are subscribers. But it is not economical. So, some calls have to be rejected and retried when the lines are being used by other subscribers. The grade of service refers to the proportion of unsuccessful calls relative to the total number of calls. GOS is defined as the ratio of lost traffic to offered traffic.

$$\text{GOS} = \frac{\text{Blocked Busy Hour calls}}{\text{Offered Busy Hour calls}} \quad ...(8.6)$$

$$\text{GOS} = \frac{A - A_0}{A_0} \quad ...(8.7)$$

where A_0 = carried traffic (equation 8.4)

A = offered traffic

$A - A_0$ = lost traffic.

The smaller the value of grade of service, the better is the service. The recommended GOS is 0.002, *i.e.* 2 call per 1000 offered may lost. In a system, with equal no. of servers and subscribers, GOS is equal to zero.

GOS is applied to a terminal to terminal connection. But usually a switching centre is broken into the following components

(*a*) an internal call (subscriber to switching office)

(*b*) an outgoing call to the trunk network (switching office to trunk)

(*c*) the trunk network (trunk to trunk)

(*d*) a terminating call (switching office to subscriber).

The GOS calculated for each component is called component GOS. The overall GOS is in fact approximately the sum of the component grade of service.

Example 8.3. *During a busy hour, 1400 calls were offered to a group of trunks and 14 calls were lost. The average call duration has 3 minutes. Find (a) Traffic offered (b) Traffic carried (c) GOS and (d) The totul duration of period of congestion.*

Given data : $n = 1400$

$h = 3$

$T = 60$, *lost calls = 14*

Sol.

(*a*) Traffic offered $A = \frac{1400 \times 3}{60} = 70 \text{ E}$

(*b*) Traffic carried $A_0 = \frac{1386 \times 3}{60} = 69.3 \text{ E}$

(*c*) $\text{GOS} = \frac{A - A_0}{A_0}$

where $A - A_0 = 70 - 69.3 = 0.7$ E (lost traffic)

$$\text{GOS} = \frac{0.7}{69.3} = 0.01$$

(*d*) Total duration = 0.01 × 3600 = 36 seconds.

There are two possibilities of call blocking. They are (*a*) Loss system and (*b*) Waiting system or delay system. In loss system, a suitable GOS is a percentage of calls which are lost because no equipment is available at the instant of call request. In waiting system, a GOS objective could be either the percentage of calls which are delayed or the percentage which are delayed more than a certain length of time.

8.2.5. Blocking Probability and Congestion

The value of the blocking probability is one aspect of the telephone company's grade of service. The blocking probability is discussed in section 5.5.2. The basic difference between GOS and blocking probability is that GOS is a measure from subscriber point of view whereas the blocking probability is a measure from the network or switching point of view. Based on the number of rejected calls, GOS is calculated, whereas by observing the busy servers in the switching system, blocking probability will be calculated. The blocking probabilities can be evaluated by using various techniques. Lee graphs and Jacobaeus methods are popular and occurate methods (See section 5.5.2). The blocking probability B is defined as the probability that all the servers in a system are busy.

Congestion theory deals with the probability that the offered traffic load exceeds some value. Thus, during congestion, no new calls can be accepted. There are two ways of specifying congestion. They are time congestion and call congestion. Time congestion is the percentage of time that all servers in a group are busy. The call or demand congestion is the proportion of calls arising that do not find a free server. In general GOS is called call congestion or loss probability and the blocking probability is called time congestion.

If the number of sources is equal to the number of servers, the time congestion is finite, but the call congestion is zero. When the number of sources is large, the probability of a new call arising is independent of the number already in progress and therefore the call congestion is equal to time congestion.

8.3. MODELLING OF TRAFFIC

To analyze the statistical characteristics of a switching system, traffic flow and service time, it is necessary to have a mathematical model of the traffic offered to telecommunication systems. The model is a mathematical expression of physical quantity to represent the behaviour of the quantity under consideration. Also the model provides an analytical solutions to a teletraffic problems. As the switching system may be represented in different ways, different models are possible. Depending on the particular system and particular circumstance, a suitable model can be selected.

In practice, the facilities of the switching systems are shared by many users. This arrangement may introduce the possibility of call setup inability due to lack of available facilities. Also in data transfer, a system has to buffer message while waiting for transmission. Here size of the buffer depends on traffic flow. As serving the number of subscribers subject to fluctuation (due to random generation of subscriber calls, variations in holding time, location of the exchange, limitation in servers etc), modelling of traffic is studied using the concepts and methods of the theory of probability.

If a subscriber finds no available server for his call attempt, he will wait in a line (queue) or leave immediately. This phenomenen may be regarded as a queueing system. The mathematical description of the queueing system characterics is called a queueing model.

Once a mathematical model is obtained, various analytical and computational tools can be used for analysis and synthesis purposes.

8.3.1. Elements of Probability

The traffic generated by a subscriber is random in telecommunication switching system. The behaviour of the network for the random request for service is random process. In this section, some elemental ideas of probability theory and proability distributions are discussed.

Probability. Probability can be defined as the relative frequency of occurance of a random event. Each event has a probability defined as the ratio of the number of times it occurs to the total number of trials. Thus the probability of the occurrence of an event A for N trials is

$$P(A) = \lim_{N \to \infty} \frac{N_A}{N} \qquad ...(8.8)$$

The probability of occurance of an event P is a positive number and that $0 \leq P \leq 1$. If an event is not possible P = 0, while if an event is certain, P = 1.

Mutually Exclusive events. Two possible outcomes of an experiment are defined as being mutually exclusive if the occurrence of one outcome precludes the occurrence of the other. If the events are A and B with probabilities P(A) and P(B), then the probability of occurrence of either A or B is written as

$$P(A \text{ or } B) = P(A) + P(B) \qquad ...(8.9)$$

For more than two mutually exclusive outcomes, say $A_1, A_2, ..., A_n$.

$$P(A_1 \text{ or } A_2 \text{ or } ... A_L) = \sum_{j=1}^{L} P(A_j) \qquad ...(8.10)$$

For example, in tossing a coin, if head occurs, occurrence of tail cannot take place.

Conditional and Joint probability. Suppose that we contemplate two experiments A and B with outcomes $A_1, A_2, ...$ and $B_1, B_2, ...$. The probability of outcome B_k, given that A_j is known to have occurred is called conditional probability given as

$$P(B_k | A_j) = \frac{P(A_j, B_k)}{P(A_j)} \qquad ...(8.11)$$

Similarly, the probability of outcome A_j, given that B_k is known to have occurred is given as

$$P(A_j | B_k) = \frac{P(A_j, B_k)}{P(B_k)} \qquad ...(8.12)$$

where $P(A_j, B_k)$ is called joint probability, that is the joint occurrence of A_j and B_k. From equation 8.11 and 8.12, $P(A_j, B_k)$ is given as

$$P(A_j, B_k) = P(B_k | A_j)\, P(A_j) = P(A_j, B_k)\, P(B_k) \qquad ...(8.13)$$

Sub 8.13 in equation 8.12 gives

$$P(A_j | B_k) = \frac{P(A_j)}{P(B_k)}\, P(B_k | A_j) \qquad ...(8.14)$$

This result is known as Bayes' theorem.

If the outcome of B_k does not depend at all on which outcome A_j accompanies it, we say that the outcomes A_j and B_k are independent. When outcomes are independent, the probability of a joint occurrence of particular outcomes is the product of the probabilities of the individual independent outcomes. It is given as

$$P(A_j, B_k) = P(A_j)\,P(B_k) \quad ...(8.15)$$

This result may be extended to any arbitrary number of outcomes. Thus

$$P(A_j, B_k, C_i, ...) = P(A_j)\,P(B_k)\,P(C_i) \quad ...(8.16)$$

Random variables and Random process. Subscribers generates calls in random manner. The call generation by the subscribers and therefore the behaviour of the network or the switching system is described as a random process. It is also referred as stochastic process. In random process, one or more quantities vary with time in such a way that the instantaneous values of the quantities are not determinable but are predictable with certain probability. The quanties are called random variables.

In telecommunication system, telephone traffic is referred as random process and the number of simultaneous active subscribers and simultaneous busy servers are assumed as random variables. The variation of traffic over a period of time (30 minutes or 60 minutes) is a typical random process. The random process may be discrete or continuous. In telecommunication, the variable representing the number of simultaneous calls is discrete. Thus, in our modelling we use discrete state stochastic processes.

Definition of statistical terms :

Mean : $$E(x) = \sum_{i=1}^{N} \frac{x_i}{N} \quad ...(8.17)$$

where x = a random variable

N = number of trials

x_i = the outcome of the individual trials.

Mean is also referred as the expectation or average of x and represented by μ.

Variance. The variance of x is

$$\text{Var}\,(x) = \frac{\sum_{i=1}^{N} (x_i - E(x))^2}{N} = \frac{\sum_{i=1}^{N} (x_i)^2}{N} - E(x)^2 \quad ...(8.18)$$

The variance of a process is a measure of how the individual outcomes differ from the mean.

Standard deviation. The standard deviation of a random variable x is given by

$$\sigma(x) = \sqrt{\text{Var}(x)} \quad ...(8.19)$$

The ratio of standard deviation to the mean of a random variable x is given by

$$\rho(x) = \frac{\sigma(x)}{E(x)} \quad ...(8.20)$$

8.3.2. Discrete Probability Distributions

While the mean and variance tell a great deal about a random variable, they do not tell us everything. The most complete information can be obtained from by the distribution of the random variable. The distribution is the probability associated with each possible outcome.

Mean and variance. The values of $P(x_1)$, $P(x_2)$, ..., $P(x_n)$ for a discrete random process $X = x_1, x_2, ..., x$ may be plotted as shown in Fig. 8.3.

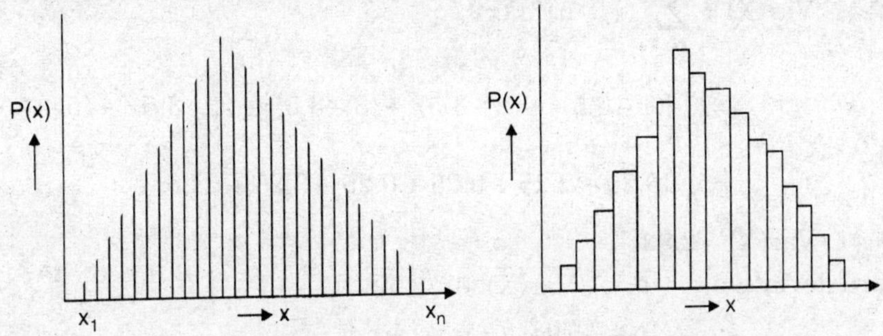

(a) Bar chart (b) Histogram

Fig. 8.3. Discrete probability distribution.

The value of x corresponding to the centre of gravity of the above diagram is called the average or mean or expectation $E(x)$

$$\mu = E(x) = \sum_{j=1}^{\infty} x_j \, P(x_j) \qquad ...(8.21)$$

$P(x_j)$ must be non-negative and must sum to 1. The mean μ is also known as the first moment of the random variable. μ_k, the kth moment of the random variable is defined as

$$\mu_k = \sum_{j=1}^{\infty} (x_j)^k \, P(x_j) \qquad ...(8.22)$$

The variance Var(X) or σ^2 is a measure of the dispersion or spread of the histogram. It is defined as the mean squared deviation of x from the mean.

$$\text{Var}(X) = \sigma^2 = \sum_{j=1}^{\infty} (x_j - \mu)^2 \, P(x_j) \qquad ...(8.23)$$

$$= \sum_{j=1}^{\infty} P(x_j)\,(x_j)^2 - \mu^2$$

The variance is generally referred as second central moment. The kth central moment is defined as

$$C_k = \sum_{j=1}^{\infty} P(x_j)\,(x_j - \mu)^k \qquad ...(8.24)$$

The standard deviation, σ if defined as the square root of the variance.

$$\sigma_x = \sqrt{\text{Var}(X)} \qquad ...(8.25)$$

Example 8.4. *The distribution for the outcomes of rolling a single die (half a pair of dice) is* $P_1 = P_2 = P_3 = P_4 = P_5 = P_6 = \frac{1}{6}$. *Find the mean, variance and standard deviation.*

Sol.

$$\text{Mean} = \sum_{j=1}^{6} x_j - P(x_j) = \frac{1}{6}[1 + 2 + 3 + 4 + 5 + 6] = 3.5$$

$$\text{Var(X)} = \sum_{j=1}^{6} (x_j - \mu)^2 P(x_j)$$

$$= \frac{1}{6}[(1-3.5)^2 + (2-3.5)^2 + (3-3.5)^2 + (4-3.5)^2 + (5-3.5)^2 + (6-3.5)^2]$$

$$= \frac{1}{6}[6.25 + 2.25 + 0.25 + 0.25 + 2.25 + 6.25]$$

$$\text{Var(X)} = 2.92.$$

Standard deviation, $\sigma = \sqrt{\text{Var(X)}} = \sqrt{2.92}$

$$\sigma = 1.708.$$

If X and Y are independent random variable, the mean and variance of sum or difference is given by

$$E(X + Y) = E(X) + E(Y)$$

$$\text{Var}(X + Y) = \text{Var}(X) + \text{Var}(Y) \qquad ...(8.26)$$

$$E(X - Y) = E(X) - E(Y)$$

$$\text{Var}(X - Y) = \text{Var}(X) + \text{Var}(Y) \qquad \text{[not minus]}$$

Bernouilli or binomial distribution. The distribution of x repetitions of an event (say head or toss) with two possible outcomes is called a binomial distribution and the numbers above are called binomial coefficients. Consider the series of trials (n) satisfies the following conditions :

(*a*) Each trial can have two possible outcomes. *e.g.* success or failure with probabilities p and $1 - p$.

(*b*) The outcome of each trial is an independent random event.

(*c*) Statistical equilibrium (*i.e.* the probabilities do not change).

The number of ways of choosing an x things out of n trial is given as

$$\binom{n}{x} = C(n, x) = \frac{n!}{x!(n-x)!} \qquad ...(8.27)$$

$C(n, x)$ is called binomial coefficient.

The probability of one particular combination of x success and $n - x$ failures is $p^x (1 - p)^{n-x}$.

Given two disjoint events with probabilities p and $(1 - p)$, the probability of first occurring x times and second occurring $(n - x)$ times in n trial, the most general form of binomial distribution is

$$P(n, x, p) = C(n, x)\, p^x (1 - p)^{n-x}$$

or simply

$$P(x) = \binom{n}{x} p^x (1 - p)^{n-x} \qquad ...(8.28)$$

The mean is $\mu = np$ ...(8.29)

The variance is $\sigma^2 = np\,(1 - p)$ ...(8.30)

Example 8.5. *Suppose that the prob of a person being on the phone 0.1 and there are 10 people in the office. Decide number of lines required in order to ensure that everyone get reasonable chance.*

Sol. For $n = 10, p = 0.1, 1{-}p = 0.9$ and for various x (no of persons), using equation (8.28), binomial distribution is tabulated in table 8.1.

Table 8.1. Binomial distribution of example 8.5.

x	$P(x) = \binom{n}{x} p^x (1-p)^{n-x}$	$P \geq x$
0	0.34868	1.00000
1	0.38742	0.65132 [$P \geq x - P(x)$ for $x = 0$]
2	0.19371	0.26390
3	0.05739	0.07019
4	0.01116	0.02180
5	0.00149	0.00164
6	0.00014	0.00015
7	0.00001	0.00001

$$\text{mean} = np = 10 \times 0.1 = 1$$

$$\text{variance} = \sqrt{np\,(1-p)} = \sqrt{10 \times 0.1 \times 0.9} = 0.95.$$

Table 8.1. illustrates that probability of 5 or more people talking is only 0.00164. Thus considerable saving can be achieved by using only 4 lines or 3 lines with certain wait :

Poisson distribution. From 8.28, we know

$$P(x) = \binom{n}{x} p^x (1-p)^{n-x}$$

When n is large, say $n \to \infty$,

$$p(x) = \frac{n^x}{x\,!}\, p^x (1-p)^{n-x}$$

As $\mu = np$, $\quad p = \mu/n, \quad$ and $\quad n = \mu/p$

$$P(x) = \frac{\mu^x}{x\,!\,.\,p^x} \,.\, p^x\,(1-p)^{n-x} \qquad (n = \mu/p)$$

$$= \frac{\mu^x}{x\,!}\left(1 - \frac{\mu}{n}\right)^{n-x} \qquad (p = \mu/n)$$

$$= \frac{\mu^x}{x\,!}\left(1 - \frac{\mu}{n}\right)^{n}\left(1 - \frac{\mu}{n}\right)^{-x} \simeq \frac{\mu^x}{x\,!}\left(1 - \frac{\mu}{n}\right)^{n}$$

By power series,

$$(1-x)^n = 1 - nx + \frac{n(n-1)}{2!}x^2 - \frac{n(n-1)(n-2)}{3!}x^3 + \ldots\ldots$$

Thus $$P(x) = \frac{\mu^x}{x!}\left[1 - n\left(\frac{\mu}{n}\right) + \frac{n(n-1)}{2!}\left(\frac{\mu}{n}\right)^2 - \frac{n(n-1)(n-2)}{3!}\left(\frac{\mu}{n}\right)^3 + \ldots\ldots\right]$$

$$P(x) = \frac{\mu^x}{x!}\left[1 - \mu + \frac{\mu^2}{2!} + \frac{\mu^3}{3!} + \ldots\ldots\right]$$

$$P(x) = \frac{\mu^x}{x!}e^{-\mu} \qquad \ldots(8.31)$$

where μ = mean and equation 8.31 is a poisson distribution. Variance $\sigma^2 = \mu$ (also).

8.3.3. Continuous Probability Distributions

The random process may be discrete or continuous. For continuous function, the histogram (Fig. 8.3(*b*)) becomes a continuous. As probability P(x) tends to zero, the probability density function $p(x)$ is plotted against x. The probability density $p(x)$ is defined by

$$p(x) = P(a \le x \le b) = \int_a^b p(x)\,dx \qquad \ldots(8.32)$$

and $$\int_{-\infty}^{\infty} p(x)\,dx = 1$$

For continuous distribution,

Mean, $$\mu = \int_{-\infty}^{\infty} x\,p(x)\,dx \qquad \ldots(8.33)$$

Variance, $$\sigma^2 = \int_{-\infty}^{\infty} x^2\,p(x) - \mu^2 \qquad \ldots(8.34)$$

For independent random variable X and Y,

mean, E(X ± Y) = E(X) + E(Y) ...(8.35)

Var (X ± Y) = Var (X) + Var (Y)

Negative exponential distribution. The probability that the interval between event T, will exceed time t is given by

$$p\,(T \ge t) = e^{-t/\overline{T}} \qquad \ldots(8.36)$$

where $\overline{T}$ is the mean interval between events.

Gaussian or normal distribution. A random variable, x is said to be normally distributed, if its density function has the form

$$n(\mu, \sigma : x) = p(x) = \frac{1}{\sqrt{2\pi\sigma^2}}\,e^{-(x-\mu)^2/2\sigma^2} \qquad \ldots(8.37)$$

where μ = mean

σ = standard deviation.

The equation 8.37 is plotted in Fig. 8.4.

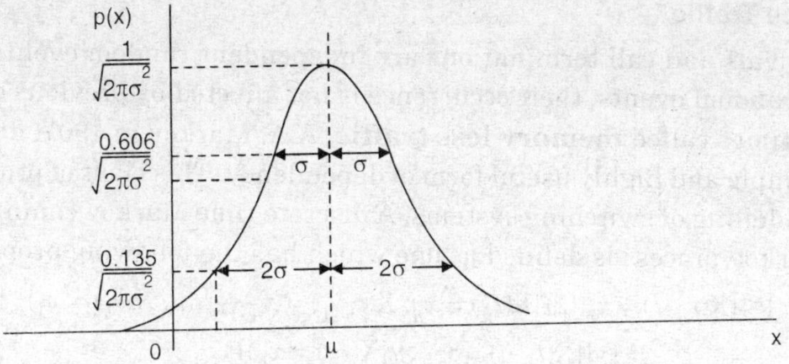

Fig. 8.4. The Gaussian density function.

The central limit theorem states that the probability density function of the sum of a large number of independent variables tends towards $n(\mu, \sigma : x)$ as 'n' increases. Substituting, $t = (x - \mu)/\sigma$ in equation 8.37, and neglecting σ

$$p(t) = \frac{1}{\sqrt{2\pi}} e^{-t^2/2} \qquad ...(8.38)$$

Equation 8.38 is called standard normal distribution.

8.3.4. Statistical Parameters

The random process may be discrete or continuous. Similarly the time index of random variables can be discrete or continuous. Thus, there are four different types of process namely (*a*) continuous time continuous state (*b*) continuous time discrete state (*c*) discrete time continuous state and (*d*) discrete time discrete state. In telecommunication switching sytstem, our interest is discrete random process and therefore for modelling a switching system, we use discrete state stochastic process. A discrete state stochastic process is often called a chain.

A statistical properties of a random process may be obtained in two ways :

(*i*) Observing the behaviour of the system to be modelled over a period of time repeatedly and thus the data obtained is called a single sample. The average determined by measurements on a single sample function at successive times will yield a **time average**.

(*ii*) Simultaneous measurements of the output of a large number of statistically identical random sources, sields a collection of sources is called as an **ensemble** and the individual noise waveforms is called the **sample function**. The statistical average made at some fixed time $t = t_1$ on all the sample functions of the ensemple is the **ensemble average.**

The above two ways are analogous to obtaining the statistics from tossing a die repeatedly (large number) or tossing one time the large number of dice.

In general, time average and ensemble average are not the same due to various reasons. When the statistical characteristics of the sample functions do not change with time, the random process is described as being **stationary**. The random process which have identical time and ensemble average are known as **ergodic processes.** An ergodic process is stationary, but a stationary process is not necessarily ergodic.

Telephone traffic is nonstationary. But the traffic obtained during busy hour may be considered as stationary (which is important for modelling) as modelling non-stationary is difficult.

8.3.5. Pure Chance Traffic

Here, the call arrivals and call terminations are independent random events. If call arrivals are independent random events, their occurrence is not affected by previous calls. This traffic is therefore sometimes called **memory less traffic.** A.A. Markov in 1907, defined properties and proposed a simple and highly useful form of dependency. This class of processes is of great interest to our modelling of switching systems. A discrete time Markov chain *i.e.* discrete time discrete state Markov process is defined as one which has the following property.

$$\begin{aligned} P\,[\{X(t_{n+1}) = x_{n+1}\}/\{X(t_n) = x_n,\ X(t_{n-1}) = x_{n-1}, \ldots\ldots X(t_1) = x_1\}] \\ = P\,[\{X(t_{n+1}) = x_{n+1}\}/\{X(t_n) = x_n\}] \end{aligned} \quad \ldots(8.39)$$

where $t_1 < t_2 \ldots\ldots < t_n < t_{n+1}$ and x_i is the ith discrete state space value.

Equation 8.39 states that the propability that the random variable X takes on the value x_{n+1} at time step $n + 1$ is entirely determined by its state value in the previous time step n and is independent of its state values in earlier time steps ; $n - 1, n - 2, n - 3$ etc.

8.3.6. The Birth and Death Process

The birth and death process is a special case of the discrete state continuous time Markov process, which is often called a continuous-time Markov chain. The number of calls in progress is always between 0 and N. It thus has N + 1 states. If the Markov chain can occur only to adjacent states (*i.e.* probability change from each state to the one above and one below it) the process is known as birth-death (B–D) process. The basic feature of the method of Markov chains is the kolmogorov differential-difference equation, for the limiting case, can provide a solution to the state probability distribution for the Erlang systems and Engset systems.

Let N(t) be a random variable specifying the size of the population at time t. For a complete description of a birth and death process, we assume that N(t) is in state k at time t and has the following properties :

1. P(k) is the probability of state k and P(k + 1) is the probability of state k + 1.

2. The probability of transition from state k to state k + 1 in short duration Δt is $\lambda_k\,\Delta t$, where λ_k is called the birth rate in state k.

3. The probability of transition from state k to state $k - 1$ in the time interval Δt is $\mu_k\,\Delta t$, where μ_k is called the death rate in state k.

4. The probability of no change of state in the time interval Δt is equal to $1 - (\lambda_k + \mu_k)\,\Delta t$.

5. The probability in Δt, from state k to a state other than $k + 1$ or $k - 1$ is zero.

Based on the above properties, birth and death process of N(t) and state transition rate diagrams are shown in Fig. 8.5.

At statistical equilibrium (*i.e.* stationary), let P_{jk} is the conditional probability, that is the probability of state increases from j to k. Similarly P_{kj} is the probability of state decrease from k to j.

The probabilities P(0), P(1), ... P(N) are called the **state probabilities** and the conditional probabilities P_{jk}, P_{kj} are called **transition probabilities.** The transition probabilities satisfy the following condition :

$$P_{jk}(t) \geq 0, \sum_{k=0}^{\infty} P_{jk}(t) = 1 \qquad ...(8.40)$$

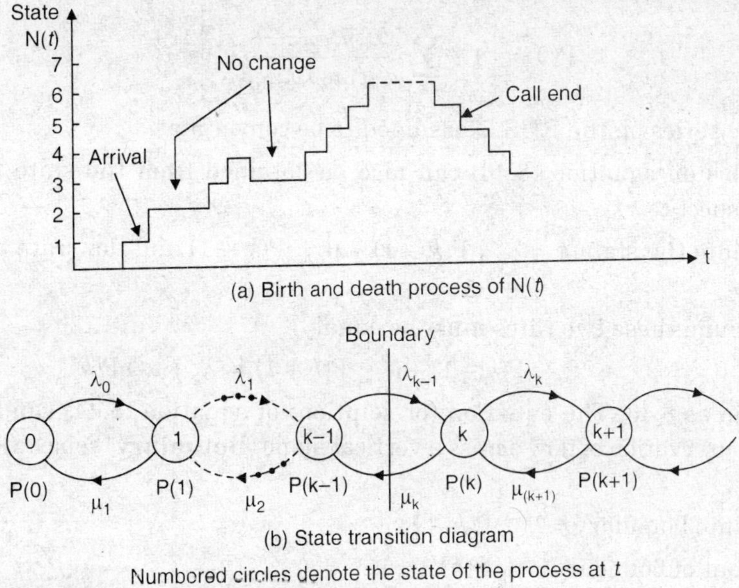

Fig. 8.5. State transition diagram of birth and death process.

Markov theorem states that for any Markov process characterized by the transition probability P_{jk}, the limit

$$\lim_{t \to \infty} P_{jk} = P(k) \qquad ...(8.41)$$

exist and does not depend on j and the probability $P(k)$.

According to Markov's,

$$\frac{d}{dt} P(k) = -(\lambda_k + \mu_k) P(k) + \lambda_{k-1} P(k-1) + \mu_{k+1} P(k+1)$$

$$k = 0, 1, 2, ..., \text{ with } \lambda_{-1} = \mu_0 = P_{-1} = 0 \qquad ...(8.42)$$

This set of differential-difference equations represents the dynamic behavior of the birth and death process. As $t \to \infty$, Equation 8.42 becomes

$$-(\lambda_k + \mu_k) P(k) + \lambda_{k-1} P(k-1) + \mu_{k+1} P(k+1) = 0 \qquad ...(8.43)$$

with $\lambda_{-1} = \mu_0 = P_{-1} = 0$.

This set of equations (8.41 and 8.42), together with the normalization condition (8.43) uniquely determines the required

$$\sum_{k=0}^{\infty} P(k) = 1 \qquad ...(8.44)$$

and state probabilities $P(k)$ as

$$P(k) = \frac{\lambda_{k-1}}{\mu_k} P(k-1) \qquad ...(8.45)$$

Consequently, $$P(k) = \frac{\lambda_0 \lambda_1 \ldots\ldots \lambda_{k-1}}{\mu_1 \mu_2 \ldots\ldots \mu_k} P(0),\ k = 1, 2, 3, \ldots\ldots \quad \ldots(8.46)$$

The probability P(0) can be determined by the equation (8.44) as

$$P(0) = \left[1 + \sum_{k=1}^{\infty} \frac{\lambda_0 \lambda_1 \ldots\ldots \lambda_{k-1}}{\mu_1 \mu_2 \ldots\ldots \mu_k}\right]^{-1} \quad \ldots(8.47)$$

where the infinite series on the RHS is assumed to be convergent.

The equilibrium equation (8.43) can also be obtained from the state transition rate diagram (with respect to k).

Flow rate into the state $k = \lambda_{k-1} P(k-1) + \mu_{k+1} P(k+1)$ and flow rate out of the state $k = (\lambda_k + \mu_k) P(k)$.

In equilibrium, these two rates must be equal

$$\lambda_{k-1} P(k-1) + \mu_{k+1} P(k+1) = (\lambda_k + \mu_k) P(k)$$

which is the same as 8.43. The equation (or solution) of equation (8.44) can be obtained by considering the conservation of flow across a vertical plane (**Boundary**) separating the adjacent states.

Flow rate into boundary = $\lambda_{k-1} P(k-1)$

and Flow rate out of boundary = $\mu_k P(k)$

At equilibrium $$\mu_k P(k) = \lambda_{k-1} P(k-1)$$

$$P(k) = \frac{\lambda_{k-1} P(k-1)}{\mu_k}$$

which is the same as 8.43.

8.4. LOSS SYSTEMS

The service of incoming calls depends on the number of lines. If number of lines equal to the number of subscribers, there is no question of traffic analysis. But it is not only uneconomical but not possible also. So, if the incoming calls finds all available lines busy, the call is said to be **blocked.** The blocked calls can be handled in two ways.

The type of system by which a blocked call is simply refused and is lost is called **loss system.** Most notably, traditional analog telephone systems simply block calls from entering the system, if no line available. Modern telephone networks can statistically multiplex calls or even packetize for lower blocking at the cost of delay. In the case of data networks, if dedicated buffer and lines are not available, they block calls from entering the system.

In the second type of system, a blocked call remains in the system and waits for a free line. This type of system is known as **delay system.** In this section loss system is described. Delay system is discussed in the section 8.5.

These two types differs in network, way of obtaining solution for the problem and GOS. For loss system, the GOS is probability of blocking. For delay system, GOS is the probability of waiting.

Erlang determined the GOS of loss systems having N trunks, with offered traffic A, with the following assumptions. (*a*) Pure chance traffic (*b*) Statistical equilibrium (*c*) Full availablity and (*d*) Calls which encounter congestion are lost. The first two are explained in previous section. A system with a collection of lines is said to be a fully-accessible system, if all the lines are equally accessible to all in arriving calls. For example, the trunk lines for inter office calls are fully accessible lines. The lost call assumption implies that any attempted call which encounters congestion is immediately cleared from the system. In such a case, the user may try again and it may cause more traffic during busy hour.

The Erlang loss system may be defined by the following specifications.

1. The arrival process of calls is assumed to be Poisson with a rate of λ calls per hour.
2. The holding times are assumed to be mutually independent and identically distributed random variables following an exponential distribution with $1/\mu$ seconds.
3. Calls are served in the order of arrival.

There are three models of loss systems. They are :

1. Lost calls cleared (LCC)
2. Lost calls returned (LCR)
3. Lost calls held (LCH)

All the three models are described in this section.

8.4.1. Lost Calls Cleared (LCC) System

The LCC model assumes that, the subscriber who does not avail the service, hangs up the call, and tries later. The next attempt is assumed as a new call. Hence, the call is said to be cleared. This also referred as blocked calls lost assumption. The first person to account fully and accurately for the effect of cleared calls in the calculation of blocking probabilities was A.K. Erlang in 1917.

Consider the Erlang loss system with N fully accessible lines and exponential holding times. The Erlang loss system can be modeled by birth and death process with birth and death rate as follows.

$$\lambda_k = \begin{cases} \lambda, k = 0, 1, \ldots\ldots N-1 \\ 0 \quad k \geq N \end{cases} \qquad \ldots(8.48)$$

$$\mu_k = \begin{cases} k\mu, k = 0, 1, \ldots\ldots N \\ 0, \quad k > N \end{cases} \qquad \ldots(8.49)$$

From 8.46, $$P(k) = \frac{\lambda_0\, \lambda_1 \ldots\ldots \lambda_{k-1}}{\mu_1\, \mu_2 \ldots\ldots \mu_k} P(0), k = 1, 2, 3$$

Substituting equation (8.48 and 8.49) in the above equation, we get

$$P(k) = \frac{1}{k!}\left(\frac{\lambda}{\mu}\right)^k P(0), k = 1, 2, 3, \ldots\ldots N \qquad \ldots(8.50)$$

From equation (8.4), the offered traffic is

$$A = \frac{\lambda}{\mu}$$

Substituting in (8.50), we get

$$P(k) = \frac{1}{k!}(A)^k P(0),\ k = 1, 2, 3, ...\ N \qquad ...(8.51)$$

The probability P(0) is determined by the normalization condition

$$\sum_{k=0}^{N} P(k) = P(0) \sum_{k=0}^{N} \frac{A^k}{k!} = 1$$

$$P(0) = \frac{1}{\sum_{k=0}^{N} \frac{A^k}{k!}} \qquad ...(8.52)$$

Substituting (8.52) in (8.51), we get

$$P(k) = \frac{A^k / k!}{\sum_{k=0}^{N} \frac{A^k}{k!}} \qquad ...(8.53)$$

The probability distribution is called the truncated **Poisson distribution** or **Erlang's loss distribution.** In particular when k = N, the probability of loss is given by

$$P(N) = B(N, A) = \frac{A^N}{N! \sum_{k=0}^{N} \left(\frac{A^k}{k!}\right)} \qquad ...(8.54)$$

where $A = \lambda/\mu$.

This result is variously referred to as **Erlang's formula of the first kind, the Erlangs-B formula or Erlangs loss formula.**

Equation 8.54 specifies the probability of blocking for a system with random arrivals from an infinite source and arbitrary holding time ditributions. The Erlang B formula gives the time congestion of the system and relates the probability of blocking to the offered traffic and the number of trunk lines.

Values from B(N, A) obtained from equation 8.54, have been plotted against the offered traffic 'A' erlangs for different values of the number of N lines in Fig. 8.6.

In design problems, it is necessary to find the number of trunk lines needed for a given offered traffic and a specified grade of service. The offered load generated by a Poisson input process with a rate λ calls per hour may be defined as

$$A = \int_0^\infty \lambda t \, dH(t) = \lambda h \qquad ...(8.55)$$

where λt = average number of increasing calls at fixed time internal

h = average holding time

The average number of occupied or busy trunks is defined as the carried load

$$A' = \sum_{k=1}^{N-1} k\, P(k) + N \sum_{k=N}^{\infty} P(k) \qquad ...(8.56)$$

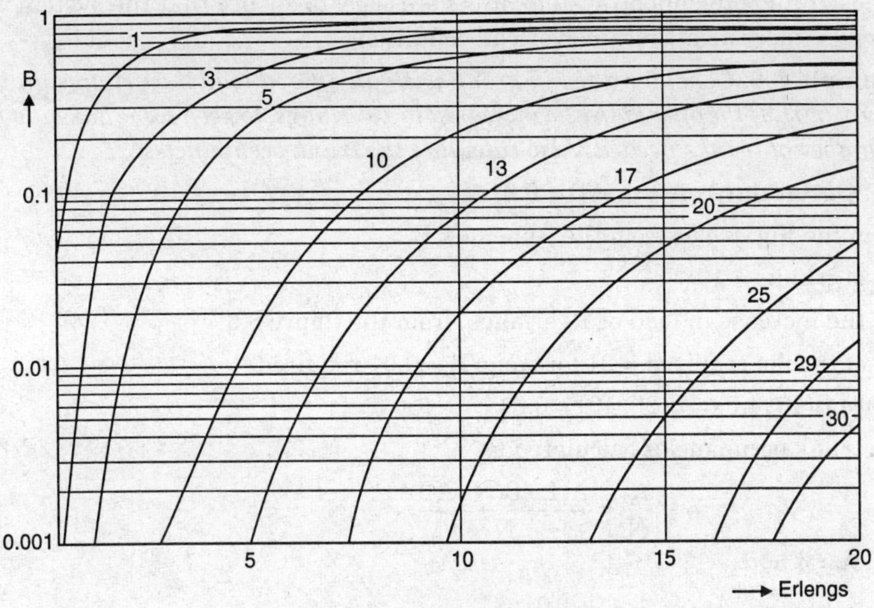

Fig. 8.6. Plot of B(N, A).

$$A' = A\,[1 - B(N, A)] \qquad ...(8.57)$$

Thus, the carried load is the position of the offered load that is not lost from the system. The carried load per line is known as the trunk occupancy. From 8.57,

$$\rho = \frac{A'}{N} = \frac{A(1-B)}{N} \qquad ...(8.58)$$

The trunk occupancy ρ is a measure of the degree of utilization of a group of lines and is sometimes called the utilization factor. Fig. 8.7 presents the output channel utilization for variuos blocking probabilities and number of servers.

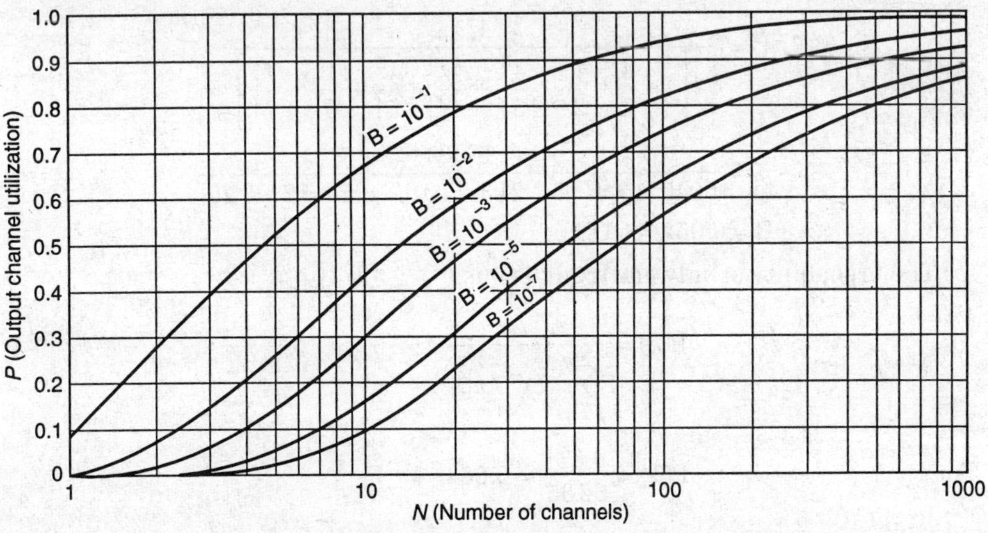

Fig. 8.7. Output channel utilization of LCC system.

In designing a telephone system, it is necessary to ensure that the system will operate satisfactorily under the moderate overload condition.

Example 8.6. *Consider a trunk group with an offered load 4.5 erlangs and a blocking probability of 0.01. If the offered traffic increased to 13 erlangs, to keep same blocking probability, find the number of trunks needed. Also calculate the trunk occupancies.*

Sol. Given data : A = 4.5, B = 0.01

From the Fig. 8.6 or from the Appendix B,

No. of trunks = 10

For the increase in load of 13 erlangs, from the figure 8.6

No. of trunks required = 21 for same B = 0.01 required

Hence B(10, 4.5) = B(20, 13) = 0.01

The trunk occupancies calculated as

$$\rho = \frac{A'}{N} = \frac{A(1 - B(N, A))}{N}$$

for N = 10, A = 4.5

$$\rho = \frac{4.5(1 - 0.01)}{10} = 0.4455$$

for N = 21, A = 13

$$\rho = \frac{13(1 - 0.01)}{21} = 0.613$$

Thus, the group of 20 trunks is more efficient than the group of 10 trunks.

Example 8.7. *A group of 7 trunks offeres 4E of traffic, find (a) the grade of service (b) the probability that only one trunk is busy (c) the probability that only one trunk is free and (d) the probability that at least one trunk is free.*

Sol. Given data : N = 7, A = 4E

From equation 8.54,

(*a*)

$$B(7, 4) = \frac{4^7}{7!\left[1 + \frac{4}{1} + \frac{4^2}{2!} + \frac{4^3}{3!} + \frac{4^4}{4!} + \frac{4^5}{5!} + \frac{4^6}{6!} + \frac{4^7}{7!}\right]}$$

$$= \frac{16384}{5040\,[1 + 4 + 8 + 21.3 + 10.6 + 8.5 + 5.7 + 3.25]}$$

$$B = 0.052 \doteq \text{GOS}.$$

(*b*) The probability of only one trunk is busy

$$P(k) = \frac{A^k / k!}{\sum_{k=0}^{N} (A^k / k!)}$$

For *k* = 1

$$P(1) = \frac{4/1!}{62.35} = 0.064$$

(*c*) The probability that only one trunk is free

$$P(6) = \frac{4^6/6!}{62.35} = \frac{5.68}{62.35} = 0.0912$$

(*d*) The probability that at least one trunk is free

$$P(k < 7) = 1 - P(7) = 1 - B = 1 - 0.052 = 0.948.$$

8.4.2. Lost Calls Returned (LCR) System

In LCC system, it is assumed that unserviceable requests leave the system and never return. This assumption is appropriate where traffic overflow occurs and the other routes are in other calls service. If the repeated calls does not exist, LCC system is used. But in many cases, blocked calls return to the system in the form of retries. Some examples are subscriber concentrator systems, corporate tie lines and PBX trunks, calls to busy telephone numbers and access to WATS lines. Including the retried calls, the offered traffic now comprise two components *viz.*, new traffic and retry traffic. The model used for this analysis is known as lost calls returned (LCR) model. The following assumptions are made to analyse the RLC model.

1. All blocked calls return to the system and eventually get serviced, even if multiple retries are required.

2. Time between call blocking and regeneration is random statistically independent of each other. This assumption avoid complications arising when retries are correlated to each other and tend to cause recurring traffic peaks at a particular waiting time interval.

3. Time between call blocking and retry is somewhat longer than average holding time of a connection. If retries are immediate, congestion may occur or the network operation becomes delay system.

Consider a system with first attempt call arrival ratio of λ (say 100). If a percentage B (say 8%) of the calls blocked, B times λ retries (*i.e.* 8 calls retries). Of these retries, however a percentage B will be blocked again.

Hence by infinite series, total arrival rate λ′ is given as

$$\lambda' = \lambda + B\lambda + B^2\lambda + B^3\lambda + \ldots\ldots$$

$$\lambda' = \frac{\lambda}{1 - B} \quad \ldots(8.59)$$

where B is the blocking probability from a lost calls cleared (LCC) analysis.

The effect of returning traffic is insignificant when operating at low blocking probabilities. At high blocking probabilities, it is necessary to incorporate the effects of the returning traffic into analysis. The effect of high blocking probability, is illustrated in the following example.

Example 8.8. *Consider a trunk group of 10 circuits serving a first attempt offered traffic load of 7 erlangs. What is the blocking probability. If the number of circuits increased to 13, what is the blocking probability. Find the blocking probability for the retries assuming random retries for all blocked calls.*

Sol. As 7E of traffic arise from a large number of subscribers, infinite source analysis (or Erlang B loss system) is justified.

(*a*) For A = 7E, N = 10, From the table (Appendix B) or from Fig. 8.6, the blocking probability B = 0.079 or approximately 8%.

If the number of server increased to 13, for the same traffic of A = 7E, the blocking probability is B = 0.02 or approximately 2%.

The difference is shown in Fig. 8.8.

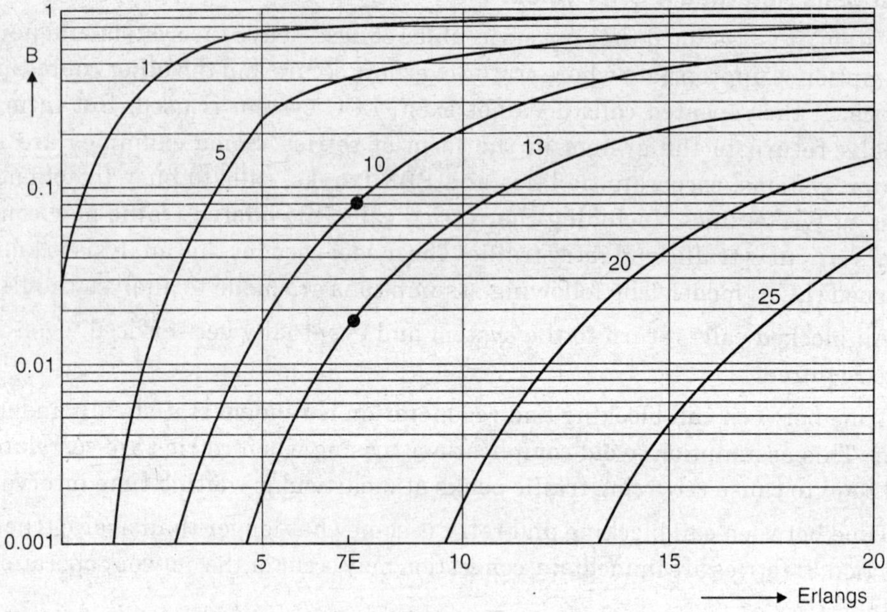

Fig. 8.8. Blocking probability of LCR for example 8.7.

(*b*) With lost call returned analysis, for the given N = 10, A = 7, the blocking probability is 0.079

We know $A = \frac{\lambda h}{T}$

where h is average holding time assumed as 3 minutes

T is duration of the observation (normally 60 minutes).

Hence, $\lambda = \text{arrival rate} = \frac{AT}{h} = \frac{7 \times 60}{3} = 140$

From equation 8.59, $\lambda' = \frac{\lambda}{1-B} = \frac{140}{1-0.079} = 152$

For $\lambda' = 152$, $A' = \frac{\lambda' h}{T} = \frac{152 \times 3}{60} = 7.6E$

Thus, the total traffic load including retries is approximately 7.6E. The corresponding B = 0.1 or 10%.

Similarly, the blocking probabilities for some more iterations are tabulated in table 8.2.

Table 8.2. LCR analysis for example 8.7

Iteration (For N = 10)	B (From table)	$\lambda = \frac{AT}{h}$	$\lambda' = \frac{\lambda}{1-B}$	$A' = \frac{\lambda' h}{T}$
1. A = 7	0.079	140	152	7.6
2. A = 7.6	0.1	152	169	8.45
3. A = 8.45	0.145	169	198	9.9
4. A = 9.9	0.215	198	252	12.6
5. A = 12.6	0.3125	252	367	18.32
6. A = 18.32	0.474	367	697	34.8

If the number of circuits in the trunk group increased to 13, the blocking probability is 0.02 or 2%. If the similar iteration procedure as in table 8.2 is made, the difference in blocking probability can be identified. Fig. 8.9 shows the blocking probability of LCR with LCC.

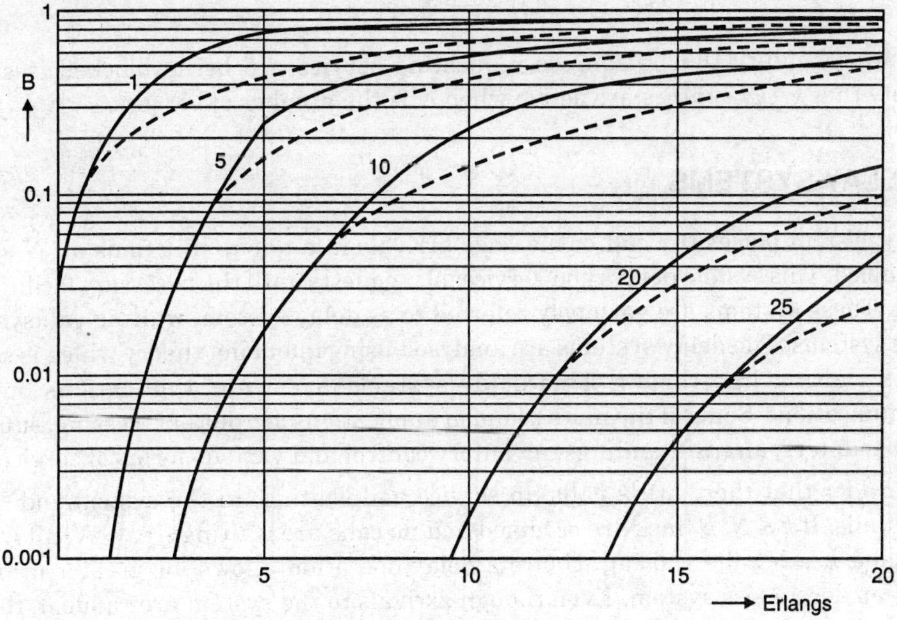

Fig. 8.9. Blocking probability of LCR with LCC.

8.4.3. Lost Calls Held (LCH) System

In a lost calls held system, blocked calls are held by the system and serviced when the necessary facilities become available. The total time spend by a call is the sum of waiting time and the service time. Each arrival requires service for a continuous period of time and terminates its request independently of its being serviced or not. If number of calls blocked, a portion of it is lost until a server becomes free to service a call. An example of LCH system is the time assigned speech interpolation (TASI) system.

LCH systems generally arise in real time applications in which the sources are continuously in need of service, whether or not the facilities are available. Normally, telephone network does not operate in a lost call held manner. The LCH analysis produces a conservative design that helps account for retries and day to day variations in the busy hour calling intensities. A TASI system concentrates some number of voice sources onto a smaller number of transmission channels. A source receives service only when it is active. If a source becomes active when all channels are busy, it is blocked and speech clipping occurs. Each speech segment starts and stops independently of whether it is served or not. Digital circuit multiplication (DCM) systems in contrast with original TASI, can delay speech for a small amount of time, when necessary to minimize the clipping.

LCH are easily analysed to determine the probability of the total number of calls in the system at any one time. The number of active calls in the system at any time is identical to the number of active sources in a system capable of carrying all traffic as it arises. Thus the distribution of the number in the system is the poisson distribution. The poisson distribution given as

$$P(x) = \frac{\mu^x}{x!} e^{-\mu}.$$

The probability that k sources requesting service are being blocked is simply the probability that $k + N$ sources are active when N is the number of servers.

8.5. DELAY SYSTEMS

The delay system places the call or message arrivals in a queue if it finds all N servers (or lines) occupied. This system delays non-serviceable requests until the necessary facilities become available. These systems are variously referred to as delay system, waiting-call systems and queueing systems. The delay systems are analysed using queueing theory which is sometimes known as waiting line theory. This delay system have wide applications outside the telecommunications. Some of the more common applications are data processing, supermarket check out counters, aircraft landings, inventory control and various forms of services.

Consider that there are k calls (in service and waiting) in the system and N lines to serve the calls. If $k \leq N$, k lines are occupied and no calls are waiting. If $k > N$, all N lines are occupied and $k - N$ calls waiting. Hence a delay operation allows for greater utilization of servers than does a loss system. Even though arrivals to the system are random, the servers see a somewhat regular arrival pattern. A queueing model for the Erlang delay system is shown in Fig. 8.10.

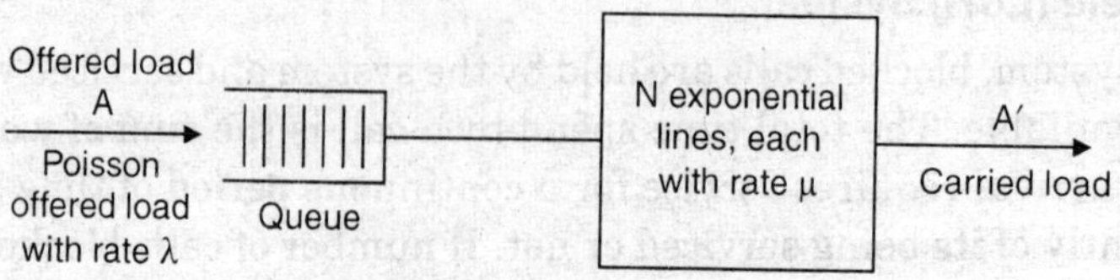

Fig. 8.10. Queueing model.

The basic purpose of the investigation of delay system is to determine the probability distribution of waiting times. From this, the average waiting time W as random variable can be easily determined. The waiting times are dependent on the following factors :

1. Number of sources
2. Number of servers
3. Intensity and probabilistic nature of the offered traffic
4. Distribution of service times
5. Service discipline of the queue.

In a delay system, there may be a finite number of sources in a physical sense but an infinite number of sources in an operational sense because each source may have an arbitrary number of requests outstanding. If the offered traffic intensity is less than the servers, no statistical limit exists on the arrival of calls in a short period of time. In practice, only finite queue can be realised. There are two service time distributions. They are constant service times and exponential service times. With constant service times, the service time is deterministic and with exponential, it is random. The service discipline of the queue involves two important factors.

1. Waiting calls are selected on first-come, first served (FCFS) or first-in-first-out (FIFO) service.

2. The second aspect of the service discipline is the length of the queue. Under heavy loads, blocking occurs. The blocking probability or delay probability in the system is based on the queue size in comparison with number of effective sources.

We can model the Erlang delay system by the birth and death process with the following birth and death rates respectively.

$$\lambda_k = \lambda,\ k = 0, 1, \ldots\ldots \quad \ldots(8.60) \qquad \text{and} \qquad \mu_k = \begin{cases} k\mu,\ k = 0, 1, \ldots\ldots N-1 \\ N\mu\ k \geq N \end{cases} \quad \ldots(8.61)$$

Under equilibrium conditions, the state probability distribution P(k) can be obtained by substituting these birth rates into the following equation (8.46).

$$P(k) = \frac{\lambda_0\,\lambda_1 \ldots\ldots \lambda_{k-1}}{\mu_1\,\mu_2 \ldots\ldots \mu_k}\,P(0)\ k = 1, 2, \ldots\ldots$$

we set

$$P(k) = \begin{cases} \dfrac{1}{k!}\left(\dfrac{\lambda}{\mu}\right)^k P(0) & 0 \leq k \leq N \\[2ex] \dfrac{\left(\dfrac{\lambda}{\mu}\right)^k}{N!\,N^{k-N}} P(0) & k \geq N. \end{cases}$$

As $A = \dfrac{\lambda}{\mu}$, we get

$$P(k) = \begin{cases} \dfrac{A^k}{k!} P(0) & 0 \leq k \leq N \\[2ex] \dfrac{A^k}{N!\,N^{k-N}} P(0); & k > N. \end{cases} \quad \ldots(8.62)$$

Under normalised condition,

$$\sum_{k=0}^{\infty} P(k) = 1 \quad \text{or} \quad \sum_{k=0}^{N-1} \frac{A^k}{k!} P(0) + \sum_{k=N}^{\infty} \frac{A^k}{N!\,N^{k-N}} P(0) = 1$$

$$\frac{1}{P(0)} = \sum_{k=0}^{N-1} \frac{A^k}{k!} + \frac{N^N}{N!} \sum_{k=N}^{\infty} \left(\frac{A}{N}\right)^k$$

$$= \sum_{k=0}^{N-1} \frac{A^k}{k!} + \frac{N^N}{N!}\left[\left(\frac{A}{N}\right)^N + \left(\frac{A}{N}\right)^{N+1} + \left(\frac{A}{N}\right)^{N+2} + \ldots\ldots\right]$$

$$= \sum_{k=0}^{N-1} \frac{A^k}{k!} + \frac{N^N}{N!}\left(\frac{A}{N}\right)^N \left[1 + \frac{A}{N} + \left(\frac{A}{N}\right)^2 + \ldots\ldots\right]$$

$$= \sum_{k=0}^{N-1} \frac{A^k}{k!} + \frac{A^N}{N!}\left[\frac{1}{1 - A/N}\right] = \sum_{k=0}^{N-1} \frac{A^k}{k!} + \left[\frac{A^N}{N!} + \frac{A^N}{N!}\left(\frac{A}{N-A}\right)\right]$$

$$\frac{1}{P(0)} = \sum_{k=0}^{N} \frac{A^k}{k!} + \frac{A^N}{N!}\left(\frac{A}{N-A}\right)$$

$$P(0) = \frac{1}{\sum_{k=0}^{N} \frac{A^k}{k!} + \frac{A^N}{N!}\left(\frac{A}{N-A}\right)} \qquad \ldots(8.63)$$

We know (from 8.51)

$$P(k) = \frac{A^k}{k!} P(0), \; k = 1, 2, \ldots\ldots, N \qquad \ldots(8.64)$$

Now, the probability of waiting (the probability of finding all lines occupied) is equal to

$$P(W > 0) = C(N, A) \qquad \ldots(8.65)$$

From 8.63 and 8.64

$$C(N, A) = \frac{A^N / N!}{\sum_{k=0}^{N} \frac{A^k}{k!} + \frac{A^N}{N!}\left(\frac{A}{N-A}\right)} \qquad \ldots(8.66)$$

$$\frac{1}{C(N, A)} = \frac{\sum_{k=0}^{N} \frac{A^k}{k!}}{\frac{A^N}{N!}} + \frac{\frac{A^N}{N!}\left(\frac{A}{N-A}\right)}{\frac{A^N}{N!}}$$

$$\frac{1}{C(N, A)} = \frac{1}{B} + \frac{A}{N-A}.$$

$$\text{Prob. (delay)} = P(W > 0) = C(N, A) = \frac{BN}{N - A(1-B)}, \text{ BA neglected} \qquad \ldots(8.67)$$

where B = Blocking probability for a LCC system

N = Number of servers

A = Offered load (Erlangs)

Equation 8.66 and 8.67 are referred as **Erlang's second formula, Erlang's delay formula or Erlang's C formula.**

For single server systems (N = 1), the probability of delay reduces to ρ, which is simply the output utilization or traffic carried by the server. Thus the probability of delay for a single server system is also.

The distribution of waiting times for random arrivals, random service times, and a FIFO service discipline is

$$(P > t) = P(> 0)\, e^{-(N-A)t/h} \qquad \text{...(8.68)}$$

where P(> 0) = probability of delay (equation 8.67)

h = average service time of negative exponential service time distribution.

By integrating equation 8.68, the average waiting time for all arrivals can be determined as

$$W(t)_{avg.} = \frac{C(N, A)\, h}{N - A} \qquad \text{...(8.69)}$$

$W(t)_{avg.}$ is the expected delay for all arrivals. The average delay of only those arrivals that get delayed is commonly denoted as

$$T_W = \frac{h}{N - A} \qquad \text{...(8.70)}$$

Example 8.9. *A message switching network is to be designed for 90% utilization of its transmission link. Assuming exponentially distributed message lengths and an arrivals rate of 10 messages per min. What is the average waiting time and what is the probability that the waiting time exceeds 3 minutes ?*

Given data : $\rho = 90\% = 0.9$

$\lambda = 10$ *messages / minute.*

Sol. Assuming single channel, N = 1

For N = 1, prob (delay) = P(> 0) = ρ = 0.9. Also A = ρ = 0.9.

The average service time $h = \frac{\text{Prob. (delay)}}{\lambda} = \frac{0.9}{10} = 0.09$

Average waiting time $W(t)_{avg} = \frac{P(>0)\, h}{N - A} = \frac{0.9 \times 0.09}{1 - 0.9} = 0.81$ min.

Prob. of the waiting time exceeding 3 minutes

$$= P(> 5) = P(> 0)\, e^{-(N-A)t/h} = 0.9 \times e^{-(1-0.09)3/0.09} = 0.032.$$

Thus 3.2% of the message experience queing delay of more than 3 minutes.

Example 8.10. *A PBX has 4 operators and receives 300 calls during a busy hour. The average holding time is 36 seconds. Assume that call arrivals are poisonian and service time is negative exponential distribution. Calculate (a) the percentage of calls on queue (b) average delay (c) percentage of calls delayed for more than 45 seconds, 30 seconds and 20 sec.*

Given data : *N = 3, n = 300 calls, h = 36 sec.*

Sol. $$A = \frac{nh}{T} = \frac{300 \times 36}{3600} = 3 \text{ Erlangs.}$$

(*a*) From equation (8.63, 8.64 and 8.67), we get

$$P(0) = 0.038$$

$$B = 0.0285$$

$$C(4, 3) = 0.105$$

i.e. 10.5% of calls have delay on answer.

(*b*) Average delay of a call

$$W(t)_{\text{avg.}} = \frac{C(N, A)h}{N - A} = 3.78 \text{ seconds.}$$

(*c*) Percentage of calls delayed for more than 45 seconds

$$P(t \geq 45) = C(N, A)\, e^{-(N-A)t/h} = 0.03$$

that is 3% calls delayed for more than 45 sec

$$P(t \geq 30) = 0.045$$

that is 4.5% calls delayed for more than 30 sec

$$P(t \geq 20) = 0.06$$

that is 6% calls delayed for more than 20 sec.

8.6. COMBINED LOSS AND DELAY SYSTEM

In the Erlang loss system, no waiting is allowed, while in the Erlang delay system, if number of sources is greater than the channels, they are placed in queue and no loss occurs. The combined delay and loss system can be modelled by the birth and death process. The following birth and death rates are considered

$$\lambda_k = \begin{cases} \lambda, & 0 \leq k < n \\ 0, & k \geq n \end{cases} \quad \text{...(8.71)}$$

and

$$\mu_k = \begin{cases} k\mu, & 0 \leq k < N \\ s\mu, & N \leq k \leq n \\ 0, & k > n \end{cases} \quad \text{...(8.72)}$$

where N = number of channels

n = number of sources

Substituting 8.71 and 8.72 in 8.46, we get

$$P(k) = \begin{cases} \dfrac{A^k}{k!} P(0) & 0 \leq k < N \\[2ex] \dfrac{A^k}{N!\,N^{k-N}} P(0) & N \leq k \leq n \end{cases} \quad \text{...(8.73)}$$

Under normalised condition,

$$\sum_{k=0}^{N} P(k) = 1 \quad \text{or} \quad \sum_{k=0}^{N-1} \frac{A^k}{k!} P(0) + \sum_{k=N}^{n} \frac{A^k}{N!\,N^{k-N}} P(0) = 1$$

$$\frac{1}{P(0)} = \sum_{k=0}^{N-1} \frac{A^k}{k!} + \frac{A^N}{N!}\left(\frac{N}{N-A}\right)\left[1-\left(\frac{A}{N}\right)^{n-N+1}\right] \quad \text{...(8.74)}$$

The probability of waiting is given by

$$P(W > 0) = \frac{A^k}{k!} P(0)$$

From 8.74, C_n (N, A) = P(W > 0) = $\frac{A^N}{N!}\left(\frac{N}{N-A}\right)\left[1-\left(\frac{A}{N}\right)^{n-N+1}\right] P(0)$ ...(8.75)

The probability of loss is equal to

$$P(n) = B_n(N, A) = \frac{A^N}{N!\,N^{n-N}} P(0) \quad \text{...(8.76)}$$

whe n = N, Equation 8.76 reduces the Erlang-B formula,

i.e. $$B_N(N, A) = B(N, A) \quad \text{...(8.77)}$$

when $n = \infty$, Equation 8.75 becomes the Erlang-delay formula

$$C_\infty(N, A) = C(N, A) \quad \text{...(8.78)}$$

ACRONYMS

B-D	—	Birth and Death process
DCM	—	Digital circuit multiplication
CCS	—	Cent call seconds
CM	—	Call minutes
CS	—	Call seconds
E	—	Erlangs
FCFS	—	First come first served
FIFO	—	First in first out
GOS	—	Grade of service
LCC	—	Lost calls cleared
LCH	—	Lost calls held
LCR	—	Lost calls returned
TASI	—	Time assigned speech interpolation

RELATED WEBSITES

http://personal.telefonica.terra.es/web/vr

http://pass.maths.org.uk/issue 2/erlang

http://rdweb.cns.vt.edu/~(gaylord/Traffic engineering

http://www.tele.dtu.dk/teletraffic

http://www.winlab.rutgers.edu/wpcs/erlang lecture.Pdf.

REVIEW QUESTIONS

1. What is the purpose of traffic engineering ?
2. Define calling rate and holding time.
3. What is traffic pattern ?
4. Write short notes on (*a*) Erlangs and (*b*) CCS
5. Define GOS.
6. Explain the two types of congestion.
7. Comment on modelling of traffic.
8. What is Erlang loss system ? Name three models of the loss system.
9. Give the expression of Erlang's-B formula.
10. Compare LCR and LCH system.
11. Draw the queueing model of delay systems.
12. Give the expression of Erlang's-C formula.
13. Comment on combined loss and delay system.
14. Derive an expression for the state probability P(*k*) using B-D process.
15. Derive an expression to obtain the Erlang's formula for the first kind of loss system.
16. A group of 10 trunks is offered 5E of traffic, find (*a*) GOS (*b*) the probability that only one trunk is busy (*c*) the probability that only one trunk is free and (*d*) the probability that at least one trunk is free.
17. Derive an expression to obtain the Erlang's second formula of delay system.

9

Telephone Network Organisation

9

Telephone Network Organisation

9.1. INTRODUCTION

A telecommunication network contains a large number of links joining different locations, which are known as the nodes of the network. These nodes may be end instruments (subscriber nodes), switching centres (switching nodes), networks providing just a link between nodes (transmission nodes) or service nodes (which provides service on demand such a voice mail boxes, stock market price announcement, sports results etc). To provide efficient communication, a telephone network should include various transmission system (for example, terestial, microwave, optical satellite communications), switching system (to identity and connect calling and called subscriber) and to exchange information between subscriber and switching systems or between interexchanges, a good signalling system required.

The calling and called subscriber should be connected almost instantly. So, as an identification, a numbering system is introduced and it varies region to region and country to country. Telephone networks require certain form of procedure to route a particular call to the destination for effective and cost effective communication. So, the telephone network should be implemented with a good routing plan. An establishment of an exchange includes heavy expenses on switching equipments, establishing trunks and links, buildings, infrastructures human resources to handle the exchange etc. These capital lost and the day-to-day expenses must be met by the exchanges through its subscribers. So, the billing and charging the subscriber calls or data transfer is a vital part of the network.

Also, introducing a new exchange, extention of the existing exchanges, upgration of the facilities and speed up the switching, changing the sales strategy based on the competition, addition of new services, management of maintenance, providing employment to the skilled peoples etc., are based on the government policies or the telephone company's business strategies. Thus, the functions of telecommunication networks is limit less and network management is an important part of any telecommunication network organization.

In this chapter, network planning, routing plan, numbering and charging are discussed.

9.2. NETWORK MANAGEMENT

The basic goal of the network management is to maintain efficient operations during equipment failures and traffic overloads. Also controlling the flow of call requests during network overload is a vital function of network management. For the effective network management the study of various services provided by the network, offered load of the network, classification of the network based on services offered, interconnection of different types of networks and network planning is important. Based on the data available for the above factors, the network management is become updated.

9.2.1. Network Planning

The planning of telecommunication system comprising a network of switching centres includes various plans. From intiating the network to the extension of the network based on the increased load, network planning plays a vital role.

Network services : The capabilities often collectively referred to "as intelligence" within the network are listed below. Depending upon the applications the network is to handle the interconnectivity with other networks. The following functions and its essential parts are included in the network. Various services are :

1. **Switching.** The process of interconnecting incoming calls or data to the appropriate outgoing channel called destination is referred or switching. Various switching methods right from manual exchange to the automated digital switching system were discussed in previous chapters.

2. **Routing.** The ability of the network to select a path to connect calling and called subscriber for telephone conversations or providing path for data transfer between source and destination is referred as routing. The network generally choses a path and somctimes user may specify it.

3. **Flow control.** It is the ability of a network to reject traffic. Managing the rate at which traffic enters a network is referred to as flow control. A network without effective flow control procedures becomes very inefficient.

4. **Security.** There are two ways of providing security of the network. First, to increase the security of operation in presence of faults. To provide adequate security, the complete network may be duplicated or triplicated. Second, preventing unauthorized access to the network and the data it carries. This may be achieved by pass words, data encryption and providing limiting factors in accessing the network.

5. **Signalling.** A signalling system link the variety of switching system, transmission system and subscriber equipments in a telecommunication network to enable the network to function as a whole. The signalling system was explained in chapter 7.

6. **Traffic management.** The ability of the network to keep track of traffic levels is referred as traffic management. Traffic management is useful both in short term and long term bases. On a short term basis, it can be used to support dynamic routing and flow control. Over a long term it can be used in network design to identify parts of the network where capacity may be productively increased or decreased.

7. **Accountability.** This includes charging, billing, accounting and inventory control. This is the ability of the network to track the users of the network.

8. **Administration.** It is related to the ability of the network to identify the load of a network and providing corresponding upgradition of parts, extention of networks facility. It also identities the sales strategy, investment planning etc.

9. **Inter networking.** It is the ability of the network to perform the functions needed to communicate with and across other networks. This includes providing routes for traffic crossing through, into and out of the network, and allocating resources such as buffers and link capacity to traffic originating other networks.

Network levels : All the above functions and some other service related functions are usually classified or grouped into different levels. This grouping ease the network and the concerned network engineers to carryout the functions efficiently. The grouping differs from network to network or divisions to divisions or between telephone companies. The ability of the network to provide these functions has a profound effect on the type of network. In particular, the nodes of the network become more complex and more expensive as their functionality increases. Fig. 9.1 shows the different levels of the network management.

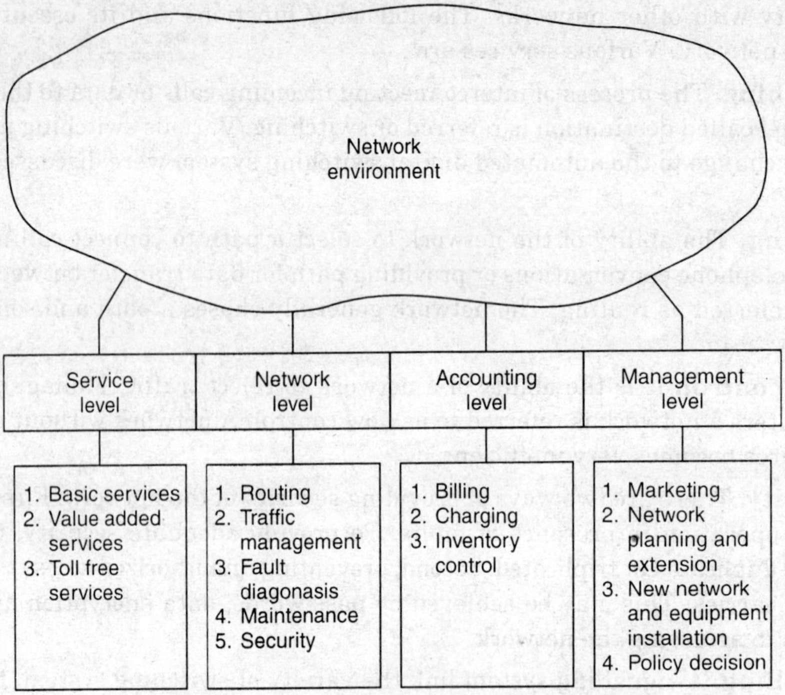

Fig. 9.1. Different levels of network management.

Variuos networking plans : A national telecommunication network is large and complex. Therefore certain plans are needed to govern the design of network. The plans are independent and are affected by the predicted (or planned) growth rate of the telecommunication system. More specific network planning are :

1. Routing plan
2. Numbering plan
3. Charging plan
4. Transmission plan
5. Signalling plan

6. Grade of service
7. Network control and network administration.

The choice of a plan for a telecommunication system generally involves comparison of the economics of various possible plans. It also involves comparison of the economy of various possible plans and involves a certain amount of human judgement.

In next sections, the routing plans, numbering plans and charging plans are discussed. The transmission plans includes various communication methods, sources of transmission, models of transmission etc. The signalling plane and grade of service are explained sufficient enough in chapter 7 and 8. The network control and network adminstration are beyond the scope of this book and not dealt.

9.2.2. Types of Networks

There are in general three networks which can be used for any services (voice or data transfer). They are :

(*a*) **Public switched network.** It allows access to the end office, connects through the long-distance network, and delivers to the end point. There are many hierarchies depends on the wish of network provider. But, the goal of the network hierarchies is to complete the call in the least amount of time and the shortest route possible.

(*b*) **Private networks.** Many companies, depending on their size and need, create or build their own networks. If their networks are underutilised, they may give their network for hire or lease. These networks employ mixture of technologics.

(*c*) **Hybrid networks.** To provide a service, if an organisation uses both private and public networks, the network is referred as hybrid network. Normally, the high-end usage services are connected via private facilities, the lower volume locations use the switched network. This usually works out better financially for the organisation because the costs can be fully justified on a location by location basis. As the private line networks are designed to suit integration of voice, data, video graphics and facisimile transmission, pressure is more on it.

The customers are generally in need of variety of services. Even though certain networks are used to perform many services, they are more effective for a particular one or two services. Hence, based on the services, the networks are classified as

1. The Public Switched telephone Network (PSTN)—for telephony.
2. The public switched telegraph network—for telex
3. Data networks—for voice and data
4. Cellular radio network—for mobile communication
5. Special service networks—to meet specialised demands.

As the above networks may be used to perform mixture of services, many authors categorised the networks into only three classes. They are generally considered as major telecommunication networks.

1. PSTN or POTS
2. Data networks
3. ISDN.

The PSTN networks are studied in earlier chapters. The data networks and ISDN networks are discussed in the chapters 11 and 12 respectively.

9.3. ROUTING PLAN

Routing planning refers to the procedures that determine which path in a network are assigned to particular connections. The switching centres may use fixed routes to each destination. Adaptive routing may be employed in which each exchange may use different routes for the same destination, depending upon traffic conditions. For effective routing of a call, some form of interconnection of switching exchanges are required. In the following pharagraph variuos forms of networking and its related routing are discussed.

9.3.1. Basic Topologies

Three basic topologies are adopted for interconnecting exchanges. Exchanges are interconnected by group of trunk lines referred as trunck groups. Two trunk groups are required between any two exchanges. Mesh, star and mixed or hierarchical are the three basic topologies. Determination of the total number of trunk circuits in any network is necessarily a function of the amount of traffic between each pair of stations or exchanges.

Mesh-connected network. This is also called fully connected topology. The advantage of mesh network is that each station has a dedicated connection to other stations. Therefore, this topology offers the highest reliability and security. If one link in the mesh topology breaks, the network remains active. A major or disadvantage of this topology is that it uses too many connections and therefore requires great deal of wiring, espeically when the number of station increases. The mesh topoly requires N(N – 1)/2 connections. For 100 stations, 4950 links required. Fig. 9.2 depicts a full connected mesh structure.

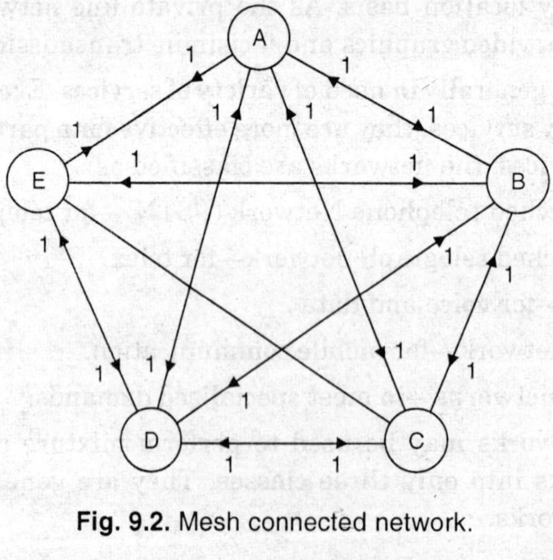

Fig. 9.2. Mesh connected network.

As an example, assume that each exchange generates 1 erlang traffic to each of the other centres. From the appendix B, it can be found that the number of trunks required is approximately four with around 1% grade of service. This is for oneway traffic. It is not necessary to have equal number of trunks for two way communication with respect to a particular station.

In this case, six trunks are sufficient for grade of service of just over 1%. Thus, for low traffic levels, (below about 10 erlangs) the use of two way, provides a significant reduction in the number of trunk circuits required.

For high traffic networks, the cost of network increases. As a compromise, generally, the circuits are divided into three groups. Two groups are used for one way trunks and third group is for both way trunks. An exchange first finds its outgoing trunk before trying one in common group.

This arrangement is practicable if number of exchanges in the network are limited and placed nearby (*i.e.* the lines are short). As the telephone network users are increasing at rapid rate, the concept of mesh network is uneconomical and the technological growth of transmission of signal by multiplexing made this technique to become completely obsolete.

Star topology. It is an alternative to the mesh arrangement. The network configuration shown in Fig. 9.3 (*a*) is called star network. In star network, the number of lines is equal to the number of stations. As shown, a star connection utilises an intermediate exchange called a tandom exchange. Through the tandom exchange (TE) all other exchanges communicate.

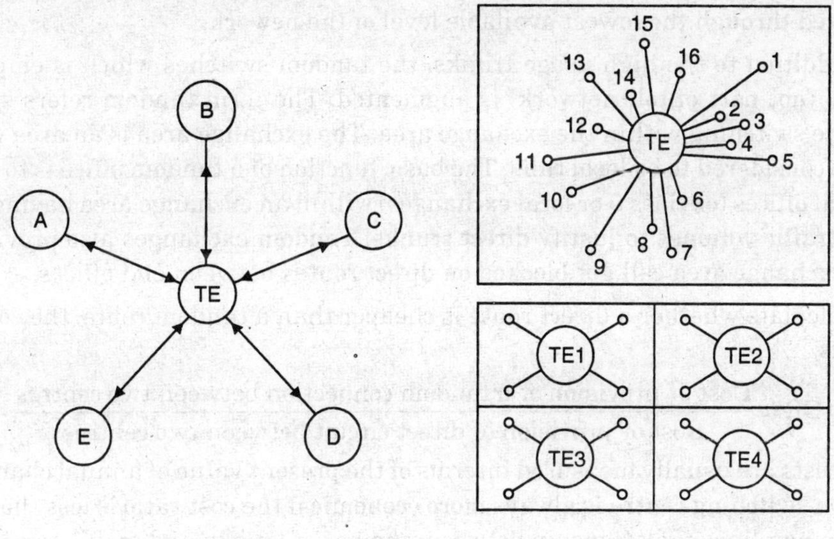

(a) Star network (b) Area with single TE and multi TE

Fig. 9.3. Star topology.

If the number of stations served by a TE increases, they are divided into smaller network, each served by its TE. Fig. 9.3(*b*) shows the star network with splitted setup. This configuration reduces the line cost but increases the exchange costs. With the star arrangement the outer centres require fewer trunk terminations, but the trunk centre has greatly increased number of terminations. So, the star arrangement reduces the design requirements on all but one of the switching centres. It needs larger, and more powerful trunk centre. As only one larger centre is required, star arrangement preferable. Note that the exchange area indicates that all the calls in that area are considered to be local calls.

Hierarchical networks. Many star networks may be inter connected by using an additional tandom exchange, leading to two level star network. An orderly construction of multilevel star networks leads to hierarchical networks. Fig. 1.11 shows the two types of hierarchical structures of AT & T and ITU–T. Hierarchical networks are capable of handling heavy traffic with minimal number of trunk groups. The hierarchical network requires more switching nodes, but achieves significant savings in the number of trunks. Determination of the total number of trunk circuits in entire network is necessarily a function of the amount of traffic between each pair of switching nodes. The efficiency of circuit utilization is the basic motivation for hierarchical switching structures.

In Fig. 1.11, it is shown that, if there is a high traffic intensity between any pair of exchanges, direct trunk groups may be established between any pair of exchanges (dotted lines or trunks of AT & T hierarchical network). These direct routes are known as high usage routes or trunks. In a strictly hierarchical network, traffic from subscriber A to B and vice versa flows through the highest level of hierarchy. Å traffic route via the highest level of hierarchy is known as the final route. Whenever high usage route exists, route is primiarily routed through them. The overflow traffic is routed through hierarchical network. Traffic is always routed through the lowest available level of the nework.

In addition to the high usage trunks, the tandom switches which is employed at the lowest level (not part of toll network) is augmented. The term tandom refers specifically to intermediate switching within the exchange area. The exchange area is an area within which all calls are considered to be local calls. The basic function of a tandom office is to interconnect those central offices (or class 5 or local exchanges) within an exchange area having insufficient interoffice traffic volumes to justify direct trunks. Tandom exchanges also provide alternate routes for exchange area call get blocked on direct routes between end offices.

To calculate whether a direct route is cheaper than a tandom route, the cost ratio (CR) is defined as

$$\text{CR} = \frac{\text{Cost of provision of a tandom connection between two centres}}{\text{Cost of provision of direct circuit between two centres}} \qquad ...(9.1)$$

The costs are usually measured interms of the present value of annual charges. Routing via a tandom switching centre is always more economic if the cost ratio is less than or equal to one. But the non-linear relationship between number of trunks and traffic carried can make tandom rather than direct routing more economic even for values of λ greater than unity. As a general rule, increasing the capacity of an existing trunk route always requires fewer additional trunks than the provision of a new direct trunk route.

9.3.2. Alternative Routing

Based on the assumption that the routing is made only by direct routing or tandom routing, it is found that to route a stream of traffic, tandom route is more economical. In fact, even greater economics are often possible if just a proportion of the traffic is routed directly. This approach is known as alternative routing.

In alternative routing, connections should use the direct trunks (referred as high usage route), because direct route provides better transmission quality and use fewer network facilities. If all the direct trunks are busy, calls are routed via a tandom exchanges or alternate routes to maintain suitably low blocking probabilities. Thus, the networks are designed to allocate a limited number of heavily utilized trunks, in the direct route and provides alternate routes for over flow.

If the high usage route consists of N tunks and the offered traffic is A erlangs, the probability of all trunks busy is given by the Erlangs–B formula (equation 8.54). The traffic carried on high usage route A_H is given by

$$A_H = A(1 - B(N, A)) \text{ erlangs} \quad \text{...(9.2)}$$

the overflow traffic is

$$A_0 = AB(N, A) \text{ erlangs} \quad \text{...(9.3)}$$

The Erlang-B formula is a good representation of the traffic on a high usage route because blocked calls are diverted to the alternative route and does not reappear. But the number of circuits required by a final route to carry the overflow traffic should not be calculated from Erlang's-B formula, because this traffic is not poissonian. The characteristic of traffic for high usage route with overflow is shown in Fig. 9.4.

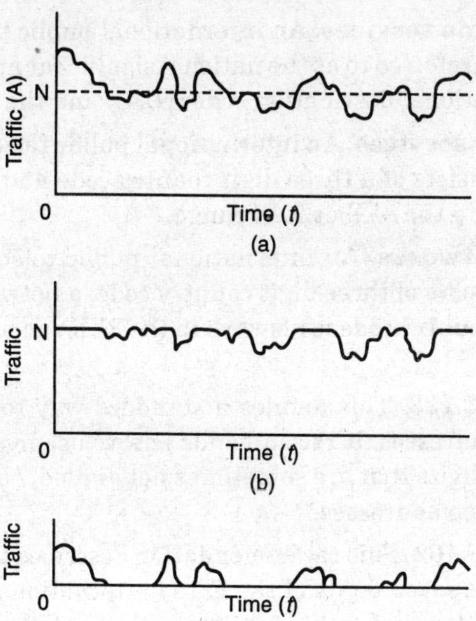

Fig. 9.4. Characteristic traffic for high usage route with overflow.

Fig. 9.4 (*a*) shows the traffic offered to high usage route. The traffic carried by the high usage routes are shown in Fig. 9.4 (*b*). Fig. 9.4 (*b*) depicts that the traffic carried is equal to the traffic offered, if it is less than or equal to number of high usage trunks. If the offered traffic is greater than the number of high usage routes, overflow occurs and the traffic carried equal to number of trunks.

Fig. 9.4 (*c*) shows the over flow traffic. The traffic offered to the final route is thus more peakly than poisonian traffic. The analysis of this traffic requires mean as well as variance. The Wilkinson equivalent Random theory is the widely used method to analyse the random overflow traffic.

9.4. NUMBERING PLAN

The numbering plan is used to identify the subscribers connected in a telecommunication network. The main objective of numbering plan by any nation is to standardise the number length wherever practical according to CCITT recommendations. Other objectives includes (*a*) to meet the challanges of the changing telecom environment (*b*) to meet subscriber needs for a meaningful and user friendly scheme (*c*) to reserve numbering capacity to meet the undefined future needs. In this section, recommendations of ITU, International and National numbering plan are discussed. The numbering plan in India is also focussed.

9.4.1. ITU Recommendations in Numberbing

Some important recommendations of ITU are discribed below :

Recommendation E.164 : It provides the number structure and functionality for three catagories of numbers used for international public telecommunication. The three catagories of numbers are :

1. **National telephone services.** An international public telecommunication number (for geographic areas) is also referred to as the national significant number (NSN). NSN consists of the country code (CC), national destination code (NDC) and the subscriber number (SN).

2. **Global telephone services.** An international public telecommunication number for global telephone service consists of a three digit country code and global subscriber number. The country code is always in the 8XX or 9XX range.

3. **International networks.** An international public telecommunication number for international networks consists of three digit country code, a network identification code and a subscriber number. The country code is alsways in the 8XX range. The identification code is one to four digits.

Recommendation E.123. This defines a standard way to write telephone numbers, email addresses and web addresses. It recommends how to use hyphen (-), space ( ), or period (•). ( ) are used to indicate digits that are sometimes not dialled, / is used to indicate alternate numbers and • is used in web addresses.

Recommendation E.162. This recommendation describes that the originating country must analyse a maximum of seven digits of the E.164 international number. When a number is being analysed, it will be done according to this recommendations.

Also, the international numbering plan or world numbering plan has been defined in recommendations E.160 ; E.161 and E.162.

9.4.2. International Numbering Plan

This plan has to be implemented irrespective of a country's national numbering plane and implemented in accordance to the recommendations of ITU. With some standard international framework, subscribers from different countries can call each other. This plan makes it possible to access all countries with the same country code any where in the world.

For the international numbering plan, the world has been divided into nine geographical area as given below. The general rule is that within each global region each country code starts with the same digit. Fig. 9.5 shows the geographical map of world numbering zones.

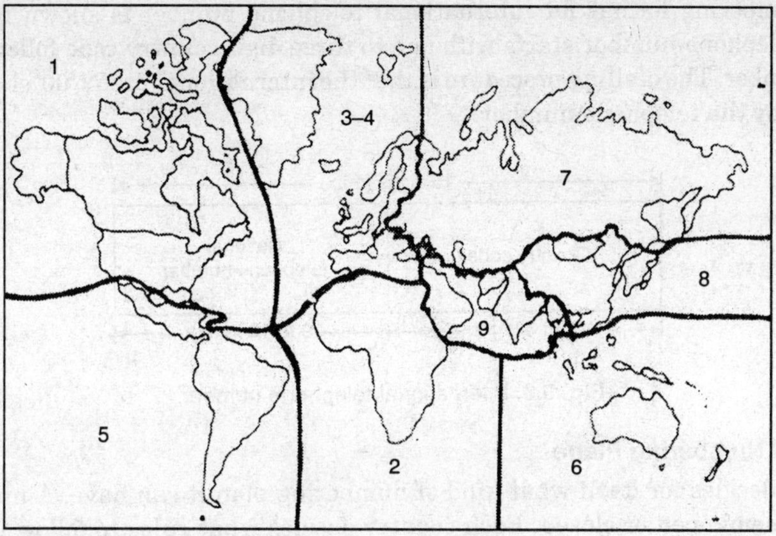

Fig. 9.5. World numbering zones.

Table 9.1 shows the zone code, Zone and two examples in each zone. Digit '0' is not used to indicate zone. Generally '0' is used as Trunk prefix and '00' is used for international prefix.

Table 9.1. World numbering zones

Zone code	Zone	Example	
		Country	Country code
1	North America, caribbean (23 countries)	Canada United States	1 1
2	African continent (61 countries)	Egypt South Africa	20 27
3	Europe (34 countries)	Italy France	39 33
4	Europe (15 countries)	Germany United Kingdom (Czech Republic)	49 44 420
5	Central and South America (28 countries)	Costarica Brazil	506 55
6	Oceania, South Pacific (32 countries)	Malaysia Singapore	60 65
7	Former USSR (2 countries)	Khzakhsten Russian Federation	7 7
8	East Asia (21 countries)	Japan International free phone service Bangladash	81 800 880
9	Middle East, South-West Asia	India Pakistan Iraq Nepal	91 92 964 977

The numbering format for international telephone number is shown in Fig. 9.6. An internation telephone number starts with one to three digit country code followed by 9 to 12 subscriber number. The dialling procedure is that the international prefix '00' should be dialled first followed by the telephone number.

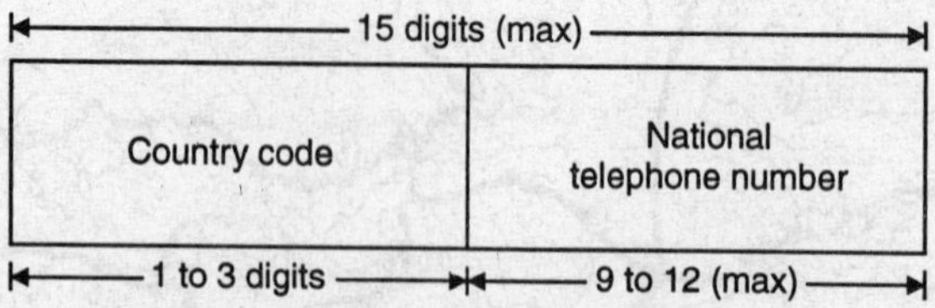

Fig. 9.6. International telephone number.

9.4.3. National Numbering Plane

Each country decides for itself what kind of numbering plan it can have. A numbering plan may be open, semi open or closed. Each country decides what rules to follow when issueing telephone numbers. Such a numbering plan is called national numbering plan.

An open numbering plane or non-uniform numbering scheme allows variations in the number of digits to be used to identify the subscriber. This plan is used in countries equipped extensively with non-director strowger switching system. This scheme is almost extinct. A closed numbering plan or uniform numbering plan refers to a numbering plan which only allows telephone numbers of a predetermined length. Special services (toll free, premium rate, etc.) are usually excluded from this rule.

A semi-open plan permits number lengths to differ by almost one or two digits. Today, this scheme is the most common and is used in many countries including India, Sweden, Switzerland and U.K.

The dialling procedure for national numbering plan are also comes in two catagories. A closed numbering plan refers to a numbering plan which requires users to dial all numbers at all times. This means that local-local calling also requires the area code to be dialled, as well as the trunk prefix. In open dialling plan local calls can be placed without the trunk prefix and area code.

National numbering format is shown in Fig. 9.7.

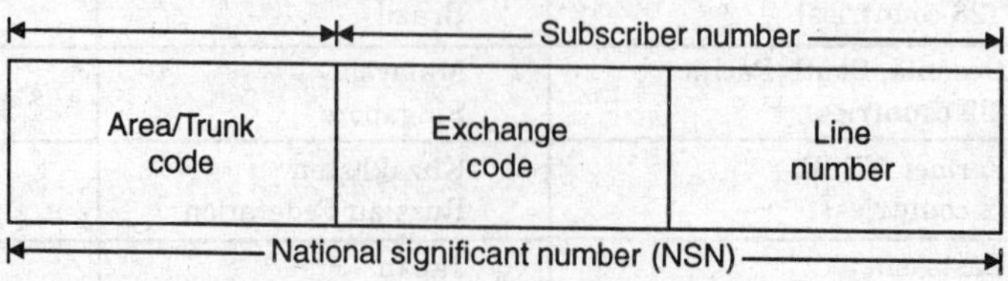

Fig. 9.7. National numbering format.

Thus, the National significant number (NSN) is the combination of trunk code, exchange code and the number. The exchange code and line number together called as subscriber number (SN). NSN length varies from country to country.

9.4.4. Numbering Plan in India

DOT India has released its national numbering plan dated April 2003. It was last reviewd during 1993. This existing numbering plan was formulated at a time when there was no competition in the basic telecom services were not available in the country. Further, the existing numbering plan was meant to address monopolistic environment in national and international long distance dialling. The new numbering plan has been formulated for a projected forecast of 50% teledensity by the year 2030 and thus making numbering space available for 75 crore telephone connections in the country comprising of 30 crore basic and 45 crore cellular mobile connections. The new national numbering plan will be able to meet the challenges of multi operator, multi service environment and will be flexible enough to allow for scalability for next 30 years without any change in basic structure. This plan is aimed at PSTN services, callular mobile services and paging services. This section focuses mainly on the PSTN services.

List of commonly used abbreviations :

BSO — Basic service operators
CAC — Carrier access code
CC — Country code
CIC — Carrier identification code
ICIC — International carrier identification code
ILD — International long distance
LDCA — Long distance number charging area
LDCC — Long distance charging centre
NDC — National destination code
NLD — National long distance
NLDO — National long distance operators
NSN — National significant number
POI — Point of interconnect
SDCA — Short distance number charging area
SDCC — Short distance charging centre
SN — Subscriber number
TAX — Trunk automatic exchange
TC — Trunk code

LDCA, SDCA, NDA, SN and NSN : Broadly, our country is divided into eight regions with each region being identified by a single digit code as shown in Fig. 9.8. The above said notations are generally used in telecommunications and thus discussed to some extent.

LDCA. Long distance charging centre comprises of one or several SDCA's. There are at present 322 LDCA's. Each LDCA has a long distance charging centre (LDCC) which is a Trunk Automatic Exchange (TAX).

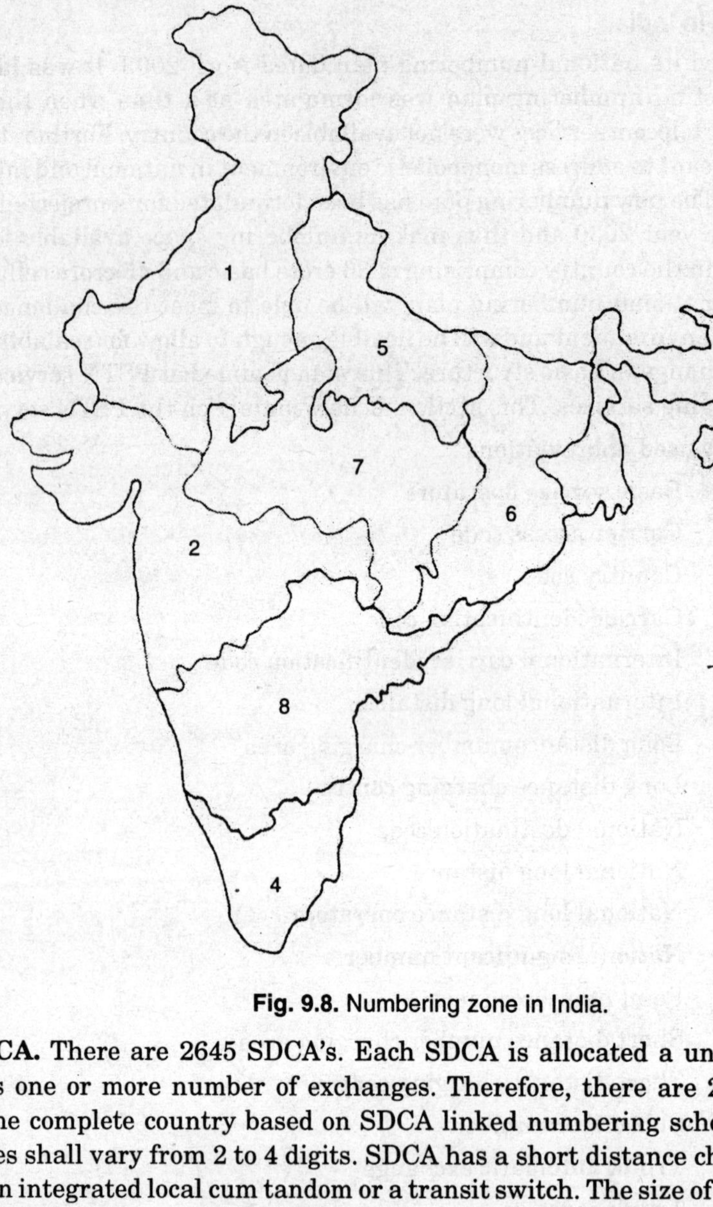

Fig. 9.8. Numbering zone in India.

SDCA. There are 2645 SDCA's. Each SDCA is allocated a unique trunk code. Each SDCA has one or more number of exchanges. Therefore, there are 2645 codes required to identity the complete country based on SDCA linked numbering scheme. The length of the trunk codes shall vary from 2 to 4 digits. SDCA has a short distance charging centre (SDCC). SDCC is an integrated local cum tandom or a transit switch. The size of SDCA generally varies between 800 sq. kms. to 2000 sq. kms.

NPA. Numbers in an Numbering Plan Area (NPA) are not duplicated and called 'subscriber numbers'. To make a call from one subscriber to another subscriber in the same NDA, only the subscriber number needs to be dialled. At present NDA is same as SDCA.

SN. Subscriber number (SN) is a 6 to 8 digit number which includes 2 to 4 digit Telephone exhcnage code.

NSN. The numbering scheme envisages the use of SDCA based linked numbering with 10 digit National Significant Number (NSN). The variants of SNS is shown in Table 9.2.

Table 9.2. Variants of SNS

SDCA code	Subscriber Number
2 digit	8 digit
3 digit	7 digit
4 digit	6 digit

National Numbering Scheme. There are ten levels of numbering schemes starting from 0 to 9. In each level, there are many sublevels as a classification of services. All the levels are defined briefly and some sublevels are explained in the following paragraph.

Level 0. Prefix codes. There are various sublevels. Some are described. Sublevel '000' as a prefix shall be used for home country direct service (Bilateral) and international toll free service (Bilateral). The format is 000 + country code + operator code '000800' is used for bilateral international toll free service. Sublevel '00' as a prefix shall be used for international dialling. The format as per E.164 recommendation is 00 + country code + NSN.

Sublevel '0' as a prefix shall be used for national long distance calls. The format is

0 + SDCA code + subscriber number

the sublevel '09' is used for cellular mobile services, satellite based services and Intelligent Network (IN) Services.

Level 1. Special services. This level is used for accessing special services like emergency services, supplementary services, inquiry and operator assisted services. The format contains 3 to *N* digit depending on service.

Level 2 to 6. PSTN subscriber number. This format contains the telephone exchange code and subscriber number.

Level 7 and 8. These two levels not being allocated and the same are reserved for new services.

Level 9. Services. The range of numbers in level '9' except '90', '95' and '96' ('96' is used in paging services) are reserved for cellular mobile services. Starting from 90 to 99, there are 9 sublevels. '90' is used as spare not allocated for any service.

Numbering format :

The PSTN numbering format shall be as per the table 9.3 given below :

Table 9.3. PSTN Numbering format

Trunk code (SDCA code)	+	Telephone exchange code	+	Last n digits of subscriber number
ABCD	+	EF	+	PQRS
ABCD	+	EFG	+	PQR
ABC	+	EF	+	PQRST
ABC	+	EFG	+	PQRS
AB	+	EFG	+	PQRST
AB	+	EFGH	+	PQRS

The trunk code is 2 to 4 digits. The telephone exchange code and last *n* digits of subscriber number together called subscriber number and is from 6 t0 8 digits. Hence national number to call a subscriber is 8 to 12 digits. To call a subscriber in another SDCA, prefix '0' must be dialled first.

SDCA code. Digit A can have any value from 1 to 8. Digit B, C and D can have any value between 0 to 9. AB codes can have any value between 11 to 89 (79 trunk codes). The code 10 is earmarked for carrier access code for NLD service and ILD service. Code 11, 20, 22, 33, 40, 44, 79 and 80 (8 codes) are used presently. 39, 50, 60, 69 and 70 (5 codes) are available for allotment to SDCA's with 3-digit code depending on the requirement. Certain three digit spare codes like 555, 666 and 888 are not to be used as SDCA codes. These are reserved for future services.

Telephone exchange code. Digit E can have any value between 2 to 6. The 0, 1, 7, 8, 9 are not allowed as '0' is used as trunk prefix, level '1' is used for special services, 7 and 8 are kept spare for future services and level '9' is used for cellular mobile services, paging services and access to adjacent areas. Digit F, G and H can have any value from 0 to 9. EFG can take value between 200 and 699 (500 exchange codes).

Last n digits of subscriber number. Digit P, Q, R, S and T can have any value from 0 to 9.

Dialling. For a call within a local area *i.e.* SDCA, subscriber number (Telephone exchange code + last *n* digits of subscriber code) only need to be dialled. Thus the number of digits needed to be dial is 6 to 8.

For calls outside the SDCA, 0 + NSN is need to be dialled (NSN = SDCA code + Telephone exchange code + last *n* digit of subscriber number). However, access to adjacent areas can also be possible by dialling '95' followed by NSN.

Note. It should be noted that the information given above are according to the matrials from the websites, magazines and reference books only. These informations are subject to change according to the change in Government policies and the change in ITU recommendations. The readers of this book are suggested to refer the related websites for updated changes in numbering plans.

9.5. CHARGING PLAN

The cost of providing a telecommunication network consists of the capital cost and the current operating expanses. The capital cost includes switching systems, buildings, lines and land. Operating cost includes staff salaries, maintenance costs, water and electricity charges and miscellaneus expenses. All of these costs must be met by the income obtained by the telecom operator from its subscribers. The telecom operator charges the subscribers for its services by the following three ways.

1. An initial charge for providing a network connection (as installation charges)
2. A rental or leasing charge
3. Call charges.

The initial costs are covered partly from installation charges and partly from rental. The operating costs of the telephone exchange are recovered through rental and call charges. According to the goverment policy, the rental may be levied on a monthly, bimonthly or by some other modes.

The quantity of equipments used, routing exchanges, switching systems, lines carrying voice/data and human involvement in establishing a connection between subscribers differs with respect to the distance between the subscribers, the time at which the call is made (at busy hour or off peak hour), the area (business or residential) etc. The charging methods for individual calls fall under two broad categories.

1. Duration independent charging
2. Duration dependent charging

Traditionally, charges for long distance calls have been proportional to distance multiplied with duration. The local calls within a numbering area are usually charged on a duration independent basis.

A meter for each subscriber counts the number of charging units based on the service providers policy decision. In India, TRAI in consultation with POT decides the charging units and also the tariff for these units. In India, recently, system integration Sai info system chose intel corporation to modernize rural telephone billing systems. The new intel based solution enables BSNL to bill not only its traditional landline subscribers, but also the rapidly growing number of cell phone users accessing local wireless loops.

As the billing procedure changes time to time according to the Govt. policies and to meet service providers expenses, the charging plans can not be explained. The readers can refer the day to day changes through websites or newspapers.

ACRONYMS

PSTN	—	Public Switched Telephone Network
POTS	—	Plain Ordinary Telephone Service
ISDN	—	Integrated Services Digital Network
TE	—	Tandem Exchange
AT & T	—	American Telephone and Telegraph Company
ITU-T	—	ITU-Telecommunications
CCITT	—	Comite Consultantif International Telegraphique et Telephonique
NSN	—	National Significant Number
CC	—	Country Code
NDC	—	National Destination Code
SN	—	Subscriber Number
TRAI	—	Telecom Regulatory Authority of India
BSNL	—	Bharat Sanchar Nigam Limited

RELATED WEBSITES

http: / / www.numbering plans.com / index

http: / / www.dotindia.com / nnp 2003.pdf

http: / / www.telecom-digest.org

http: / / www.teletechnics.co.nz / reference

http: / / www.bsnl.co.in / service / basic tariff.

REVIEW QUESTIONS

1. What is the basic goal of the network management ?
2. What are the various network planning ? Explain each planning in brief.
3. With neat diagram, illustrate the different levels of network management.
4. Write short notes on three types of networks.
5. Explain in detail, the basic topologies of the routing plan.
6. Define cost ratio.
7. With necessary equations explain the concept of alternative routing.
8. List and explain briefly the ITU recommendations related to numbering.
9. Explain the international telephone numbering format.
10. Explain the national numbering format.
11. Tabulate the PSTN numbering format followed in India.
12. What is charging plan ?

10

Transmission Networks

10

Transmission Networks

10.1. INTRODUCTION

As digital switching networks are highly sophisticated and complex, demands for improved and increasingly sophisticated services that have large bandwidth, better performance monitoring facilities and greater network flexibility are increased. These feature were based on high order multiplexing. The process of combining two or more communication paths into one path is referred as multiplexing. The fundamental types of multiplexing are space division multiplexing (SDM), frequency division multiplexing (FDM) and Time division multiplexing (TDM). These three multiplexing were described in section 3.6.

The TDM can be implemented in two ways. They are synchronous TDM and asynchronous TDM. The multiplexing which involves light signals transmitted through fiber optic channels is called wave length division multiplexing (WDM). In the section 10.2, synchronous TDM, asynchronous TDM and WDM are explained in detail.

Telephone companies have high speed digital networks to handle communications between their central offices. The link between the subscriber and local exchange is called local loop or subscriber loop. This local loop is analog line (twisted copper pair cable) with a potential bandwidth of 1 MHz or more. The actual wiring of the local loop may be considered to be a lossy transmission line. In the section 3.5, the subscriber loop characteristics of loop design and limiting factors were described.

The digital subscriber line (DSL) is the technology used between a customer premises and telephone companies, enabling more bandwidth over the already installed copper cabling that user have traditionally had. In the section 10.3 various types of DSL are explained briefly. The most popular form of DSL technology referred as Asymmetric Digital Subscriber Line (ADSL) is explained in detail.

The development of optical fiber transmission and large scale integrated circuits made more complex standards possible. The multiplexing technique allowed for combining of slightly nonasynchronous rates lead to the term plesiochronous digital hierarchy (PDH). PDH dominated 20 years with maximum of 140 Mbps.

SONET and later SDH are high speed optical network and dominates the digital communication networks. The SONET is explained in detail and SDH is explained in brief in the section 10.4.

10.2. DIGITAL MULTIPLEXING

Multiplexing is a technique that allows the simultaneous transmission of multiple signals across a single data link. These links can carry large numbers of voice and data transmission simultaneously using multiplexing. Fig. 10.1 explains the concept of multiplexing.

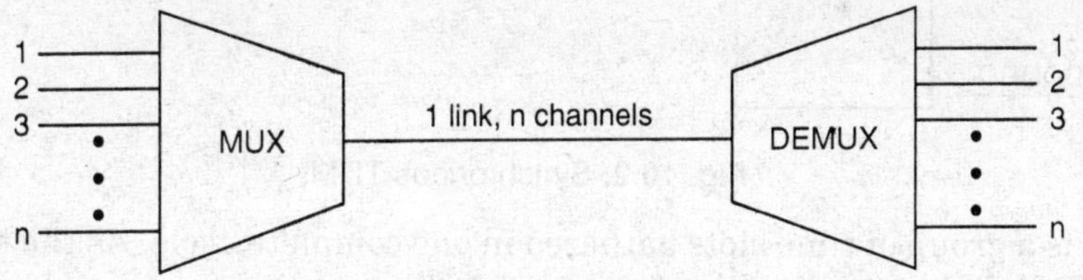

Fig. 10.1. Multiplexing.

There are three high speed multiplexing techniques. They are FDM, TDM and WDM. Early telephony multiplexing systems used FDM (which is also the standard multiplexing technique used for cable television). TDM has become the dominant method used by carriers for accomodating multiple data streams on a single cable. TDM is used extensively in the circuit switched world to send voice traffic. Some examples of TDM are T1 (1.5 Mbps) and T3 (45 Mbps) circuits, which are used for voice traffic as well as for data. Another technology that relies heavily on TDM is SONET. TDM have certain limitations over high speed transmission especially in SONET network.

WDM is a method of transmitting data from different sources over same fiber optic link. It maximise the use of fiber optic without the bottle necks and complexity. In this section TDM and WDM are discussed in detail.

10.2.1. Time Division Multiplexing (TDM)

TDM is a digital process that can be applied when the data rate capacity of the transmission medium is greater than the data rate required by the sending and receiving devices. The TDM concept is explained sufficiently in the section 3.6. The TDM can be implemented in two ways. They are Synchronous TDM and Asynchronous TDM. These two multiplexing techniques are explained in the following paragraphs.

Synchronous TDM. In a synchronous TDM based network the channels are scanned in sequence and each channel is given access to the data link at a particular time slot. The TDM network inserts a frame slot at the beginning and end of a group of channels. One pass across all channels results in a frame. For n devices (or channels) there are n time slot in each frame. Fig. 10.2 shows the synchronous TDM multplexing process.

In Fig. 10.2, there are four channels connected to TDM. These channels are scanned in sequence. In synchronous TDM, the multiplexer allots exactly the same time slot to each device or channel. Each time its allocated time slot comes up, a channel has the opportunity to send a portion of its data. If a device is unable to transmit or does not have data to send, its time slot remains empty. An empty slots get filled with bits to keep frames a consistent size, thereby frames arrive at predictable intervals.

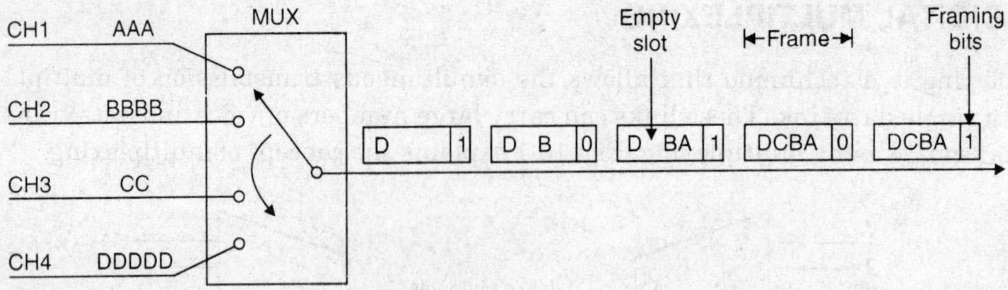

Fig. 10.2. Synchronous TDM.

A frame is a group of time slots gathered in one complete cycle. As the time slot order in synchronous TDM does not vary from frame to frame, very little overhead information needs to be included in each frame. As the time slots are in the same order, there is no need of additional bits needed to indicate path. But one or more synchronization bits are usually added to the beginning of each frame. These bits are called framing bits. In most cases, this synchronization information consists of one bit per frame (shown in Fig. 10.2).

The switch in TDM MUX shown moves from one device to device at a constant rate and in a fixed order. This process is called interleaving. Each device sends different message. The multiplexer interleaves the different messages and forms them into frames before putting them onto the link.

Data rate. If a device creates 250 characters per sec with 8 bit per character, the data rate is 250 × 8 = 2000 bits per sec or 2 kbps. If all the input devices sharing a link or transmitting at the same data rate, each device has one time slot per frame. However, it is possible to connect devices of different data rates to a synchronous TDM.

Bit stuffing. The number of slots in a frame and the input lines to which they are assigned remain fixed throughout a given system. The data rates of different devices control number of those slots. The device may have one slot, other may have two or three. But as the time slot is fixed, the different data rates of the device must be integer multiples of each other. If a device is 3 times faster than other, three slots can be alloted. But if that device has 2.5 times faster than the other, we cannot allot $2\frac{1}{2}$ time slot. So, the remaining space should be filled with additional bits, to make it as integer multple of other device.

By the technique called bit stuffing, the multiplexer adds extra bits to a device's source stream to speed relationships among the various devices into integer multiples of each other.

At receiver, exactly reverse operation in performed. The demultiplexer removes the message from frame and pass it to the approximate receiving device.

Asynchronous time division multiplexing. This type of TDM is also called statistical TDM as it avoids the redundant bits requirement. In synchronus TDM, the total speed of the input lines can be greater than the capacity of the link. In synchrons TDM, for n channel input, there are atleast n time slots. In asynchronous TDM, for n input channel, the number of time slots is less than n (say m). Hence with the same link, asynchronous TDM can support more devices than synchronous TDM. The number of asynchrons TDM frame (m) is based on a statistical analysis of the number of input lines that are likely to the transmitted at any given time. The operation of asynchronous TDM is explained with two cases.

Case (*i*). Fig. 10.3, shows the multiplexing of asynchorous TDM where there are 5 input lines and of five only three input channels are sending data.

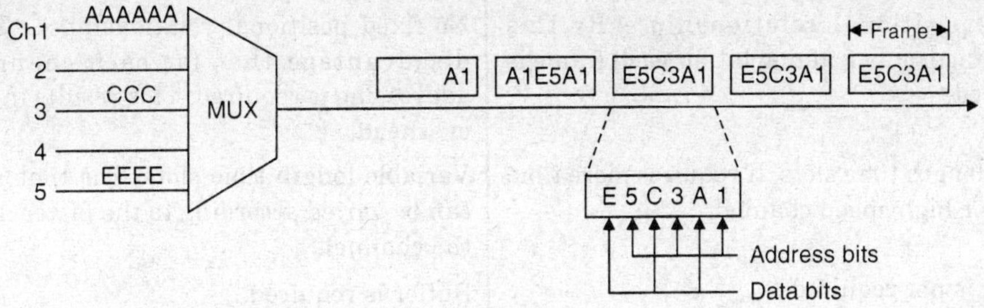

Fig. 10.3. Case (*i*) Asynchronous TDM.

In the above figure, 3 time slots are used in each frame. For the first three frames, the input is symmetrically distributed among all the channels. In fourth frame 3rd channel not having data and hence channel 1 data (A) is filled in 3rd time slot. This causes the absence of fixed positional relationships.

Hence, each time slot must carry a bit or bits to send data to a particular output line. This additional bit is attached by the multiplexer and discarded by the demultiplexer once it has been read. This is called addressing. This causes the system to become inefficient and additional overhead. For this case, in a synchronous TDM system, six frames (max data of A in channel 1) and five time slots (for 5 channels) requires 30 time slots. But only 13 time slots would have been filled in asynchronous system.

Case (*ii*). Fig. 10.4 shows the multiplexing of asynchronous TDM where there are 5 input lines and of five, four channels are sending data.

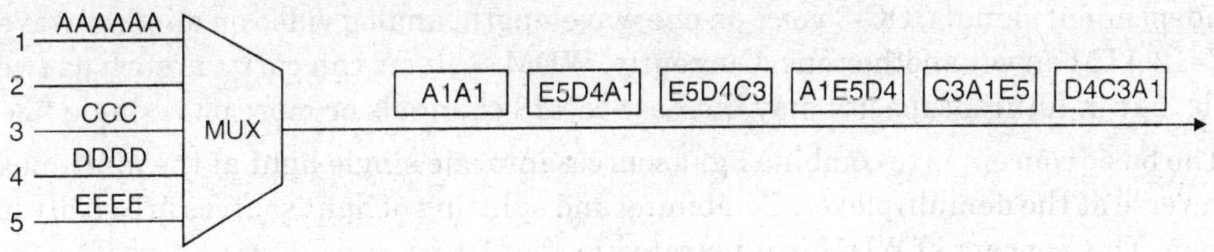

Fig. 10.4. Case (*ii*) Asynchronous TDM.

From figure, it can identified that, when the number of active senders does not equal to the number of slots in a frame, the time slots are not filled symmetrically. For example channel one data occupies the first slot in first frame, second slot in second frame and so on.

If the second channel is also active, and for example if it has 3 data inputs, the channel one data occupies the first slot of first frame, third slot of second frame and no slot in third frame and soon. Hence if the speed of the line is equal to three times of the input lines, then the data to be transmitted will arrive faster than the multiplexer can put it on the link. In this case a buffer is needed to store data until the multiplexer is ready for it.

Table 10.1 shows the comparison of synchronous TDM and asynchroquos TDM.

Table 10.1. Comparison of synchronous TDM and Asynchronous TDM

Synchronous TDM	Asynchronous TDM
1. Fixed positional relationship – By this advantage, only framing bit for each frame is required.	No fixed positional relationships – This is the disadvantage that for each channel, input addressing is required. This results in additional overhead.
2. Fixed length time slots. It requires more time slots for high speed channel.	Variable length time slots. The time slot length can be varied according to the faster data rate of the channel.
3. Buffer is not required.	Buffer is required.

TDM have certain limitations, particularly when dealing with the optical network such as SONET/SDH. TDM devices are required to convert data from light waves to electronic signals and back again. This conversion process add complexity to the system. Another related concern with TDM based system is scalability. Currently, the common top speed for TDM is 10 G bits/sec. So the circuit TDM equipments can not handle the faster networks.

There is an another method for expanding fiber capacity beyond TDM without the bottlenecks and complexity called as wave division multiplexing (WDM). In the next section the WDM and densed WDM (DWDM) are described.

10.2.2. Wavelength Division Multiplexing (WDM)

WDM is a method of transmitting data from different sources over the same fiber optic link at the same time whereby each data channel is carried on its own unique wavelength. WDM divides the light travelling through fiber into wavelengths. As a wavelength is inversely proportional to frequency, WDM is logically equivalent to FDM. WDM wave lengths can each carry independent signals, OC-3 voice on one wavelength, analog video on another wavelength and OC–12 ATM on yet another one. Currently, WDM systems can carry as such as two dozen channels, but in future, capacity may increase to 128 channels or more on a single fiber.

The basic concept is to combine light sources into one single light at the multiplexer and do the reverse at the demultiplexer. Combining and splitting of light sources are easily handled by a prism. This concept of WDM is illustrated in Fig. 10.5.

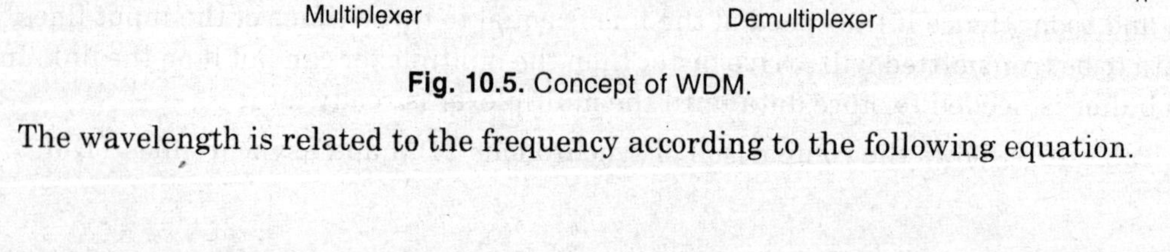

Fig. 10.5. Concept of WDM.

The wavelength is related to the frequency according to the following equation.

$$\lambda = \frac{c}{f} \quad ...(10.1)$$

where λ = wavelength measured in meters

c = velocity of light = 3×10^8 m/sec

f = frequency in Hertz:

Systems with extremely close WDM channel spacing (on the order of 0.04 nm) are sometimes referred to as FDM systems. Systems that use electronic multiplexing to produce electrical FDM signals for modulation of a single optical carrier are referred to as subcarrier multiplexing systems. Because the individual channels of these systems are typically close together in frequency and hence in wavelength, passive separation is usually infeasible.

All WDM device share the common characteristics of being purely passive and being reversible so any paritcular device can perform either mutliplexing or demultiplexing. This passive property of WDM devices is a dominant attraction for some applications such as fiber to home. The need for remote power is the disadvantage of active components. Thus passive WDM devices are preferred. The use of WDM in this application is sometimes referred to as the passive photonic loop (PPL).

An alternative approach for demultiplexing involves power splitting of the received signal followed by wavelength filtering to extract individual channels. This approach is passive and functionally identical to diffractive separation. Its main advantage is that power splitting can be implemented as passive taps distributed along the fiber route. The disadvantage is that there is a waste of optical fiber. It has applications where distances are less important than flexibility (like LAN). WDM demultiplexing with passive diffraction does not inherenty cause a loss of signal power in the individual channels.

10.2.3. Dense Wavelength Division Multiplexing (DWDM)

DWDM is an extension of WDM. This technology enables the service provider to accomodate consumer demand even for increasing amount of bandwidth. DWDM is a fiber optic transmission technique that employs light wavelength to transmit data parallel by bit or send by character. DWDM combines multiple optical signals so that they can be amplified as a group and transported over a single fiber to increase capacity. DWDM allows the transmission of email, video, multimedia, data and voice carried in internet protocol (IP), ATM, and SONET/SDH.

An example of a I generation DWDM system is the multiwave 1600 system from ciena corporation. It was introduced in 1996. This system provided 16 channels spaced 0.8 nm apart in the region of 1550 nm. 40 channel availability was announced in 1998. A 160 channel commercial system, each with a 10 Gbps data rate was announced by Nortel in 1999.

DWDM systems are enabled by Erbium doped fiber amplifier (EDFA), that transparently amplify all wavelength in the band and by fiber Bragg gratings fabricated into glass fiber for demultiplexing and filtering in a receiver. DWDM combines signals at a different rate (OC–3/12/24, etc.) and in different format (SONET, ATM, data, etc). Fig. 10.6, shown illustrates the concept of DWDM.

DWDM merges the optical fiber traffic into common fiber. This allows the high speed, high volume transmission optical amplifiers operate in a specific band of frequency spectrum and are optimized for operation with existing fiber, making it possible to boost lightwave signals and thereby expand their reach without converting them back into electrical form.

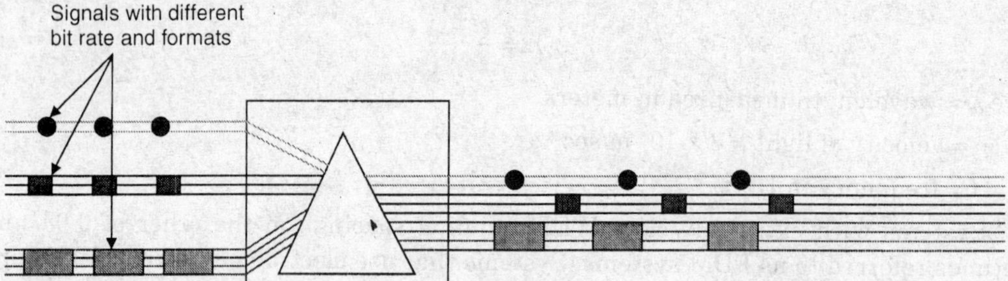

Fig. 10.6. Concept of DWDM.

Features of DWDM :

1. DWDM carries different signals of different bit rate and different formats over a single fiber.

2. DWDM gives service providers the flexibility to expand capacity in any portion of their networks. This advantage is not offered by any other technologies.

3. The fiber optic amplifier component of the DWDM system enables a service provider to save costs by taking in and amplifying optical signals without converting them to electrical signals.

4. DWDM uses decreased numer of amplifiers between source to destination. For example, DWDM system multiplexing upto 16 wavelength on a single fiber carriers can decrease the number of amplifiers by a factor of 16 at each regenerator site. Using fewer regenerators in long-distance networks results in fewer interruptions and improved efficiency.

5. Combined with optical wavelength add/drop (OWAD–wavelength or added or dropped to or from a fiber, without requiring a SONET terminal) and DWDM, the optical cross-connect (OXC) will offer service providers the ability to create a flexible, high capacity, efficient optical network with full optical bandwidth management.

6. Automatic adjustment of the optical amplifiers when channels are added or removed can be achieved with DWDM.

7. DWDM offers standards–compliant maintenance interface. Standard transaction language 1 interfaces are widely available for DWDM systems.

Limitations. The limitation in DWDM is due to the wavelength routing. The concept of wavelength routing is as follows. DWDM system uses optical cross connects (OXC). OXC's are switches with a certain number of input and output ports. The signals received at a input port are demultiplexed and switched to any one of its output port where they are remultiplexed. In this way, a virtual link created comprising a single wavelength routed through several OXC's. This is called wavelength routing.

By wavelength routing two limitation occurs. First, a signal with a wavelength is transmitted essentially unchanged after a power loss. The second one is that two signals modulating the same wavelength arriving on different ports cannot be transmitted on the same output port. The second limitations can be overcome by the wavelength converters that convert an input wavelength into a different wavelength while being transparent to the modulating data. The first limitations can be overcome by using powerloss compensation system.

10.3. DIGITAL SUBSCRIBER LINE (DSL) TECHNOLOGY

xDSL is a generic abbreviation for the many variations of digital subscriber line technology. DSL refers to the technology used between a customer premises and the telephone company enabling more bandwidth over the already installed copper cabling. Thus xDSL is a technology backed by telephone companies to provide next generation high bandwidth services to the home and business using the existing telephone cabling infrastructure. This technology accomplishes high speed delivery of data, voice, video and multimedia. The G.992.1(G.dmt), G992.2 (G.lite) standards by ITU, ANSI T1.413–1998 from ANSI, ADSL metallic interface from universal ADSL working group are the important standards provided by the standards organization. Other standards group for XDSL are European Telecommunication Standards Institute (ETSI), Internet Engineering Task Force (IETF), ADSL MIB work group and ATM Forum.

In this section, the comparison of xDSL over other existing technologies, types of XDSL technologies are discussed. The Asymmetric Digital Subscriber line (ADSL) which is the most popular form of xDSL technology is also described in detail.

10.3.1. xDSL Comparison with Other Technologies

xDSL is compared with the existing technologies like cable modems, ISDN, T1, voice band modem and wireless technologies based on the network formation and speed.

Cable TV network is a broad band technology provides services using cable modems. In this network, each subscriber in an area receives the same signals as all others in that area. The cable modem has potential bandwidth in the range 30 Mbps from the service provider to subscribers. Cable networks require two paths, one for down stream and one for upstream. XDSL is circuit oriented. Thus, each subscriber connection is independent of other. It uses existing telephone lines and does not require separate arrangement for upstream and downstream like cable TV network.

ISDN technology provides digital services in increments of 64 kbps channels. ISDN requires the phone company to install services within their phone switches to support their digitally switched connection service.

T1 (E1 is the European near equivalent) line is a 1.544 Mbps PCM system comprised of 24 TDM channels of 64 kbps each. T1 lines have been installed for end users who require dedicated high speed bandwidth between their home and work (or internet). T1/E1 uses have been used in voice and data networks throughout the world where highly available, high capacity networks needed to be built.

Voice modems or simply modems allows digital data to flow over the telephone company's traditional telephone network by performing digital to analog conversions for transmission on to the network and vice versa on the receiving end. Modems are limited to telephone company's voice bandwidth service (hence the name voice modem). With the bandwidth of 3000 Hz, typical speed is 56 kbps. Wireless access technology provides access to large number of subscribers in a relatively large area. Bandwidth ranges from few kbps to many Mbps.

xDSL is a newer modem technology that uses the existing telecommunication networks backed by telephone companies to provide next generation high speed bandwidth services to home and business. Typical speed of XDSL technology starts at about 128 kbps and goes up to

8.192 Mbps for most home user and to 50 Mbps for some installations with high capacity networks. The speed variations depends on the equipment used, distance involved, cabling quality, encoding techniques and frequency spectrum available.

10.3.2. Various Types of XDSL

Several forms of xDSL are designed to suit specific application, achieve specific goals and satisfy the needs of subscribers. XDSL may best be categorised within the modulation methods used to encode data and that uses the POTS splitter or not. A POTS splitter uses a low pass filter to separate the low end frequencies of the telephone audio spectrum from the higher frequencies of the xDSL signals. The splitter allows the traditional voice service that subscribers accustomed to. A splitter is required at both the customer premises and at the telephone exchange. xDSL that does not use a POTS splitter on customer premises is termed "splitter less xDSL". In this case splitter function is performed at the service provider end. Brief summary of well known xDSL's are given below.

1. **ADSL.** Asymmetric Digital Subscriber Line (ADSL) is the most popular form of XDSL technology. Its upstream and downstream bandwidth is asymmetric or uneven. The ADSL can provide upstream (user to provider) data rate from1.5 Mbps to 9 Mbps. Typical downstream (provider to user) speed range from 64 kbps to 1.5 Mbps. In practice, the bandwidth of down stream is high and is the high speed path.

2. **ADSL Lite or G.lite.** It is a low rate version of ADSL. It was proposed as an extension to ANS1 standard. T1.413 by UAWG (Universal ADSL working group) led by Microsoft, Intel and compaq. This is known as G.992.2 in the ITU standards. It uses the same modulation scheme as ADSL (DMT) ; but eliminates POTS splitter at the subscriber premises. It results in lower available bandwidth.

3. **VDSL.** The very high bit rate digital subscriber line (VDSL) is similar to ADSL and uses the DMT modulation technique. It is proposed for shorter local loops perhaps from 300 to 1800 meters.

4. **HDSL.** High Bit-rate Digital subscriber Line (HDSL) is generally used as a substitute for T1/E1. It was designed by Bellcore (now Telcordia). T-1 line uses AMI encoding, which is very susceptible to alternation at higher frequencies. This limits the length of aT–1 line to 1 km and requires repeaters every 6000 ft to boost the signal strength.

HDSL uses 2BIQ encoding. HDSL is becoming popular as it provides full duplex symmetric data communications at rates up to 1.544 Mbps (2.048 Mbps in Europe). The 2BIQ encoding is less susceptible to attenuation and hence the good rating can be achieved without repeaters upto a distance of 3.6 km. HDSL 2 was designed to transport T1 signalling at 1.544 Mbps over a single copper wire. HDSL 2 uses overlapped phase Trellis-code interlocked spectrum (OPTIS).

5. **SDSL.** Symmetric Digital Subscriber Line (SDSL) is a 2 wire implementation of HDSL. It achives the same data rate of HDSL. It supports T1/E1 on a single pair to a distance of 3.2 km.

6. **IDSL.** ISDN based DSL (IDSL) uses 2BIQ line coding and typically supports data transfer rates of 128 kbps. This technology is similar to ISDN, but uses the full bandwidth of two 64 kbps bearer channels plus one 16 kbps delta channel.

7. **G. Shdsl.** It is a ITU standard and offers a rich set of features (*e.g.* rate adaptive) and offers greater rate than many current standards. This technology is able to replace T1, E1, HDSL, SDSL, HDSL 2, ISDN, IDSL technologies.

8. Other popular proprietary DSL includes (*a*) Consumer installable DSL (CiDSL) is a splitterless DSL proposed by Globespan's (*b*) Etherloop proposed by Nortel. (*c*) Multi rate digital subscriber line (MDSL) and (*d*) Rate adaptive DSL (RADSL) designed by Glopes pan semiconductor). Etherloop is half duplex and hence capable of generating the same bandwidth rate in either the upstream or downstream. It is proposed for the speed of 1.5 Mbps and 10 Mbps depending on line quality and distance limitations. It uses carrierless amplitude and phase modulation (CAP).

Table 10.2 summarises the speeds characteristics of the DSL technologies.

Table 10.2. Summary of various DSL technologies

DSL Techniques	Speed		Operation
	Down stream	Up stream	
ADSL–Asymmetric DSL	1.5-8.192 Mbps	16-640 kbps	One pair of wire
RADSL–Rate adaptive DSL	64 kbps to 8.192 Mbps	16-768 kbps	One pair of wire
CDSL–Consumer DSL	1 Mbps	16-128 kbps	Now ratified as DSL-lite No splitters one pair of wire
HDSL–High bit rate	1.544 Mbps in North America	1.544 Mbps	Symmetrical services
	2.048 in rest of the world	2.048 Mbps	Two pair of wires
VDSL–Very high speed DSL	13-52.6 Mbps	1.5-6.0 Mbps	Fiber needed and ATM network probably used
IDSL–ISDN DSL	14.4 kbps (64 + 64 + 16) as BRI	144 kbps (64 + 64 + 16) as BRI	Symmetrical operation one pair of wires
SDSL–Symmetric DSL	1.544 Mbps 2.048 Mbps	1.544 Mbps 2.048 Mbps	One pair of wire
SHDSL–Single pair high bit rate DSL	2.312 Mbps	192-384 kbps	One pair of wire targetted the residential subscribers

10.3.3. Principle of Operation of xDSL

xDSL techniques uses greater range of frequencies over telephone cable than the traditional telephone services used. By using frequency range above the telephone bandwidth (300 Hz to 3400 Hz generally), XDSL can encode more data to achieve higher data rates. In order to utilize the frequencies above the voice audio spectrum, xDSL equipment must be installed on both ends. The loading coils used in copper wires between customer premises and local exchange must be removed or avoided to enable the copper wire to pass higher frequencies over the entire route. The loading coils are the inductors added in series with the copper wires to compensate the parallel capacitance of the line. This coil limits the bandwidth in the copper line.

In transmission, the bit rate may differ for upstream and downstream. The upstream is transfer of information from subscriber to local telephone exchange and the down stream is transfer of information from local telephone exchange to the customer premises. If the transfer of information in bit rate is same for upstream as well as down stream, it is referred as symmetric. If the bit rate varies for upstream and downstream, it is referred as asymmetric or rate adaptive or uneven. As the subscriber usually want to receive high volume files quickly, but usually have small files to send, XDSL one mostly asymmetric, that is down stream bit rates are higher than upsteram. XDSL with symmetric bit transfer is also available (IDSL).

10.3.4. Encoding and Modulation

Various methods of encoding and modulation techniques are available for encoding data onto XDSL circuits. They are Quadrature amplitude modulation (QAM), Pulse code modulation (PCM), Pulse amplitude modulation (PAM), V.90, Carrierless amplitude and phase (CAP) modulation, Discrete multitone (DMT) modulation etc. The common are QAM, V.90, CAP and DMT. In this section, CAP and DMT are summarised. The QAM and V.90 were discussed in section 3.7.

CAP. Carrierless amplitude and phase (CAP) modulation is similar to QAM but more complex than QAM and is not standardized. Both CAP and QAM are single carrier signal techniques. It is a proprietary standard implemented by Globespan semiconductor. While the name specifies that the modulation is "carrierless" an actual carrier is imposed by the transmit band shaping filter through which the outband symbols are filtered. Hence CAP is algorithmically identical to QAM. The upstream symbol rate is 136 K baud on a 113.2 kHz carrier, while the downstream symbol rate is 340 K baud on a 435.5 kHz carrier, 680 K baud on a 631 kHz carrier, or 952 K baud on a 787.5 kHz carrier.

Fig. 10.7 shows the spectral use of CAP.

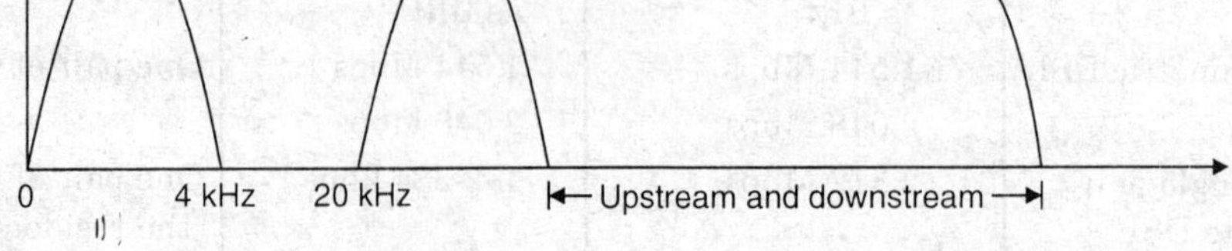

Fig. 10.7. Spectral use of CAP.

CAP uses the entire loop bandwidth (excluding 4 kHz) base band analog voice channel to send the bits all at once. CAP has some benefits over DMT and DMT has some benefits over CAP. One advantage of CAP is a lower peak-to-average signal. Power ratio relative to DMT. This means that the drivers and receivers may operate at lower power than DMT. This is so because CAP does not peak signal capacity as required is DMT circuitry. Most of the RBO's started using CAP but have since moved on to DMT.

Discrete Multitone (DMT) modulation. The use of DMT for ADSL was first proposed in 1991. In 1992, ANSI committee TIE1.4 began work toward a standard for ADSL, defined a set of requirements. In March 1993, the DMT system was chosen to be the best of the standard. ADSL uses DMT encoding methods which use QAM to divide the bandwidth of the channel into multiple subchannels, with each channel transmitting information using QAM modulation. DMT uses the frequency spectrum from 26 Hz to 1.1 MHz for broadband data. For POTS, it uses the frequency spectrum from 0 to 4 kHz.

Fig. 10.8 shows the spectral use of DMT.

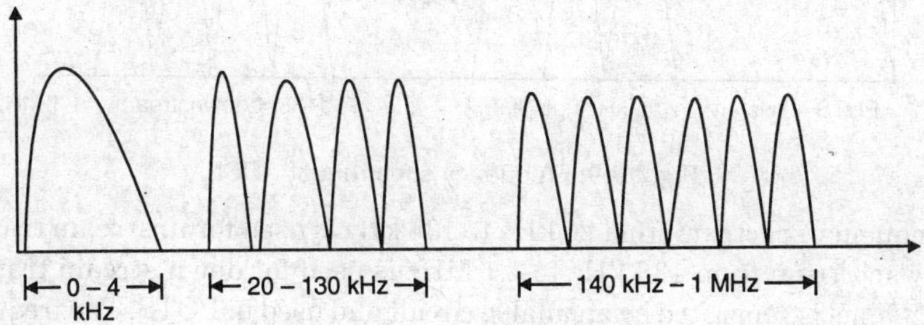

Fig. 10.8. Spectrum of DMT.

DMT modulation separates usable frequency range into 256 frequency bands (or channels). These are intimately connected to the FFT algorithm with DMT used as its modulator and demodulator. By dividing the frequency spectrum into multiple channels DMT is thought to perform better in the presence of interference sources such as AM radio transmitters. The channels 6–31 used for upstream and 32–250 used for downstream. The number of bits per symbol within each channel may be independently selected allowing the modem to be rate adaptive.

10.3.5. ADSL

The DSL Forum was formed in December 1994 to promote the DSL concept and facilitate developement of DSL system architectures, protocols, and interfaces for major DSL applications. The ANSI, working group TIE1.4, approved the first ADSL in 1995. It supported data rates upto 6.1 Mbps. The ETSI contributed an annex to T1.413 to reflect European requirements. The ITU-T standards are most commonly referred to as G.lite (G.992.2) and G.dmt (G.992.1). The ATM Forum has recognised ADSL as a physical layer transmission protocol for unshielded twisted pair media.

ADSL, a modem technology, converts existing twisted pair telephone lines (subscriber loop) into access paths for multimedia and high speed data communications. ADSL can transmit up to 6 Mbps to a subscriber, and as mush as 832 kbps or more in both directions.

ADSL Frquency spectrum. ADSL divides the bandwidth of a twisted pair cable into three bands. The twisted pair cable used in telephone wire has a frequency spectrum of ADSL.

Fig. 10.9 shows the frequency spectrum of ADSL.

ADSL uses various encoding methods to divide the available bandwidth of the channel into multiple subchannels. Earlier, FDM or Echo cancellation are used to divide the available channels. Presently, ADSL uses DMT encoding methods, which use QAM to divide the bandwidth of the channel into multiple subchannels with each channel transmitting information using QAM modulation. DMT uses the frequency spectrum from 26 kHz to 1.1 MHz for broad band data.

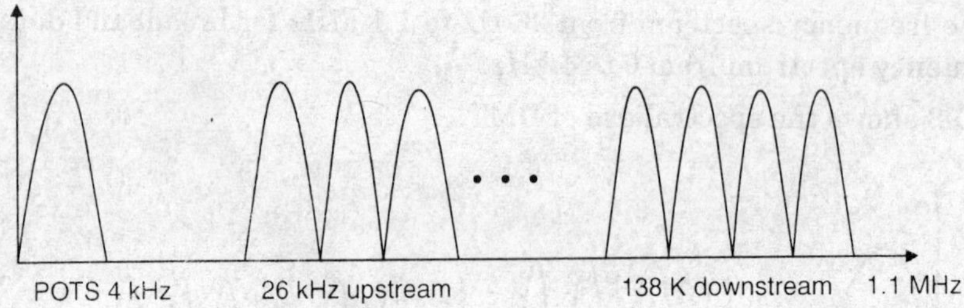

Fig. 10.9. Frequency spectrum of ADSL.

The frequency spectrum from 26 kHz to 138 kHz is used for upstream transmission, and the frequency spectrum from 138 kHz to 1.1 MHz is used for down stream transmission. The lower 4 kHz channel is separated by an analog circuit and used in POTS. The frequency spectrum above 26 kHz is divided into 249 independent subchannels, each containing 4.3 kHz bandwidth. 25 channels are used for upstream transmissions and 224 channels are used for downstream transmissions.

Topology For ADSL System. ADSL modem is connected to each end of twisted pair, one at the subscriber end and other at the central office. The ADSL modem at the exchange is called ATU-C (ADSL-Terminal Unit Central office) and the ADSL modem at the subscriber end is called ATU-R (ADSL-Remote). Fig. 10.10 shows the ADSL connection between exchange and subscriber. ADSL modems with various speed ranges and capabilities available. Fig. 10.10 shows the modem with 1.5 Mbps to 9 Mbps downstream bit rate and 16 to 640 kbps duplex channel.

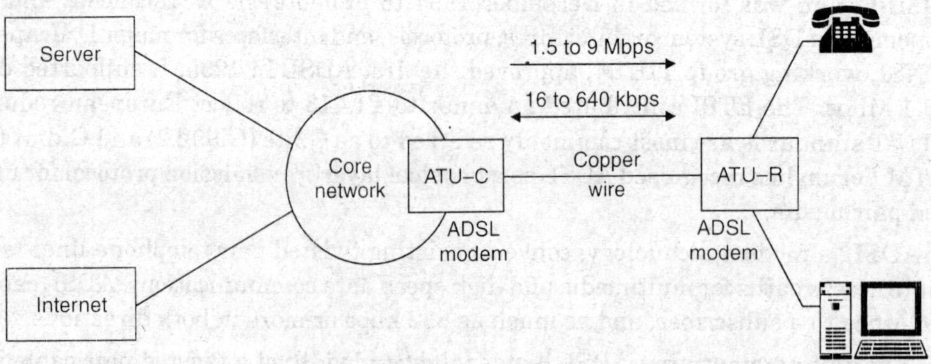

Fig. 10.10. ADSL modem connection.

Fig. 10.11 shows the topology of an ADSL system. To the access node different types of services such as Digital broadcast, broadband network, Narrow band network, Network management etc enteres. The access node provides interfacing of broadband services to ATU-C (ADSL–terminal unit central office).

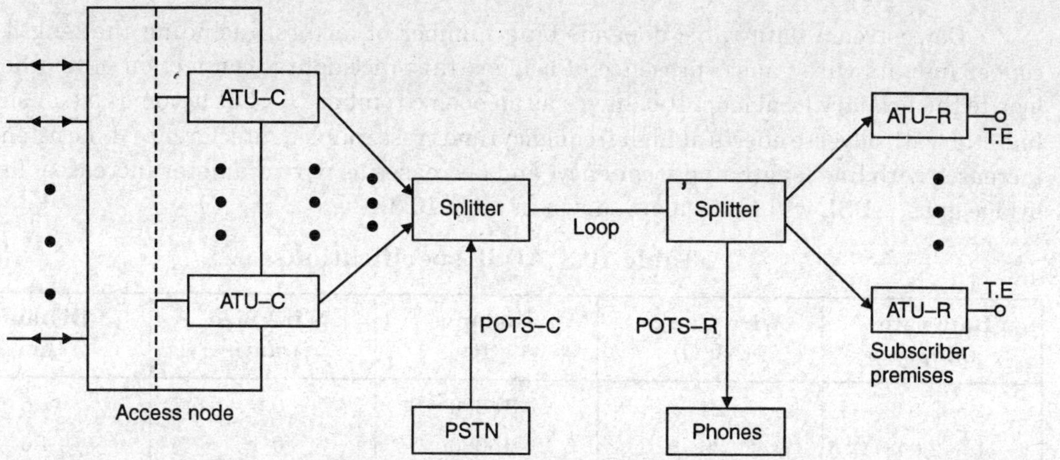

Fig. 10.11. Topology of an ADSL system.

ATU-C converts the data into ADSL format. The ADSL format fed to the splitter multiplexes them onto a single loop line. Telephone connection from the PSTN enter the system at the splitter level and are added to POTS-C (POTS–central office) area of the ADSL spectrum. The splitter in POTS-R (POTS–Remote) demultiplexes and transfers phone calls to the phones. The ADSL formats are transferred to ATU-R (ADSL-Terminating unit remote) which in turn converts to the orignal format and supplies to the terminal ends (TE).

ADSL Frame. The transport of data packets over ADSL requires link layer protocols. There are two ANSI stardards.

(*a*) Point-to-point protocol (PPP) variable length data units within and HDLC framed structure (RFC 1662).

(*b*) The ATM Forum's standard for ATM Frame UNI, also within an HDLC framed structure.

Fig. 10.12 shows the ADSL frame format. The frame begins with the standard HDLC flag (7E in hex) followed by a PPP address code field of FF and 03 hex. Two bytes of protocol ID identifies the payload type and the possible protocol that has been encapsulated in it. The frame check sequence (FCS) field uses CRC-16 for error detection and the frame ends with another 7E flag.

7E flag	FF	03	PPP protocol ID	PPP protocol data unit (Pay load)	FCS CRC–16	7E flag

Fig. 10.12. ADSL frame format.

ADSL Capabilities. ADSL will play a crucial role over the next ten or more years. Thus the service providers shall enter new markets for delivering information in video and multimedia formats. By bringing movies, television, video catalogs, remote CD-ROMs, Corporate LANs, and the internet into homes and small businesses, ADSL will make these markets viable, and profitable for telephone companies. ADSL modems available with various speed ranges and capabilities.

Downstream data rates depends on a number of factors, including the length of the copper line, its wire gauge, presence of bridged taps (accidental connection of another local loop to the primary local loop. It behaves as an open circuit at DC, but becomes a transmission line stub with adverse effects at high frequency) and cross coupled interference. Line attenuation increases with line length and frequency, and decreases as wire diameter increases. Ignoring bridge gaps, ADSL will perform as shown in table 10.3.

Table 10.3. ADSL specifications

Data rate (Mbps)	Wire Gauge (AWG)	Distance (ft)	Wire Size (mm)	Distance (km)
1.5–2-048	24	18000	0.5	5.5
1.5–2.048	26	15000	0.4	4.6
6.3	24	12000	0.5	3.7
6.3	26	9000	0.4	2.7

Advantages of ADSL. It provides many advantages to telecom companies and users. Some of them are :

1. It provides a simple, affordable mechanism to get more bandwidth to end users, both residential and small to medium businesses.

2. The high speed downstream is increasingly important for internet access, remote access to corporate server, integrated voice/data access and transparent LAN interconnection.

3. It enables carrier to ofter value added, high speed networking services.

10.4. SONET/SDH

10.4.1. Introduction

Synchronous Optical Network (SONET) is a high speed optical carrier using fiber-optic cable. SONET was originally proposed by Bellcore and standardised by ANSI. Later, the ITU had set a standard for SONET called Synchronous Digital Hierarchy (SDH). Throughout the world SDH was accepted, yet in U.S., the term SONET is continued. Although the international and U.S. versions of SDH/SONET are very close, they are not exactly identical. The differences that exist between them also addressed in this section.

The existing Digital Hierarchy which carries digitized voice over twisted wire is asynchronous at DS3 (digital stream) and lower levels. This digital hierarchy suffered from a deficiency of lot of overhead information requirement, addition of large number of multiplexers and digital cross connects. The most significant characteristics or advantages of SONET are listed below.

1. SONET uses byte multiplexing at all levels.

2. As SONET is a synchronous network, a single clock handles the timing of transmissions and equipment across the entire network.

3. Establishes a standard multiplexing format using some number of STS-1 signals as building blocks.

4. SONET/SDH contains recommendations for the standardization of fiber optic transmission system (FOTS) equipment sold by different manufacturers.

5. It defines multiplexing formats for carrying existing digital signals of the asynchronous multiplexing hierarchy (DS1, DS1C, DS2, DS3). It defines a DS0 (identifiable mapping format for DS1 signals.

6. SONET/SDH supports CCITT (ITU–T) digital signal hierarchy (E_1, E_2, E_3, E_4).

7. The flexibility of SONET accomodates applications such as ISDN with variety of transmission rates.

8. SONET provides extensive operation, administrations, maintenance and provisioning (OAM & P) functions for network managers.

9. It has enhanced network reliability, availability and universal connectivity.

SONET Signal Hierarchy. SONET defines a hierarchy of signalling levels called synchronous transport signals (STS). The lowest level, referred to as STS-1 or OC-1 (optical carrier level 1). Multiple STS-1 signals can be combined to form as STS-N (or OC-N) signal. The signal is created by interleaving bytes from N STS-1 signals that are mutually synchronised. Each STS level (STS-1 to STS-192) supports a certain data rate, specified in Mbps.

The STS is an electrical signal. The physical links defined to carry each level of STS are called optical carriers (OC). OC levels describe the conceptual and physical specifications of the links required to support each level of signalling. The most popular implementations are OC-1, OC-3, OC-12 and OC-48.

The ITU–T recommendation of SDM defines the hierarchy of signalling levels called synchronous transport module (STM). The lowest level is STM–1 has the lowest rate 155.52 Mbps. This is exactly equal to STS-3 rate. The reason for the descrepancy is that STM-1 is the lowest rate signal that can accomodate an ITU-T level 4 signal (139.264 Mbps).

Table 10.4 shows the SONET and SDM signal rates.

Table 10.4. Data rates for STS and STM signals

STS for SONET	OC	STM for SDH	Date rate (Mbps)
STS–1	OC–1		51.84
STS–3	OC–3	STM–1	155.52
STS–9	OC–9	STM–3	466.56
STS–12	OC–12	STM–4	622.08
STS–18	OC–18	STM–6	933.12
STS–24	OC–24	STM–8	1244.16
STS–36	OC–36	STM–12	1866.23
STS–48	OC–48	STM–16	2488.32
STS–96	OC–96	STM–32	4976.64
STS–192	OC–192	STM–64	9953.28

The above table reveals the relationship between different STS. For example, STS–3 rate is 3 times STS–1. STS 18 is 18 times STS 1. The meaning is 18 STS–1 channels can be multiplexed into one STS–18. Similarly six STS–3 channels can be multiplexed into one STS–18.

SONET componets. Fig. 10.13 shows SONET's components. SONET transmission relies on three basic devices.

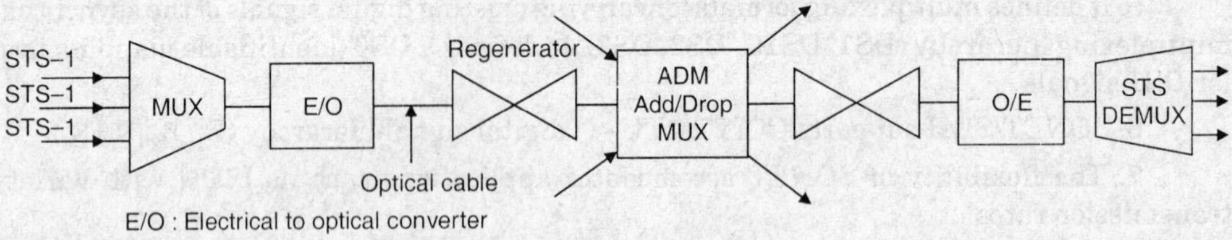

Fig. 10.13. SONET components.

STS multiplexer/demultiplexer. The function of an STS MUX is to multiplex electrical input signals to a higher data rate and then convert the results to optical signals. The STS demultiplexers convert and demultiplex optical signals to electrical signals for the users. Both multiplexer and demultiplexer are described in section 10.4.4.

Add/Drop multiplexer (ADM). It adds or removes the lower rate signals from or into high rate multiplexed signals. It does this extraction or insertion and redirect it without demultiplexing the entire signal. These multiplexers use the header information such as addresses and pointers to identify individual streams. Add/drop MUX is also explained in section 10.4.4.

Regenerator. The regenerator performs the functions of a repeater. It receives the optical signal and regenerates it. If the optical cable is longer than standard, the regenerator will be used to receive the optical signal and then to regenerate the optical signal.

Comparison of SONET and existing digital signals.

Table 10.5 shows a comparison of SONET with the existing digital signals and voice channels.

Table 10.5. Comparison of SONET and existing digital signals

SONET	Date rate (Mbps)	Equivalent capacities	
		DS–1 (1.54 Mbps)	DS–3 (44.736 Mbps)
STS–1	51.84	28	1
STS–3	155.52	84	3
STS–9	466.56	252	9
STS–12	622.08	336	12
STS–18	933.12	504	18
STS–24	1244.16	672	24
STS–36	1866.24	1008	36
STS–48	2488.32	1344	48

The design of the frame and signalling for SONET makes it compatible with the traditional, existing networks. The above table explains the equivalent capacities of digital signal (DS) service.

In the following sections, the MUX/DEMUX, ADM, SONET layers, SONET network, SONET Frame structure, virtual tributions, SONET topologies and SONET applications are described.

10.4.2. SONET Network and Layers

Fig. 10.14 shows the layered architecture of SONET network. In a simple SONET of two STS multiplexers linked directly to each other, the section, line and path are the same. Figure shows a typical SONET network with 5 STS MUX/DEMUX, 5 REG and a ADM.

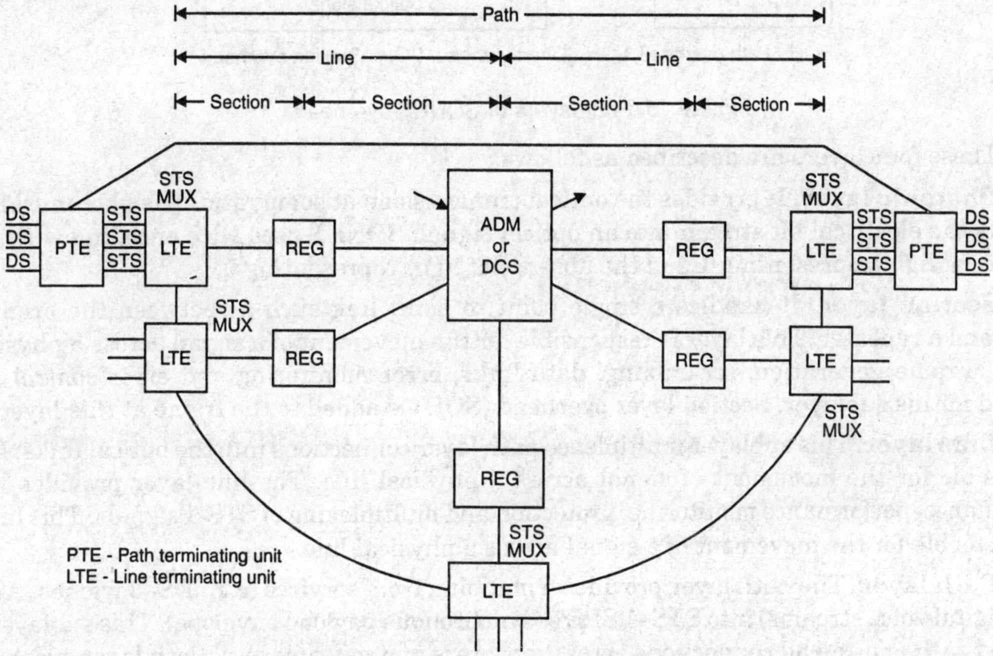

Fig. 10.14. SONET network and layers.

Section is the portion of the transmission facility or the optical link connecting two neighbour devices. It may be multiplexer to multiplexer, multiplexer to regenerator, or regenerator to regenerator. Line is the transmission medium (optical fiber) associates equipment required to transport information between two consecutive network elements. So, a line is a portion of the network between two STS multiplexers, two add/drop multiplexers or STS multiplexers to add/drop multiplexers. Path is the end-to-end portion of the network between two STS multiplexers. Path at a given rate is a logical connection between a point at which a standard frame for the signal is assembled and a point at which the standard frame is disassembled.

Fig. 10.15 shows the four layers of SONET overhead. The four layers of the structure are the path, line, section and photonic layers. The photonic layer is the lowest and performs physical layer activities. The section, line, and path layers correspond to the OSI model's data link layer.

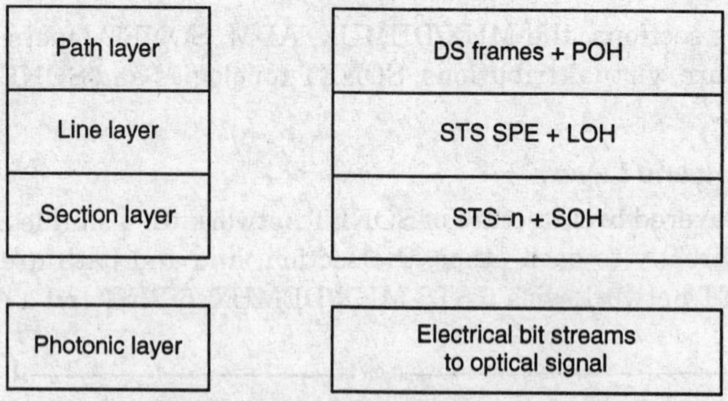

POH - Path overhead, LOH - Line overhead, SOH - Section overhead

Fig. 10.15. Four layers of SONET overhead.

These four layers are described as follows :

Photonic layer. It provides the optical transmission at some ratio. The photonic layer converts the electrical bit stream into an optical signal. SONET uses NRZ encoding with the presence of light representing 1 and the absence of light representing 0.

Section layer. It handles a single point to point link such as between the original source and a repeater. This layer is responsible for the movement of a signal across a physical section. Frame generation, scrambling, data links, error monitoring and error control are handled in this sublayer. Section layer overhead (SOH) is added to the frame at this layer.

Line layer. This sublayer multiplexes path layer connections into the optical links. It is responsible for the movement of signal across a physical line. The line layer provides line maintenance, performance monitoring, protection and multiplexing of STS-1 signals. This layer is responsible for the movement of a signal across a physical line.

Path layer. The path layer provides a mapping from services (*e.g.*, DS–3 frames, ATM cells, digital voice streams) into STS–1 SPE's (synchronous payload envelope). This sublayer is involved with end-to-end connections across repeaters and multiplexers. Path layer overhead is added at this layer. This overhead carriers path information which includes a signal label, a path status and path trace. This layer provides control signalling and error monitoring between the end points in the network.

Data received (for example DS–3 frames) from an electronic interface such as T–1 line is encapsulated in a frame at the path layer and overhead (POH) is added. The DS–3 plus POH are mapped into an STS–1 SPE (synchronous pay load envelope), and given to the line layer. The line layer may multiplex several different payloads from the path layer and adds line layer overhead (LOH). In section layer additional overhead SOH is added. It provides framing and scrambling prior to transmission by the photonic layer. The photonic layer converts the electrical bit stream from the section layer into an optical signal. Thus in SONET, overheads are inserted at a variety of locations in the middle of the frame.

The SONET network have three important devices. They are STS multipexer, regenerator and Add/drop multiplexer which are studied in the previous section. These devices and the four layers are related. STS multiplexer is a four layer device. An add/drop multiplexer is a three layered device. A regenerator is a two layered device. In the following section, the frame format is described.

10.4.3. SONET Frame Format

A SONET frame format is also called a synchronous payload envelope (SPE). The basic transmission signal is STS–1. The frame format of the STS–1 signal is shown in Fig. 10.16. In general, the frame is divided into a transport overload and the synchronous payload envelope (SPE). The SPE further divided into STS path overhead (POH) and the pay load. Once the pay load is multiplexed into synchronous payload envelope, it can be transported and switched through SONET without having to be examined and possible demultiplexes at intermediate nodes. Thus, SONET is said to be service independent or transparent.

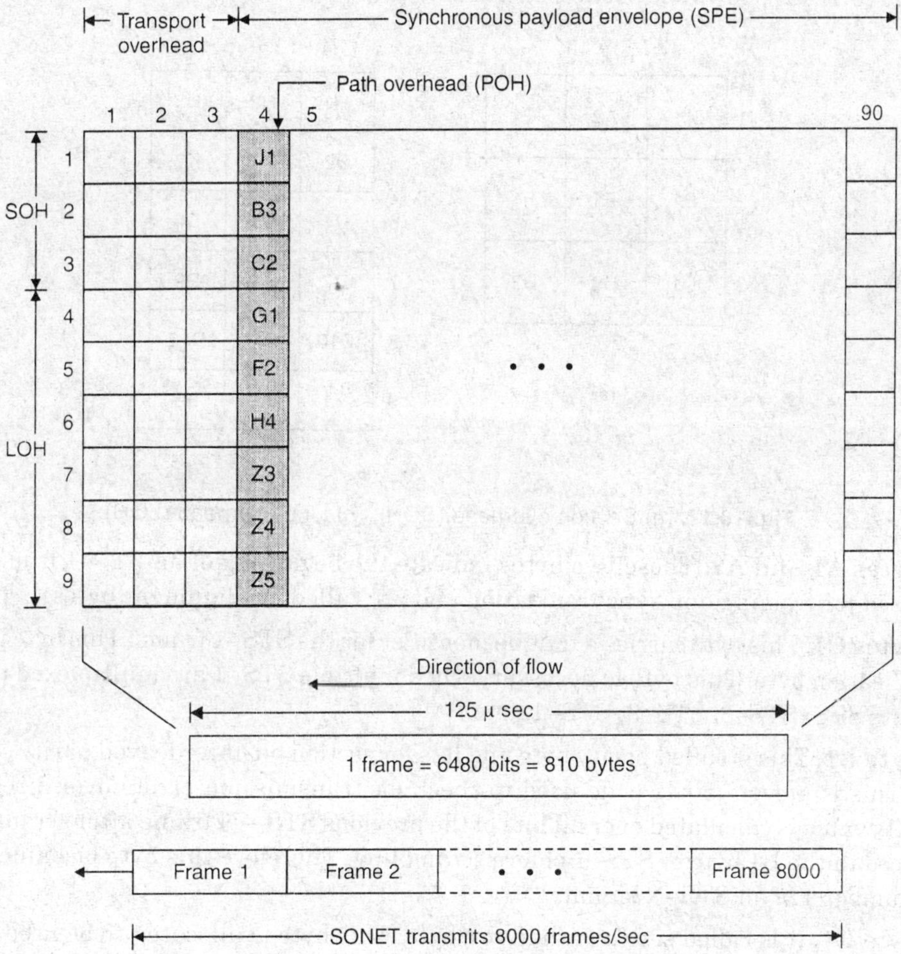

Fig. 10.16. Frame format STS-1 signal.

Each STS–1 frame is made up of 9 rows and 90 columns bytes. Thus the frame size is $9 \times 90 = 810$ bytes or 810 octets or 810×8 bits $= 6480$ bits. A frame is 125 μ sec in duration, corresponding to the 8 kHz voice sample. This gives the STS—1data rate 51.84 Mbps (*i.e.* 8000 frames/sec $= 8000 \times 6480$ bits/sec $= 51.84$ Mbps).

The upper three rows of the first three columns are used for section overhead (SOH). The lower six rows of first three columns are used for line overhead (LOH). This first three columns of the frame are used as a transport overhead. Column 4 to 90 are used for SPE. Of 87 columns of SPE, one column is used for path overhead POH (usually the first column). The path load modules end-to-end tracking information. The STS–1 frame is transmitted left to right, top to bottom.

Section over head (SOH). It contains information about frame synchronization (informing the destination of an incoming frame), carries information about OAM, handles frame alignment, and separates data from voice. SOH contains 9 bytes of the transport overhead accessed, generated and processed by section terminating equipment.

Fig. 10.17 (*a*) shows the section overhead.

	1	2	3
1	A1	A2	C1
2	B1	E1	F1
3	D1	D2	D3

(a)

	1	2	3
4	H1	H2	H3
5	B2	K1	K2
6	D4	D5	D6
7	D7	D8	D9
8	D10	D11	D12
9	Z1	Z2	E2

(b)

Fig. 10.17. (*a*) Section overhead (SOH), (*b*) Line overhead (LOH).

Bytes A1 and A2. These two bytes indicate the beginning of an STS–1 frame. A1 and A2 are used for framing and synchronization and are called as alignment bytes.

Bytes C1. This byte carries a unique identifier for the STS–1 frame. This byte is referred as identification byte. This byte is necessary when multiple STS–1 are multiplexed to create a higher rate STS (STS–3, STS–9, STS–12 etc.).

Byte B1. This is called parity byte and it is for section bit-inter leaved parity code (BIP–8) byte. This is a even parity code used to check for transmission errors over a regenerator section. Its value is calculated over all bits of the previous STS—N frame after scrambling and then placed in the B1 byte of STS--1 before scrambling. Therefore this byte is defined only for STS–1 number 1 of an STS–N signal.

Byte E1. It is called section orderwire byte. This byte is allocated to be used as a local orderwire channel for voice communication between regenerators, hubs and remote terminal locations.

Byte F1. It is called user's byte. This byte is set side for the user's purposes. The F1 bytes in consecutive frames form a 64 kbps channel that is reserved for user needs at the section level. It terminates at all section terminating equipment within a line. It can be read and written to at each section terminating equipment in that line.

Byte D1, D2 and D3. These bytes are called management bytes or section data communication channel (DCC) bytes. These bytes together form a 192 kbps message channel providing a message based channel for OAM & P between sections The channel is used from a central location for alarms, control, monitoring, administration and other communication needs. It is available for internally generated, externally generated or manufacturer specific messages.

Line overhead (LOH). Fig. 10.17 (*b*) shows the line overhead. Line overhead carries the payload pointers to specify the location of SPE in the frame and provides automatic switching (for standby equipment). It separates voice channels and provides multiplexing, line maintenance and performance monitoring. The byte descriptions of the LOH are given below.

Byte H1 and H2. These bytes are called pointer bytes. Two bytes are allocated to a pointer that indicates the offset in bytes between the pointer and the first byte of the STE SPE. The pointer bytes are used in STS-1s within an STS–N to align the STS–1 transport overhead in the STS–N and to perform frequency justification.

Byte H3. This is a pointer action byte and is allocated for SPE frequency justification purposes. The H3 byte is used in all STS–1s within an STS–N to carry the extra SPE byte in the event of a negative pointer adjustment.

Byte B2. It is line bit interleaved parity code (BIP–8) byte. This parity code byte is used to determine if a transmission error has occurred over a line. It is even parity and is calculated over all bits of the line overhead and STS–1 SPE of the previous STS–1 frame before scrambling.

Bytes K1 and K2. K1 and K2 bytes are called automatic protection switching (APS channel) bytes. These 2 bytes are used for protection signalling between line terminating entities for bidirectional automatic protection switching and for detecting alarm indication signal (AIS–L) and remote defect indication (RD1) signals.

Bytes D4 to D12. These bytes are called as line data communication channel (DCC) bytes. These 576 kbps message channel provides the same service as the D1–D3 bytes (OAM). They are available for internally generated, externally generated and manufacturer specific messages. A protocol analyzer is required to access the line DCC information.

Bytes Z1 and Z2. These bytes are referred as growth bytes. The Z1 byte is located in the second through Nth STS-1s of an STS-N (3 < = N < = 48) and one allocated for future growth. Z2 byte is located in the first and second STS-1s of an STS-3 and the first, second and fourth through the Nth STS-1s of an STS-N (12 < = N < = 48). These bytes are allocated for future growth.

Byte E2. This is a orderwrite byte. It provides a 64 kbps channel between line entities for an express orderwire. It is a voice channel for use by technicians and will be ignored as it passes through the regenerators.

Path overhead. (POH) Path overhead is part of SPE and contains following information. Performance monitor of synchronous Transport signal, path trace, parity check and path status. POH contains a bytes. The labels are shown in Fig. 10.17. The labels and functions are given below.

J1. Path trace byte — J1 byte sends a continuous 64 bit string to verify the connection. The choice of the string is left to the application program.

B3. Path parity byte.

C2. Path signal label byte — It is used to identify different protocols used at higher levels (such as FDD1 or SMDS).

G1. Path status byte.

F2. Path user channel byte.

H4. Virtual tributory indicator — Indicates payloads that cannot fit into a single frame.

Z3, Z4 and Z5. Growth bytes — reserved for future use.

Virtual tributories. SONET is designed to carry broad band payloads. In order for the SONET to carry lower data rate frames such as DS-1 and DS-2, the lower data rate frame is mapped to the STS-1 payload and is called a virtual tributary (VT). A virtual tributory is a partial payload that can be inserted into an STS-1 and combined with other partial payloads to fill out the frame. The basic frame for the SONET is STS-1 with a data rate of 51.84 Mbps. The payload of STS-1 is made up of 86 columns and grows. Instead of using all 86 payload columns of an STS-1 frame for data from one source, we can subdivide the SPE and call each component a VT. Fig. 10.18 shows the VTs in the STS-1 payload.

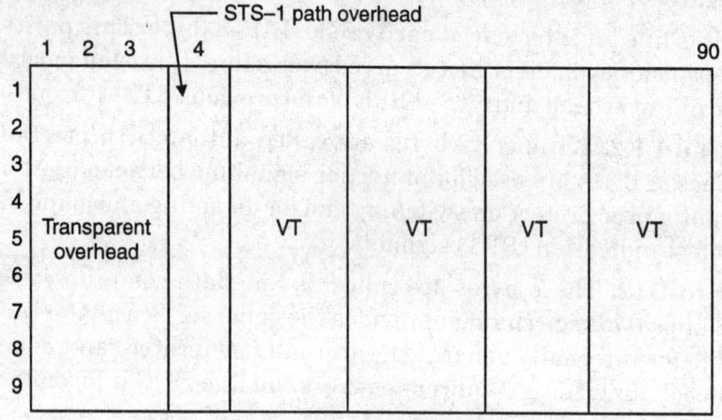

Fig. 10.18. Virtual tributories.

There are four types of VTs that map into the STS-1 payload.

1. VT 1.5 is a frame of 27 bytes that is made up of 3 columns and 9 rows. The data rate of VT 1.5 is calculated as follows.

Data rate VT 1.5 = 27 bytes × 8 bits × 8000 frames/sec = 1.728 Mbps.

The VT 1.5 is used for transmission of DS-1 with a data rate of 1.54 Mbps. The STS-1 payload can transmit 28 VT 1.55.

2. VT 2 is a frame of 36 bytes that is made up of 4 columns and 9 rows. It is used for transmission of an European E-1 line (can E-line can carry 30 voice channels). The data rate of VT 2 is

$$= 36 \text{ bytes} \times 8 \text{ bits} \times 8000 \text{ frames/sec}$$

$$= 2.304 \text{ Mbps.}$$

Fig. 10.19 shows VT 2 framed into STS-1 frame.

3. VT 3 is a 54 byte frame (made up of 6 columns and 9 rows) used for transmission of a DS-1 C frame with a data rate of 3.152 Mbps.

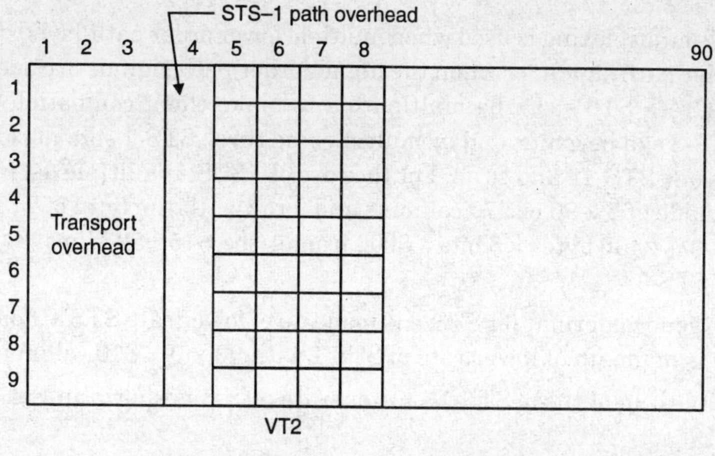

Fig. 10.19. VT2 mapped into STS-1 frame.

4. VT 6 is a 108 byte frame that is made up of 12 columns and 9 rows. It is used for transmission of a DS-2 frame with a data rate of 6.312 Mbps.

Within an STS-1 frame, each VT occupies a number of columns. Within the STS-1, VT groups can be mixed together to form an STS-1 payload. When two or more tributories are inserted into a single STS-1 frame, they are interleaved column by column. SONET provides mechanisms for identifying each VT and separating them without demultiplexing the entire stream. For example a VT group may contain one VT-6, 2, VT-2s or 4 VT 1.5s. To synchronize the various low speed signals to a common rate before multiplexing, bit stuffing is used. All services below DS-3 rate are transported in the VT structure.

10.4.4. SONET Multiplexing

Multiplexing enables one physical medium to carry multiple signals. A transmission network is a set of a links between sites. To put more than one call on each link is to give each call a time slot and transmit several calls simultaneously. This process is known as TDM. In SONET, individual SONET channels are merged into a higher level channel using TDM. In each scheme, each channel gets a specific time slot in the transmission. With SONET, multiplexing is TDM, but not statistical. Hence, there is no concept of congestion or priority in SONET. The traffic flowing in and out of a node is exactly equal. There is no peak rate.

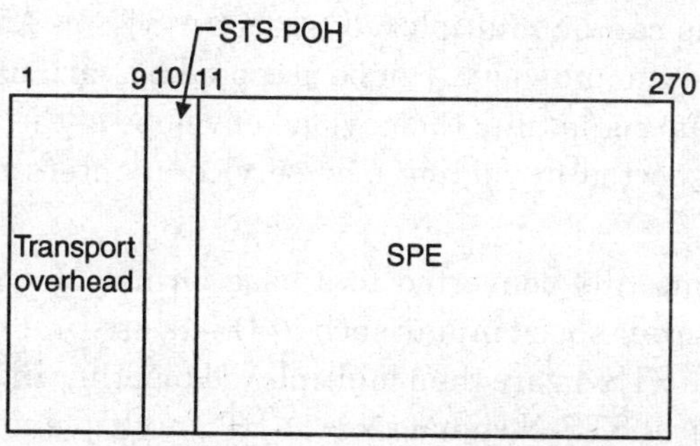

Fig. 10.20. Frame format of STS-3.

In SONET, multiplexing is used when multiple lower order path-layer signals are adapted into a higher-order path signal, or when the higher order path signals are adapted into the line overhead. Lower rate STS's can be multiplexed to make them compatible with higher rate systems. The STS-3 can be generated by multiplexing three STS-1 signals, four STS-3's can be multiplexed into one STS-12 and so on. For the case of 3 STS-1 multiplexed into one STS-3, the STS-3 frame is made of 3 × 90 or 270 columns and 9 row with the total of 2430 bytes. Thus the data rate of STS-3 is 2430 bytes × 8 bits × 8000 frames/sec = 155.52 Mbps. Fig. 10.20 shows the frame format of STS-3.

Hence the general format for STS-n is made up of lower rate STS's. For example, general format of STS-9 is made up of lower rate of STS 1 with 90 × 9 = 810 columns and 9 rows.

The multiplexing of three STS-1's to generate STS-3 is shown in Fig. 10.21.

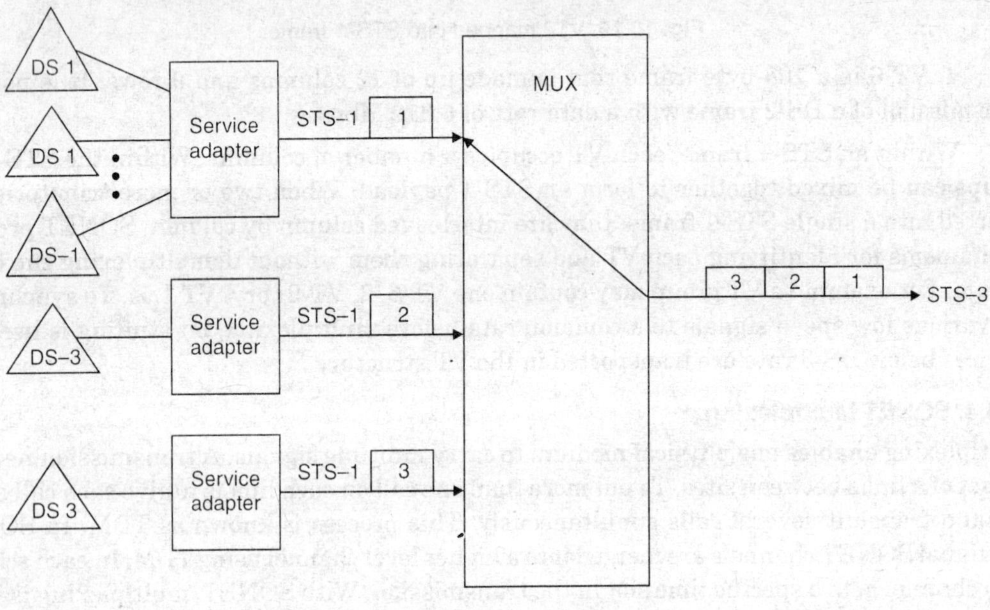

Fig. 10.21. SONET Multiplexing.

In the Fig. 10.21 three STS-1 signals are multiplexed to obtain one STS-3 signals. Similarly 3 STS-3 signals can be multiplexed to obtain STS-9. Any type of service, ranging from voice to high speed data and video, can be accepted by various types of service adapters. A service adapter maps the signal into the payload envelope of the STS-1 or VT. New services and signals can be transported by adding new service adapters at the edge of the SONET network.

All inputs are eventually converted to a base format of a synchronous STS-1 signal (51.84 Mbps or higher). Lower speed inputs such as DS-1s are first bit or byte multiplexed into VTs. Several synchronous STS-1s are then multiplexed together in either a single or two stage process to form an electrical STS-N signal ($N \geq 1$). STS multiplexing is performed at the byte interleave synchronous multiplexer. Basically the bytes are interleaved together in a format such that the low-speed signals are visible. No additional signal processing occurs except a direct conversion from electrical to optical to form an OC-N signal.

Add/drop Multiplexer (ADM). ADMs are used for extracting or inserting lower rate signals from or into higher rate multiplexed signals without completely demultiplexing the SONET signals. SONET does not restrict manufacturers to providing a single type of product, not require them to provide all types. For example, one vendor might offer an ADM with access at DS-1 only, whereas another might offer simultaneous access at DS-1 and DS-3 rates. Fig. 10.22 shows the ADM.

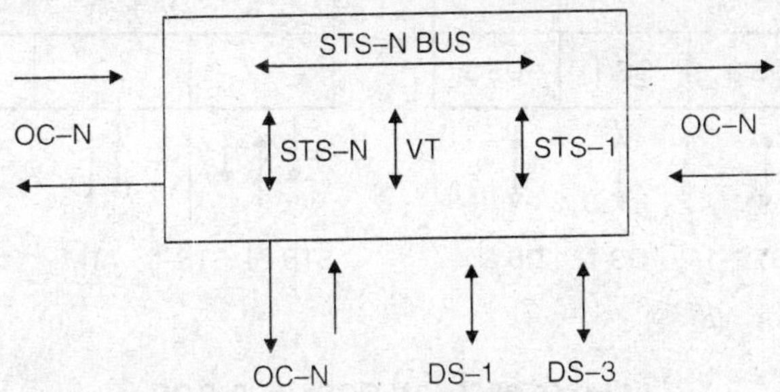

Fig. 10.22. Add/Drop Multiplexer.

At an ADM site, three possible functions are performed. They are listed below :

1. At ADM, only those signals that need to be accessed are dropped or inserted without demultiplexing the entire signals. ADM uses header information such as addresses and pointers to identify individual streams.

2. ADMs can also be configured as a survivable ring. SONET enables drop and repeat or drop and continue. With drop and repeat, a signal terminates at one node is duplicated (repeated) and is then sent to the next and subsequent nodes. The ring survivability has telephony and cable TV applications. It also provides alternate routing, if the connection cannot be made through one of the nodes. For example, when transporting video, it can be delivered (dropped) at a node and repeated for delivery to the next node-channels not treminating at a node can be passed through without physical intervention to other nodes.

3. The ADM provides interfaces between the different network signals and SONET signals. At this site, it can drop lower rate signals to be transported on different facilities, or it can add lower rate signals into the higher rate STS-N signal. The rest of the traffic simply continues straight through.

Digital cross connects (DCS). DCS is ideally used at a SONET hub. It accepts various optical carrier rates, accesses the STS-1 signals and switches. One major difference between a cross connect and an add/drop multiplexer is that a cross connect may be used to interconnect a much larger number of STS-1s. There are two types of DCS. They are wideband digital cross connects (W-DCS) and Broad band digital cross connects (B-DCS). Both W-DCS and B-DCS are shown is Fig. 10.23.

The W-DCS terminates SONET and DS-3 signals and has the basic functionality of VT and DS-1 level cross connection. It is the SONET equivalent to the DS-3/DS 1 DCS and accepts optical OC-N signals as well as STS-1s, DS-1s and DS-3s. In W-DCS, the switching is done at

VT level. As W-DCS can automatically cross connect VTs and DS-1s, the W-DCS can be used as a network management system. Thus W-DCS is ideal for grooming at a hub location.

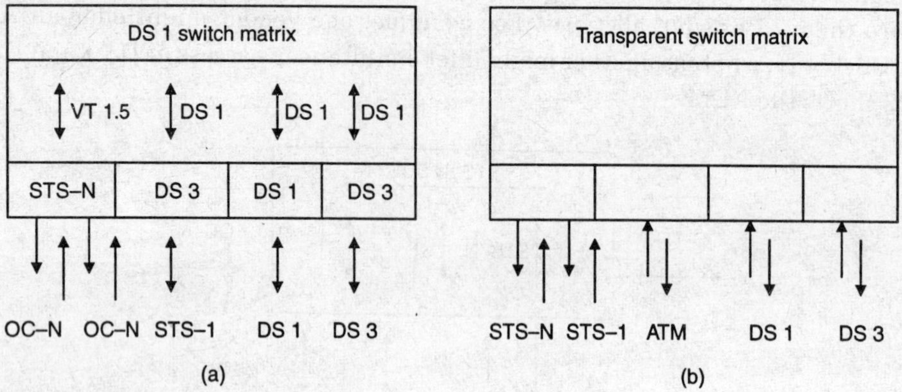

Fig. 10.23. (*a*) W-DCS, (*b*) B-DCS.

B-DCS interfaces various SONET signals and DS-3s. It accepts STS-1 and switches at this level. It make two way cross connections at the DS-3, STS-1 and STS-N levels. It is best used as a SONET hub, where it can be used for grooming STS-1s for broadband restoration purposes, or for routing traffic.

10.4.5. SONET Topologies

Different topologies are used in SONET network using various multiplexers. Popular topologies are 1. Point to point 2. Point to multipoint 3. Hub and spoke and 4. Ring. Each topologies are discussed briefly in this section.

Point to point. This configuration involves two terminal multiplexers linked by fiber with or without a regenerator in the link. In this configurtion, the SONET path and the service path (DS-1 or DS-3 links, end to end) are identical. Fig. 10.24 shows the point-to-point configuration.

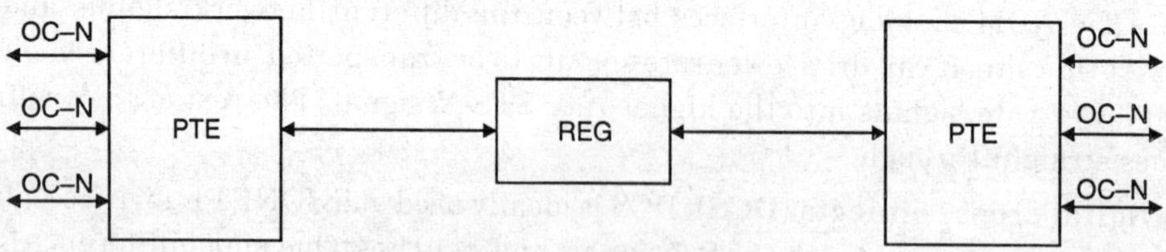

Fig. 10.24. Point-to-point configuration.

This point-to-point multiplexing arrangement can act as a stand-alone environment and not to have to interface with the public switched networks.

Point to multipoint. This configuration includes ADM circuits along the way. ADM is placed along a SONET link to facilitate adding and dropping tributary channels at intermediate points in the network. Fig. 10.25 shows point to multipoint architecture.

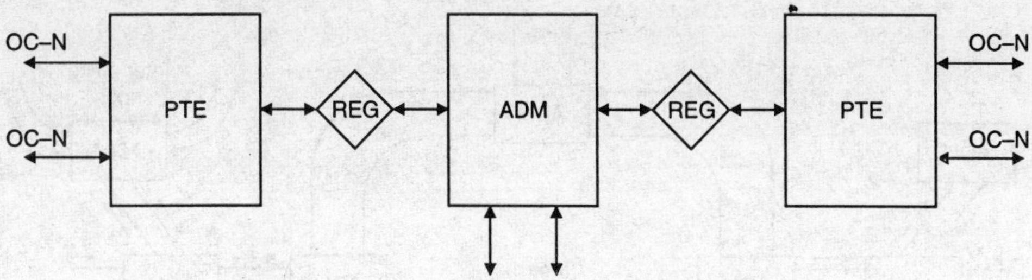

Fig. 10.25. Point to multipoint configuration.

Hub and smoke. The hub or star network provides some added flexibility in the event of unpredicted growth or changes in the architecture of the network. A hub concentrates traffic at a central site and allows easy reprovisioning of the circuits. A SONET multiplexer can be hubbed into a DCS where it is concentrated and then forwarded to the next node as shown in Fig. 10.26.

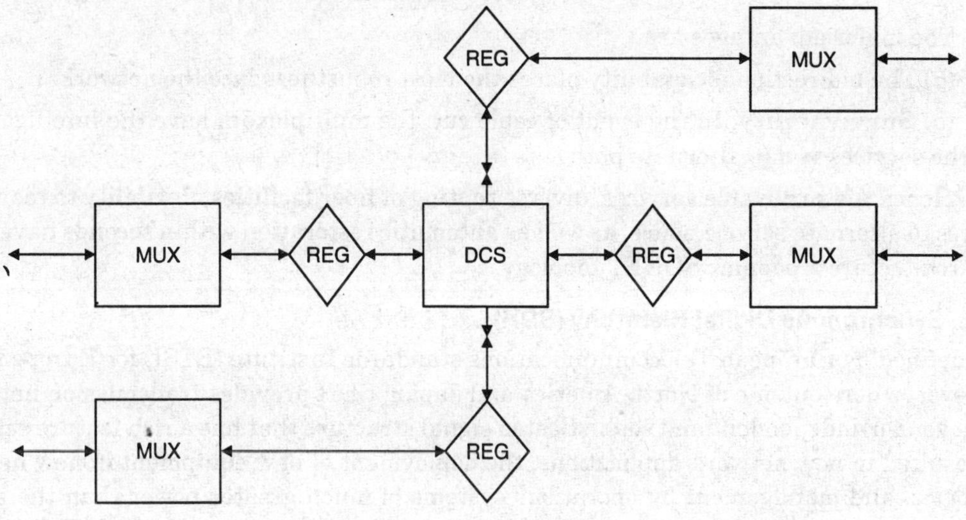

Fig. 10.26. Hub network.

Hub system is used in many larger organizations where regional offices are located and district or branch offices are tied into the network through the hub. The flexibility in accepting new loads and the development of blocking or non-blocking network are the features of this topology.

Ring network. Multiple ADMS can be put into a ring configuration for either bidirectional or unidirectional traffic. In ring architecture, where SONET automatic protection switching (APS), is employed. Fig. 10.27 shows the ring architecture of SONET multiplexers.

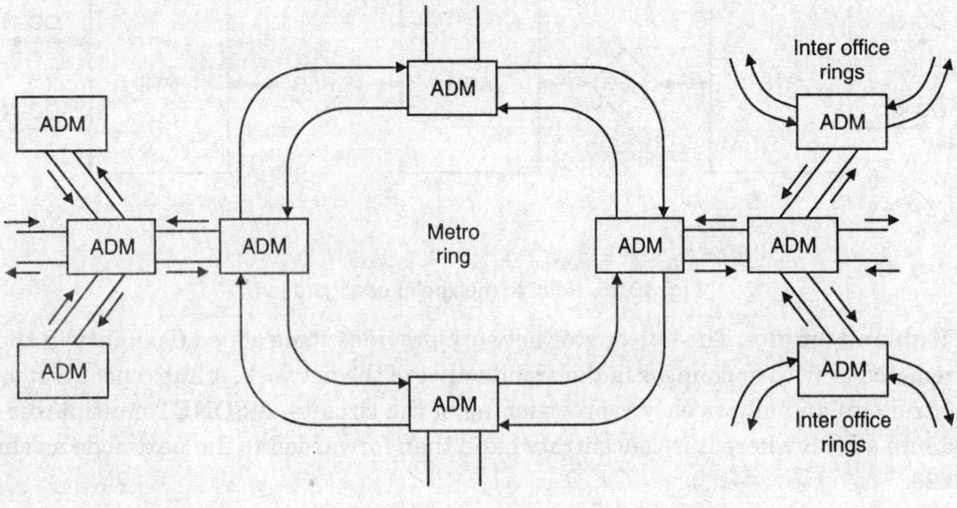

Fig. 10.27. Ring network.

The major advantages are :

(*a*) The bidirectional capability places the most robustness into the network.

(*b*) **Survivability.** In the event of cable cut, the multiplexers have the intelligence to send the services via an alternate path.

Hence the survivable services, diverse routing of fiber facilities, flexibility to rearrange services to alternate service nodes, as well as automatic restoration within seconds have made ring architecture a popular SONET topology.

10.4.6. Synchronous Digital Hierarchy (SDH)

SDH defined by European Telecommunications standards Institute (ETSI) for Europe is now used everywhere outside of North America and Japan. SDH provides transmission networks with a vendor-independent and sophisticated signal structure that has a rich feature set. This has resulted in new network applications, the deployment of new equipment in new network topologies, and management by operations systems of much greater power than the earlier transmission networks. The original SDH standard defined the transport of 1.5/2/6/34/45/140 Mbps within a transmission rate of 155.52 Mbps and is being developed to carry other types of traffic such as ATM and IP, within the rates that are integer multiples of 155.52 Mbps.

SDH was designed to allow for flexibility in the creation of products for electronically routing telecommunication traffic. The key products are optical line system, radio-relay system, terminal multiplexers, ADM, hub multiplexers and digital cross-connect switches.

SDH Frame. The SDH forms a multiplexing rate based on the STM-N (synchronous transmission module) frame format. The frame has a repetitive structure with a period of 125 μ sec. The basic STM-1 frame consists of 270 columns and 9 rows (2430 octets). 9 columns and 9 rows are for section overhead (81 octets). The remaining 2349 octets create the pay load.

Higher rate (STM-N) frames are derived from multiples of STM-1 according to the value of N. The first nine columns contain the section overhead (SOH) for transport-support features such as framing, management operations channels and error monitoring, with the first segment containing the frame word for demultiplexer alignment. The remaining columns can be assigned in many ways to carry lower bit rate signals. The administrative unit (AU) is the unit of provision for bandwidth in the main networks. Its capacity can be used to carry a high bit rate signal. (AU-2-45 Mbps, AU-4-140 Mbps). AU can be further divided to carry lower rate signals.

Figure 10.28 shows the STM-1 frame formats.

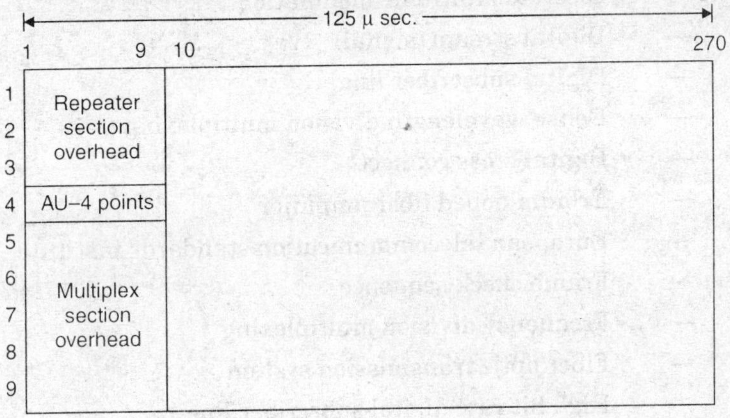

Fig. 10.28. STM-1 frame format.

Typical rates and speed of STS and STM are compared in table 10.6.

Table 10.6. Comparison of STS and STM rates

STS	STM	Speed
STS–1	STM–0	51.84 Mbps
STS–3	STM–1	155.52 Mbps
STS–9	STM–3	466.56 Mbps
STS–12	STM–4	622.08 Mbps
STS–18	STM–6	933.12 Mbps
STS–24	STM–8	1.2446 Gbps
STS–36	STM–12	1.866 Gbps
STS–48	STM–16	2.488 Gbps

Almost all new fiber transmission systems now being installed in public networks use SDH or SONET.

ACRONYMS

ADM	—	Add drop multiplexer
ADSL	—	Asymmetric digital subscriber line
AIS	—	Alarm indication signal
ATU-C	—	ADSL Terminal unit-central office
APS	—	Automatic protection switching

ATU-R	—	ADSL Terminal unit-Remote office
CAP	—	Carrier amplitude and phase modulation
CiDSL	—	Consumer installable DSL
DCC	—	Data communication channel
DCS	—	Digital cross connect
DMT	—	Discrete multitone modulation
DS	—	Digital stream (signal)
DSL	—	Digital subscriber line
DWDM	—	Dense wavelength division multiplexing
DXC	—	Digital cross connect
EDFA	—	Erbium doped fiber amplifier
ETSI	—	European telecommunication standards institute
FCS	—	Frame check sequence
FDM	—	Frequency division multiplexing
FOTS	—	Fiber optic transmission system
HDSL	—	High bit rate digital subscriber line
IDSL	—	ISDN based DSL
IETF	—	Internet engineering task force
LOH	—	Line over head
MDSL	—	Multirate DSL
OAM & P	—	Operation, Administrations, Maintenance and provisioning
OC-1	—	Optical carrier level-1
OPTIS	—	Overlapped phase trellis-code interlocked spectrum
OWAD	—	Optical wavelength add/drop
OXC	—	Optical cross connect
PDH	—	Plesiochronous digital hierarchy
POH	—	Path over head
PPL	—	Passive photonic loop
PPP	—	Point to point protocol
QAM	—	Quadrature amplitude modulation
RADSL	—	Rate adaptive DSL
RDI	—	Remote deflect indication
SDH	—	Synchronous digital hierarchy
SDM	—	Space division multiplexing
SDSL	—	Symmetric digital subscriber line
SOH	—	Section over head
SONET	—	Synchronous optical network

SPE	—	Synchronous payload envelope
STS	—	Synchronous transport signals
TDM	—	Time division multiplexing
UAWG	—	Universal ADSL working group
VDSL	—	Very high bit rate digital subscriber line
VT	—	Virtual Tributory
WDM	—	Wavelength division multiplexing
HDLC	—	High level data link control

RELATED WEBSITES

http: / / www.cis.ohimo-state.edu / ~jain /

http: / / www.networkmagazine.com / article

http: / / www.catv.org / frame / bbb.html

http: / / www.ciscu.com / univercd / cc / td / doc

http: / / www.atomicfrog.com / archiever / phreak

REVIEW QUESTIONS

1. Explain the concept of multiplexing using a simple diagram.
2. Explain and distinguish the synchronous TDM and asynchronous TDM.
3. With neat diagram, explain the concept of WDM.
4. With neat diagram, explain the concept of DWDM.
5. List the features of DWDM.
6. What is wavelength routing ?
7. What are the limitations and its remedial of DWDM system.
8. Compare XDSL technology with the existing system.
9. List and briefly explain various types of XDSL.
10. State the principle of operation of XDSL.
11. Write short notes on the following modulation techniques.
    (*a*) CAP (*b*) DMT.
12. Explain ADSL and its frequency spectrum.
13. Give brief explanation on ADSL topology system.
14. With neat diagram, explain the ADSL frame format.
15. List the advantages of ADSL technology.
16. What are the important freatures of SONET.
17. With neat diagram, explain the SONET components.
18. Explain the SONET network and its layers with neat diagrams.
19. Explain the STS-1 SONET frame format. Based on this prinicple ; draw the STS-3 SONET frame format.

20. Define SOH, LOH and POH.
21. Define VT.
22. Explain the four types of VT.
23. Explain the SONET multiplexing.
24. Write short notes on ADM and DCS.
25. List the popular topologies of SONET and compare them with neat diagrams.
26. What are the major advantages of ring network ?
27. What is SDH ?
28. How the SDH format differs from SONET ?
29. How the higher STM frames can be derived ?
30. Tabulate the comparison of STS and STM rates.

11

Data Networks

11

Data Networks

11.1. INTRODUCTION

Communication between two computers over the telephone lines begun around 1968. Previously, to transfer information between equipments or computers, a vendor or a company had to rent the interface equipment from the telephone company. As this causes the delayed procedure and led to monopolizing of telephone company, the Federal communications commission (FCC) produced a rule 67, by which a vendor can directly connect an equipment to telephone company network. The device Modem was developed which converts the serial digital data form produced by a computer to a form of analog signal that can be sent through the telephone voice circuits. Thus, a communication networks enable users to transfer information in the form of voice, video, E-mail and computer files.

When a network is to transfer a stream of data from a source to destination, it must assign to the data stream a route, that is, a sequence of links or channels connecting the source to the destination. The network should also allocate the data stream a portion of the capacity or BW in each channel along the route to be used. These decisions are performed by switches (or sometimes routers) in telephone exchanges. The process is called switching. There are three types of switching namely message switching, circuit switching and packet switching. The circuit and message switching were explained in the section 4.7. The packet switching is explained in the section 11.3.

In 1978, the International Organization for Standards (ISO) was asked to come up with a solution that would allow the transparent communication and data transfer between and among systems regardless of manufacturer. The model developed by ISO is the OSI reference model. OSI stands for the open systems interconnect reference model. This model allows any two different systems to communicate regardless of their underlying architecture. The purpose of the OSI model is to open communication between different systems without requiring changes to the logic of the underlying hardware and software. In this chapter, the OSI model is described in the section 11.4.

Till 1980's, OSI model was widespread and dominated the entire networking, commercially as well as by architecture. In 1990's TCP/IP has become firmly established as the dominant commercial architecture. Now the TCP/IP is the protocol of choice in many LAN-to-WAN environments. The concepts of TCP/IP is explained in the section 11.5. LAN technologies become almost a necessity for small offices. Experts predicts that within a few years, LAN setups will find their way into homes. Various LAN technologies are explained in the section 11.6.

Asynchronous transfer mode (ATM) is a high performance, cell oriented switching and multiplexing technology that uses fixed length packets to carry different types of traffic. ATM is defined as a transport and switching method in which information does not occur periodically with some reference (hence the name asynchronous). The ATM concept is described in the section 11.7.

The internet is a social as well as a technological phenomenon. Internet is the world's largest computer network. It was created nearly twenty five years ago as ARPA net. Its goal was to create a method for widely separated computers to transfer data efficiently even in the event of nuclear attack. The internet is a network formed by the co-operative interconnection of various computing networks.

11.2. DATA TRANSMISSION IN PSTN

The transmission medium is the physical foundation for all the data communications. The amount of data carried across the networks crossed the voice traffic level. The data is growing at a rate of 30 percent per year. It will take 12 years to double the amount of voice carried on the network at the current growth rate, whereas the data is doubling approximately every 90 days. In the beginning, data transmission was organized using telegraph or telex networks as they could carry digital signals directly. But teletype machines were slow, noisy and consumed large amounts of power. The speed was limited to 110 bauds. (Baud rate is a measure of the rate at which binary data are transmitted and received). But data rates for transmission have been on the rise. With public switched telephone network, there is a possibility of carrying signals at higher speeds. Public switched telephone networks and electronic PABX's are designed to carry analog voice signals. They can be used for data transmission by employing suitable interfaces.

11.2.1. Data Rates in PSTN

Baud rate. The maximum rate of signal transitions that can be supported by a channel is known as baud rate. Baud rate is a close measure of information throughput, or the effective information data transfer rate from sender to receiver. Thus, baud rate is one that can be supported in a noiseless channel.

We know, a voice channel in a PSTN is bandlimited with a nominal bandwidth of 3.1 kHz. A maximum data rate that a noiseless or ideal voice channel can support can be obtained from the Nyquist theorem

$$D = 2\,B \log_2 L \text{ bps} \qquad \ldots(11.1)$$

where D = Maximum data rate (in Baud or bps)

B = Bandwidth of the channel

L = Number of discrete levels in the signals.

For a 3 kHz channel, and a binary signal, the maximum data rate is 6000 bps, if the signal level is two.

For higher data rates, we translate information rate into symbols per second. A symbol is any element of an electrical signal that can be used to represent one or more binary data bits. The rate at which symbols are transmitted is the symbol rate. This rate may be represented as a systems baud rate. Fig. 11.1 illustrates the pulse representation of the binary numbers used to code the samples (Fig. 11.1 (*a*)) and representation by voltage levels (symbols) rather than pulses (Fig. 11.1 (*b*)).

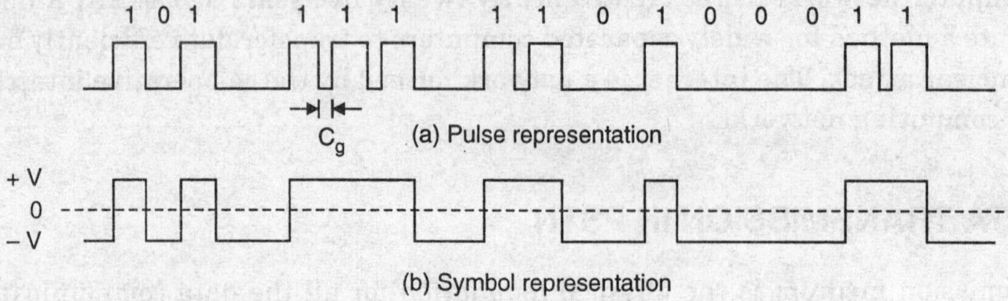

Fig. 11.1. Binary bits represented as (*a*) pulse and (*b*) symbol.

In Fig. 11.1 shown, each three digit binary number that specifies a quantized sample value is called a word. C_g is called guard time between pulses.

Bit rate. In the noisy channel, there is an absolute maximum limit for the bit rate. This limit arises because the difference between two adjacent signal levels become comparable to the noise level when the numner of signal level is increased. For noisy channel, data rate is calculated by

$$D_b = B \log_2 (1 + S/N) \qquad ...(11.2)$$

Where D_b = Data rate in noisy channel (in bps)

B = Bandwidth of the channel

S/N = Signal to noise ratio.

For S/N of 30 dB and 3 kHz Bandwidth noisy channel, D_b is 30000 bps.

Relation between baud rate (or symbol rate) and bps : The baud rate and bit rate are related as

$$D_b = D \times n \qquad ...(11.3)$$

where n = number of bits required to represent signal levels.

In the example considered for baud rate explanation n is assumed as one. Hence baud rate is equal to bps. Fig. 11.2 illustrates the relation between baud and bit rates.

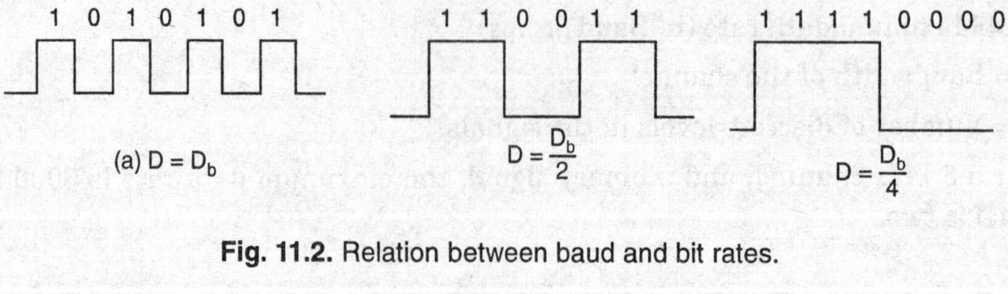

Fig. 11.2. Relation between baud and bit rates.

Fig. 11.2 (*a*) shows the baud rate equal to bit rate. Fig. 11.2 (*b*) and (*c*) shows the baud rate equal to one-half and one-fourth of bit rate respectively. It is proved that up to 2400 bauds may be transmitted reliably through a PSTN voice channel. By increasing the signal levels, the effective bit rate increases.

For low-speed applications, the difference between baud and bit rate are insignificant. Thus 300 and 1200 bps modems orginally used with personal computers were frequently referred to as 300 or 1200 baud modems.

11.2.2. Data Communications Link

In order to communicate from a terminal, computer or any equipment, the following six parts have to be put together in proper order.

1. The transmission medium that carries the traffic between source and destination.
2. Data communication equipment or data circuit terminating equipment (DCE).
3. Data terminal equipment (DTE).
4. Communication protocols and software.
5. Terminal devices.
6. Interface.

Fig. 11.3 shows the typical arrangement of the communication link for the data communication. Data link refers to the process of connecting or linking two stations together.

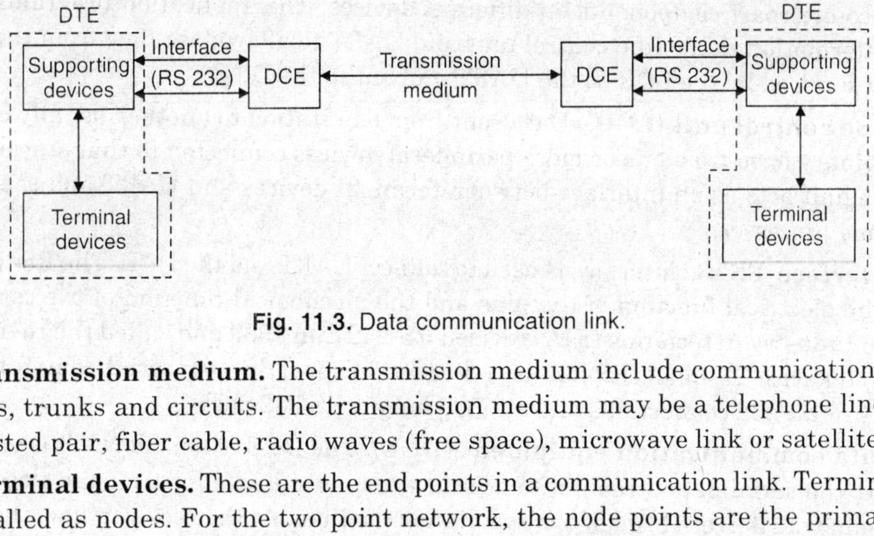

Fig. 11.3. Data communication link.

Transmission medium. The transmission medium include communication channels, path, links, trunks and circuits. The transmission medium may be a telephone lines, coaxial cable, twisted pair, fiber cable, radio waves (free space), microwave link or satellite link.

Terminal devices. These are the end points in a communication link. Terminal devices are also called as nodes. For the two point network, the node points are the primary station and the remote or secondary station. A primary station is responsible for establishing and maintaining the data link between it and a secondary station. The terminal devices includes main frame computer, personal computer, peripherals such as printers, keyboards, FAX machines and data display terminals.

Data terminal equipment (DTE). The terminal devices, communication station, UART, and line control unit (LCU) grouped together and named as DTE. Fig. 11.4 shows typical arrangement of DTE.

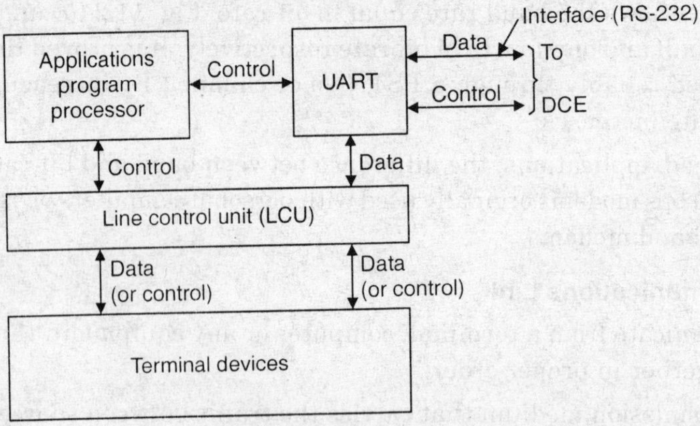

Fig. 11.4. Data terminal equipment.

UART. The universal asynchronous receiver transmitter (UART) and the universal synchronous/asynchronous receiver transmitter (USART) performs the parallel to serial conversion (and vice versa at the receiving station).

Application programme processor. An application program used by the DTE, called a protocol, defines a set of rules that determine requirements for the successful establishment of a data link and the transfer of actual information between stations. Protocols are key components of communication architectures. Protocols provide the rules for communication between counterpart components on different devices. The application programs also direct control information to the line control unit and UART to allow data flow from the peripheral currently serviced by the LCU to the UART and out to the DCE.

Line control unit (LCU). Data sent from one station to another usually originates in parallel binary form from one or more peripheral devices connected to that station through a LCU. The unit acts as an interface between terminal devices and UART and the application programme processor.

Interface. RS 232 interface is used to connect UART and the DCE. The RS-232 interface defines the electrical function of the pins and the mechenical function of the connector. The Electrical Industry Association (EIA) revised RS-232 C in 1989 and called the revision RS-232 D (connector with 25 pins). RS-232 is a standard connection for serial communication. All modems use RS 232 connections and all PCS have a RS 232 port.

Data communication equipment (DCE). The DCE is a modem. This device is used to convert the serial data stream into a form which is suitable for transmission. This serial data stream transferred through a transmission medium. At the receiver side, the serial data stream are converted back to digital and sent to DTE. DCE may be a modem or a computer based node in a data network.

Some of the standards defined by CCITT known as V series are :

V.5 — Standardisation of data signalling rates for synchronous data transmission in PSTN.

V.24 — DTE-DCE interface and control signals.

V-28 — DTE-DCE electrical characteristics for unbalanced double current interchange circuits.

V-53 — Limits the maintenance of telephone type circuits used for data transmission.

Some of the important V-series modem standards are given in Table 11.1.

Table 11.1. V-series modem standards

V-series	Speed (bps)	Modulation	Application
V-22 bis	2400 (1200 bps fall back)	16 ary QAM	FD, D-UP, 2-W
V-29	9600	16 ary QAM	FD, 4-W
V-32	up to 9600 bps	32 ary QAM	FD, D-UP, 2-W
V-33	14,400	128 ary AM-PM	FD, 4-W

FD—Full duplex, D-up = dial up connection, 4-W-4 wire based circuit.

11.3. PACKET SWITCHING

There are three types of switching used in PSTN network. The circuit switching and message switching were explained in the section 4.7. Circuit switching was designed for voice communication. Circuit switching creates temporary (dialed) or permanent (leased) dedicated links that are well suited to this type of connection. The circuit switching also limits the flexibility and not suitable for connecting variety of digital devices. More efficient utilization of the network requires greater control channel bandwidth and increased call processing capacities in the switches. But the circuit switching not providing these capabilities. Message switching overcomes the limitations of circuits switching. This switching stores the incoming messages into a computer memory and forward it to the destination when available. This causes delay in switching. The packet switching overcomes all the limitations of message and circuit switching. Thus it is highly suitable for the data communication.

The first packet switching system Arpanet, was developed by the U.S. Defence Advanced Research Projects Agency (DARPA) in 1969. The system used PDP-8 minicomputers made by Digital Equipment Corporation as packet switches, which were connected by dedicated 50 kbps telephone lines. Since then, many private and public packet switching networks, notably the X-25 system with speeds varying from 56/64 kbps to 1.5/2 Mbps (T1/E1) have been deployed.

11.3.1. Packet Switching Principles

The datastream originating at the source is divided into packets of fixed or variable size. The data communication system typically have bursty traffic. Thus, the time interval between consecutive packets may vary, depending on the burstiness of the data stream. A typical upper bound on packet length is 512 octets (bytes). Each packet contains a portion of the user's data plus some control information. As the bits in a packet arrive at a switch or router, they are read into a buffer. When the entire packet is stored, the switch routes the packet over one of its outgoing links. The packet remains quenched in its buffer until the outgoing link becomes idle. This technique is called store and forward technique. Fig. 11.5 illustrates the flow of packet switching.

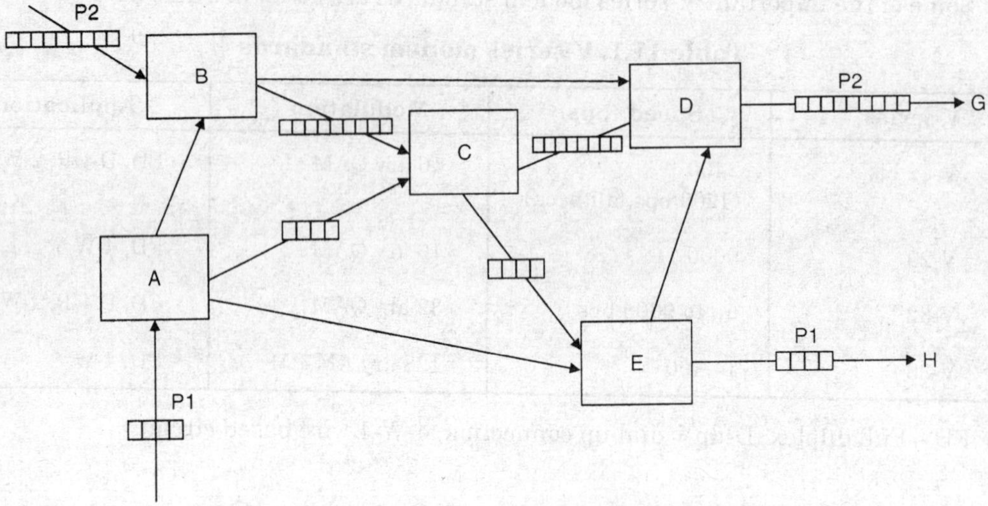

Fig. 11.5. Packet switching principle.

In Fig. 11.5, two packets (P1, P2) are entering at station A and B. The packet (P1) entering at A is targetted to the station H and the packet (P2) entering at B is targetted to the station G. Depends on the path availability the packet P1 choses the path as station A–C–E–H. Similarly the packet P2 choses the path as station B–C–D–G. In each station, the packets are stored in a buffer and forwarded to the next station after the availability.

11.3.2. Routing Control

From the previous section, it is clear that in packet switching, messages are broken into packets and sends one at a time to the network. Routing control decides how the network will handle the stream of packets as it attempts to route them through the network and deliver them to the intended destination. The routing decision is determined in one of two ways. They are

1. Datagram and
2. Virtual circuit.

Datagram. In datagram, each packet within a stream is independently routed. A routing table stored in the router (switch) specifies the outgoing link for each destination. The table may be static or it may be periodically updated. In the second case, the routing depends on the router's estimate of the shortest path to the destination. Since the estimate may change with time, consecutive packets may be routed over different links. Therefore each packet must-contain bits denoting the source and destination. Thus may be a significant overhead.

Fig. 11.6 shows a simple communication network where the concept of datagram is explained. The circled one are called the switching nodes whose purpose is to provide a switching facility that will move the data from node to node until they reach the destination. The squared one are called the stations. The stations may be computers, terminals, telephones or other communication devices. These stations also referred as end devices or the communi-cating devices.

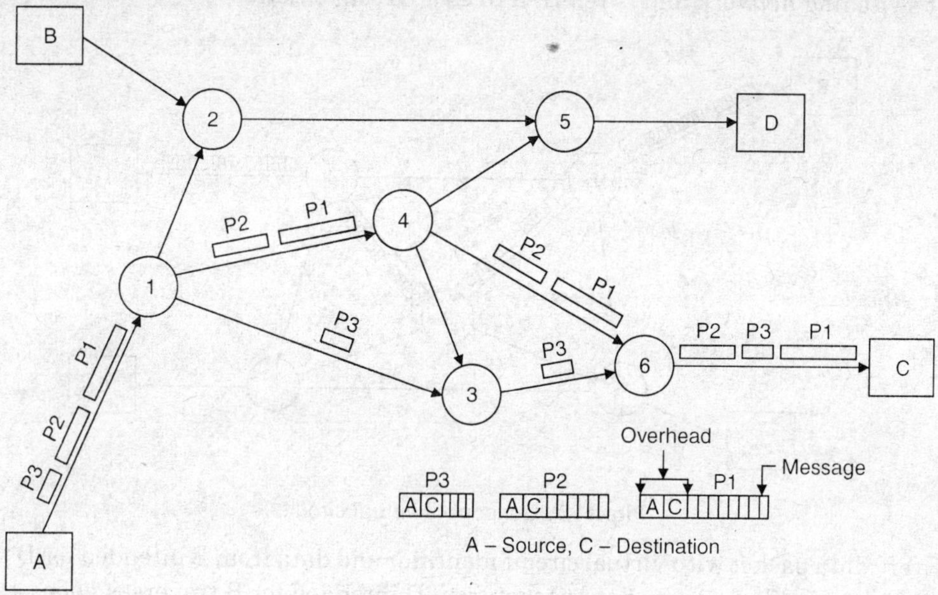

Fig. 11.6. Concept of datagram.

In the datagram approach, each packet is treated independently. In the Fig. 11.6. shown, the station A is assumed to send three packets of message namely P1, P2 and P3 (for explanation purpose named so). At first, A transmits these packets to node 1. Node 1 makes decision on routing of these packets. Node 1 finds node 4 as shortest compared to node 3. Thus it passes P1 and P2 to node 4. Accidently, if node 4 is not accessible, node 1 finds node 3 as shortest and sends packet P3 to node 3.

Node 3 and 4 sends its received messages to the destination C through node 6. It is shown that the order of the packet is changed due to the different routing of the packets. Thus in datagram, it is the responsibility of destination station to reorder the packets in proper sequence. Also if a packet crashes in a switching node, the destination C may not receive, all packets. In such a case also, it is the responsibility of station C to recover the lost packet.

Virtual circuit. In virtual circuit, a fixed route is selected before any data is transmitted in a call setup phase similar to circuit switched network. All packets belonging to the same data stream follow this fixed route called a vitrual circuit. Packet must now contain a virtual circuit identifier. This bit string is usually shorter than the source and destination address identifiers needed for datagram. Once the virtual circuit is established, the message is transmitted in packets. Fig. 11.7 shows the concept of virtual circuit.

In the Fig. 11.7 shown, suppose that end station B has two messages to send to the destination D. First B sends a control packet refered as call-request packet to node 2, requesting logical connection to D. Node 2 decides to route the request and the subsequent message packets through node 3 and 4 to destination D. If D prepared to accept the connection, it sends a call-accept packet to node 4. Node 4 sends the call-accept packet to B through node 3 and 2. Because the route is fixed for the duration of the logical connection, it is somewhat similar to a circuit switching network and is referred to as a virtual circuit.

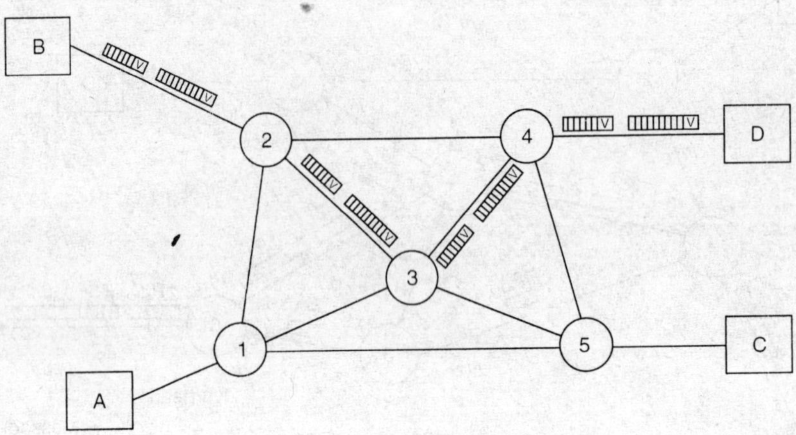

Fig. 11.7. Concept of virtual circuit.

Every data packet with virtual circuit identifier and data from B intended for D traverses node 2, 3 and 4. Similarly every data packet from D intended for B traverses nodes 4, 3 and 2. Any station can terminate the connection with a clear-request packet. At any time, each station can have more than one virtual circuit to any other station and can have virtual circuits to more than one station. As all packets follow the same route, they reach the destination in the same order. So there is no need of reordering work for destination station. Error control is the additional advantage of virtual circuit. Error control is a service that assures error free reception. For example, if a packet in a sequence from node 3 to 4 fails to arrive at node 4, or arrives with an error, node 4 can request a retransmission of that packet from node 4. As there is no necessity of routing decision during transition of packets from source to destination, with virtual circuit, packets transit rapidly. Currently available packet switching networks make use of virtual circuits for their internal operations.

Packet size. If an organization has large amounts of data to send, then the data can be delivered to a packet assembler/disassembler (PAD). The PAD (software package) receives the data and breaks it down into manageable packets. In the data communication, a packet can be a variable length. Usually upto 128 bytes of data is in one packet. X-25 services have created packets upto 512 bytes, but the average is 128. The 128 byte capability is also referred to as fast select. There is a significant relationship between packet size and transmission time. The process of using more smaller packets (for example 30 byte information may be sent as a single packet with header of 3 byte or two packets with 15 byte each plus the header in each packet or 5 packets with 6 bytes plus header) enables the speed of transmission.

11.3.3. Comparison of Circuit Switching and Packet Switching

There are two types of approaches in packet switching. Datagram and virtual circuit. The circuit switching is compared with these two approaches.

Datagram switching achieves higher link utilization than circuit switching especially when the traffic is bursty. No dedicated path is required as circuit switching. But the datagram have the disadvantage over virtual circuit. They are:

1. End to end delay may be so large or so random as to preclude applications that demant guaranteed delay.

2. The overhead due to source and destination identifiers and bits needed to delimit packets may waste a significant fraction of the transmission capacity if the packet are very short.

3. A datagram switch does not have the state information to recognize if a packet belongs to a particular application. Hence the switch can not allocate resources (bandwidth and buffers) that the application may require.

Virtual circuits are more advantageous and currently the packet switching network uses the virtual circuit approach. The overhead is comparable to circuit switching. As the packets arrives in sequence, no resequencing is needed. Statistical multiplexing of packets at the router or switch can achieve better utilization than in circuit switching. Since packets contains their virtual circuit identifiers (VCI), the switch can allocate resources depending on the VCI. During the connection setup phase, the switches may be notified that a particular VCI should be given extra resources.

11.3.4. Packet Formats

The format of a packet in packet switching network can vary significantly from one network to another. Generally header includes all related control information. In some cases, control information is communicated through a special control packets. Fig. 11.8 shows typical packet format.

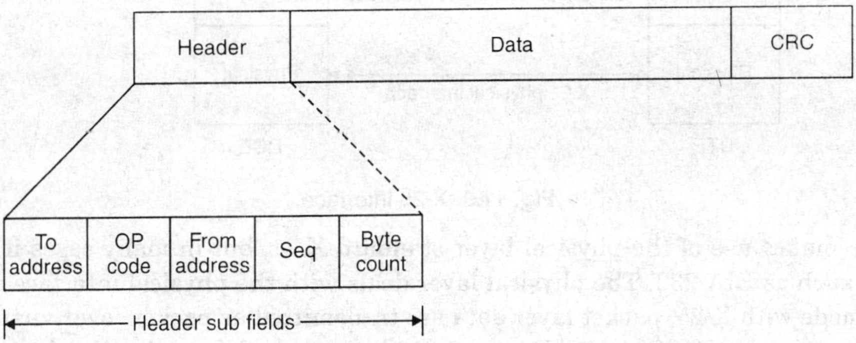

Fig. 11.8. Typical packet format.

A packet contains 3 major fields.

1. **Header.** It contains sub fields in addition to the necessary address fields. Other than the to and from address field, the following are the useful control information.

(*a*) **Op code.** It designates whether the packet is a message (text) packet or control packet.

(*b*) A **sequence number (Seq)** to reassemble messages at the destination node, detect faults and facilitates recovery procedures.

(*c*) **Byte count.** Used to indicate the length of a packet.

2. **Data.** A portion of a data stream to be transferred in the data field. Some packets may not contain a message field if they are being used strictly for control purposes.

3. **CRC.** The cyclic redundancy checks (CRC) field contains a set of parity bits that cover overlapping fields of message bits. The fields overlap in such a way that small numbers of errors are always detected. The probability of not detecting the occurance of 2 large number of errors is 1 in 2^M, where M is the number of bits in the cheek code.

11.3.5. X-25

X-25 is an ITU standard, well known and most widely used protocol established in 1976. The X-25 is subsequently revised in 1980, 1984, 1988, 1992 and 1993. The standard specifies an interface between a host system and a packet switching network. X-25 standard for packet switching is a lower three layer equivalent of the OSI model. This protocol based on a physical layer, a link layer, and packet layer. The data link layer of X-25 is link access procedure balanced (LAPB) using high level data link control (HDLC). HDLC is a bit oriented protocol based on the synchronous data link protocol (SDLC) established by IBM for Synchronous Network Architecture (SNA) networks. Fig. 11.9 shows the X-25 interface.

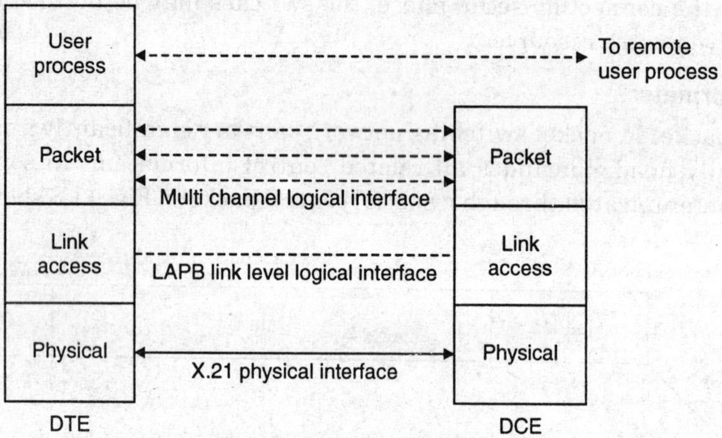

Fig. 11.9. X-25 interface.

X-25 makes use of the physical layer standard X-21, but in many cases it uses other standards such as EIA 232. The physical layer deals with the physical interface station and switching node with X-25, packet layer data are transmitted as packets over virtual circuits. The link layer provides for the reliable transfer of data across the physical link by transmitting the data as a sequence of frames. The link layer standard is known as LAPB. User data are passed down to X-25 level 3, which appends control information as a header, creating a packet. The entire X-25 packet is then passed down to LAPB entity, which appends control information at the front and back of the packet, forming an LAPB frame. The basic format of an HDLC packet is shown in Fig. 11.10.

The opening flag and closing flag are made up of 8 bit information. Packets are delimited by a starting and an ending flag (01111110). The address field is typically 8 bits long, but can be extended in increments of 8 bits (Here 16 bits).

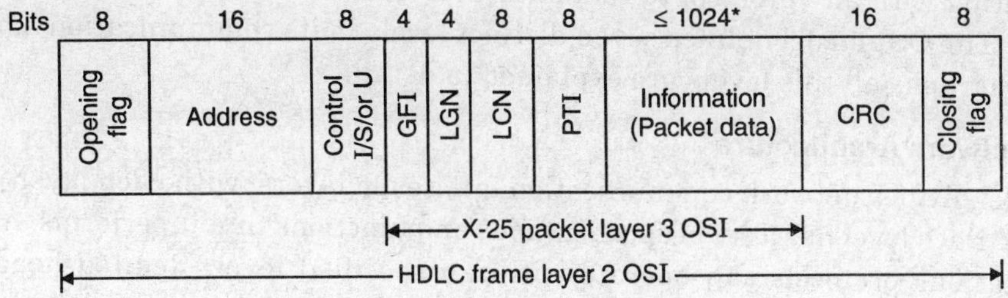

Fig. 11.10. HDLC frame.

Control I/S/ or U : Control information consists of 8 bits of data describing the type of HDLC frame.

- **Information (I).** Used to transfer data across the link at a rate determined by the receiver and with error detection and correction.
- **Supervisory (S).** Used to determine the ready state of the devices receiver is ready (RR), receiver is not ready (RNR) or reject (REJ).
- **Unnumbered (U).** Used to dictate parameters, such as set modes, disconnect and so on.

Packet information : The packet information consists of GFI, LGN and LCN.

- **General format identifier (GFI).** 4 bit of information that describes how the data in the packet is being used, from/to an end user, from/to a device controlling the end user device and so on.
- **Logical channel group number (LGN).** 4 bits of information that describe the grouping of channels.
- **Logical channel number (LCN).** 8 bit of information of the actual channel being used. Theoretical number of logical channels available is 2048. But most organizations implement significantly fewer ports or channels.

Packet type identifier (PTI). It is an 8 bit sequence that describers the type of packet being sent across the network. Six types of packets are used in X-25 switching network. They are call request, call accept, clear request, interrupt request, reset request and restart request.

Variable data field. 128 byte field is the standard implementation. But it can be as much as 512 bytes.

Cyclic Redundancy Check (CRC). The 16 bit sequence is used for error detection and/or correction.

11.4. OSI MODEL

The Open System Inter connection (OSI) model was developed by the International organization for standardization (ISO). The ISO developed OSI for networking. An open system is a set of protocols that allows two computers to communicate with each other regardless of their design, manufacturer or CPU type. Open system architectures are flexible structures set into fixed frame works. The concept of an open system approach to networking allows any device or system operating with any protocol to communicate with another device or system using its own protocol. The OSI model defines seven distinct levels in its communication model. In the following paragraphs all this levels are explained in detail.

11.4.1. OSI Network Architecture

The OSI model divides network communication into seven layers, with each layer performing specific tasks. Each level has a set of specifications and functions that it performs. Any number of communications protocols can operate within a specified level. Related header, trailer information, error detection capabilities and other overhead type are added to the message. The entire message with its overhead denoted as payload. The pay load is then encapsulated into the data portion of the next layer's message format and transported using that level's protocol rules. In the table 11.2 shown, the specification/functions of the layer are given briefly. The detailed explanation of each layer is given in the following sections.

Table 11.2 OSI layer specifications

LAYER	SPECIFICATIONS
7	**APPLICATION LAYER:** Performs information processing such as file transfer, e-mail and teletext. Detailed and application specific information about data being exchanged.
6	**PRESENTATION LAYER:** Defines the format of data to be sent : ASCII, data encryption, data compression and EBCDIC.
5	**SESSION LAYER:** Management of connections between programs. Sets up a session between two applications by determining the type of communication such as duplex, half duplex, synchronization etc.
4	**Transport layer:** Delivery of sequence of packets. Ensures data gets to destination. Manages error control, flow control and quality of service.
3	**NETWORK LAYER:** Format of individual data packets. Sets up connection, disconnects connection, provides routing and multiplexing.
2	**DATA LINK LAYER:** Manages framing, error detection, and retransmission of message. Access to and control of transmission medium.
1	**PHYSICAL LAYER:** Medium and signal formed of raw bit information. Electrical interface (type of signal), Mechanical interface (type of connector), converts electrical signal to bits, transmits and receives electrical signals.

The seven layers of OSI are grouped into three layers. The layer 1, 2 and 3 are called network support layers. The layers 5, 6 and 7 are called user support layers and layer 4 is transport layer. Fig. 11.11 shows the OSI network architecture.

Let computer A sends a data stream of bits to computer B. Communication must move from higher layer down through the lower layers on computer A. Each layer in sending machine adds its own information to the message. In receiving computer B, communication must move from lower layer up through the higher layers. At the receiving machine, the message is unwrapped layer by layer.

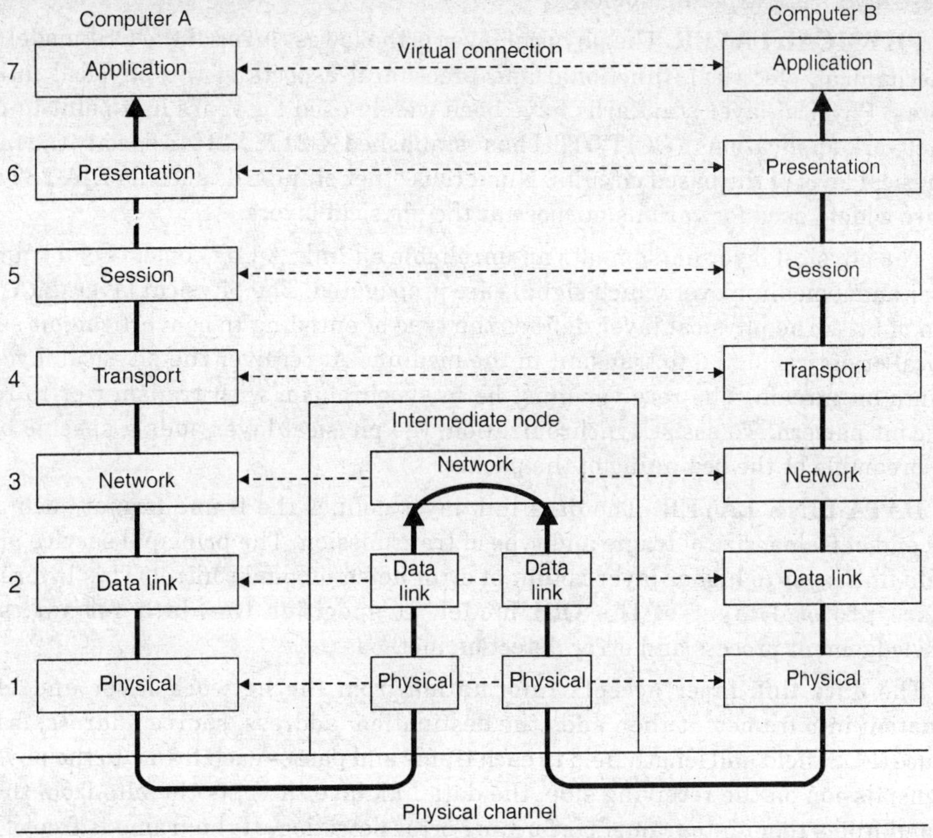

Fig. 11.11. OSI network architecture.

The information added to the message in layers of the sending computer is in the form of headers and trailers. Headers are added to the message at layers 6, 5, 4, 3 and 2. A trailer is added at layer 2. In receiving computer, each layer removes the data meant for it and original message is recovered by the receiving computer B.

Between each layers, there is an **interface** (not shown in Fig. 11.11). The passing of the data and network information down through the layers of the sending computer and back up through the layers of the receiving computer is made possible by this interface. Each interface defines what information and services a layer must provide.

Between computers, layer x on one computer communicates with layer x on another machine. This communication is governed by certain protocols. The process on each machine at a given layer are called pear to pear processes.

In the following sections, all the layers are explained under three headings as network support layers, transport layer and support layers.

11.4.2. Network Support Layers

The physical layer, data link layer and network layer are referred as network support layers. All these layers are explained below.

PHYSICAL LAYER. The physical layer is the lowest layer of the OSI model. It defines the mechanical, electrical, functional and procedural aspects of the physical link between networks. Physical layer standards have been widely used for years in a point-to-point wide area network applications. (CCITT/ITU has established X-21 & X-24 to specify the functions at the physical level of the based circuits. Numerous other standards such as EIA-232 and V-21–V-34 are widely used for various purpose at the physical layers.

The physical layer implements an unreliable bit link. A link consists of a transmitter, a receiver and a medium over which signals are propagated. The physical layer data consists of stream of bits. The physical layer defines the type of encoding to convert the bit stream into electrical or optical signal to transmit in the medium. At receiver the physical layer converts back into bit stream. The receiver must be in synchronism with transmitter to receive the specific bit pattern. To assist synchronization, the physical layer adds a specific bit pattern called preamble at the beginning of the packet.

DATA LINK LAYER. The data link layer defines the frame format such as start of frame, end of frame, size of frame and type of transmission. The principal service provided by the data link layer to higher layer is that of error detection and control. This layer is the first software protocol layer of the OSI model. It specifies the data format, sequence, acknowledgement process and error detection methods.

The data link layer accepts information from the network layer and breaks the information into frames. It then adds the destination address, source address, frame check sequence (FCS) field and length field to each frame and passes each frame to the physical layer for transmission on the receiving side, the data link layer accepts the bits from the physical layer and froms them into frame, performing error detection. If the frame is free of error, the data link layer passes the frame up to the network layer. It perform frame synchronization, that is, it identifies the beginning and end of each frame.

Existing protocols for the data link layer are :

1. **Synchronous Data Link Control (SDLC).** Developed by IBM as link access for System Network Architecture.
2. **High Level Data Link Control (HDLC).** It is a version of SDLC modified by the ISO for use in the OSI model.
3. **Link Access Procedure Balanced (LAPB).** The modified version of HDLC is LAPB.

The data link layer also performs the flow control and access control. By flow control, if the rate at which the data are absorbed by the receiver is less than the rate produced in the sender, the data link layer imposes a flow control mechanism to prevent overwhelming the receiver. By access control, when two or more devices are connected to the same link, the protocol of data link layer determines which device has control over the link at any given time.

Two sub layers defined in the data link are the media access control (MAC) and the logical link control (LLC) layer. MAC performs address management function. LLC manages flow and error control, automatic requests for retransmission (APQ) and handshake processes.

NETWORK LAYER. If two systems are attached to different networks (links) with connecting devices between the networks (links), there is often a need for the network layer to accomplish source to destination delivery. Thus the function of the network layer is to perform routing. The network layer checks the logical address of each frame and forwards the frame to the next router based on a look up table. The network layer is responsible for translating each logic address (name address) to a physical address (MAC address).

There are two types of virtual circuits used in the network layer. Connectionless and connection oriented. In connection oriented service, the network layer makes a connection between source and destination, then the transmission starts. Connectionless circuits are also known as "bandwidth on demand" circuits, which establish a connection when they are needed. The source transmits information regardless of whether the destination is ready or not. A common example of this type of service is e-mail.

11.4.3. Transport Layer

The transport layer provides a mechanism for the exchange of data between end systems. It optimizes the use of network services with providing a requested quality of service to session entities. The size and complexity of a transport protocol depend on how reliable or unreliable the undelying network and network layer services are. Essentially, this layer is responsible for the reliable data transfer between two end nodes and is sometimes referred to as host-to-host layer.

The transport layer decomposes messages into packets and combines packets into messages, possibly after resequencing them. This is required, because the size of the message may be larger than the size of the packets accepted by the network layer. To perform this packetization function, the transport layer numbers the packets belonging to the same message. This process is called segmentation and reassembly.

The transport layer controls the flow of packets to prevent the source from sending packets faster than the destination can handle them. Same function takes place in data link layer also. But here, the flow control is performed end to end rather than across a single link.

The transport layer delivery of messages is either connection oriented or connectionless. A connection oriented transport layer deliveres error free messages in the correct order. Such a transport layer provides the following services. CONNECT, DATA and DISCONNECT. A connectionless transport layer deliveres messages one by one possibly with errors with no guaranteed order of the messages. The services in UNIT–DATA, the connectionless delivery of single message.

Other processes at transport layer are :

1. Error detection and recovery to minimize data loss and timeloss due to retransmission of frames.

2. Sets quality of service (QOS) for network layer packets to assure end-to-end message integrity.

3. A process called blocking is employed when requested by sending station. Blocking is used to prevent a specific frames to reach a particular node.

11.4.4. User Support Layers

The session layer, presentation layer and application layer are considered as support layers. All these layers explained in the following paragraphs.

SESSION LAYER. The session layer establishes a logical connection between the applications of two computers that are communicating with each other. The session layer concerns file management and overall networking functions. Access availability and system time allocations are included in this layer. The session layer can partition a transfer of a large number of messages by inserting synchronous points. Specific responsibilities of the session layers are authentication of user access, fault recovery if a break in service occurs, permitting multiple applications to share a virtual circuit, connection and disconnection of any node from the network.

PRESENTATION LAYER. The presentation layer receives information from the application layer and converts into ASCII or unicode or encrypts or decrypts. This layer is concerned with the syntax and semantics of the information exchanged between two systems. The three basic forms of protocols used in the presentation layer are

1. Virtual terminal protocol, which is used to allow different types of terminals to support different applications.
2. Virtual file protocol, which handles code coversions within files, file communication and file formating.
3. Job transfer and manipulating protocols which controls the structure of jobs and records.

A separate protocol function known as abstract syntax notation (ASN) specifies file data structure.

APPLICATION LAYER. This layer enables users to access the network with applications such as e-mail, FTP and Telnet. It provides user interfaces and support for various services. Specific services are (*a*) Network virtual terminal (*b*) File transfer access and management (FTAM), (*c*) Mail services (*d*) Directory services (*e*) Specific User Service Element (SUSE) — deals with actual user requirements for access and use of the network (*f*) Common application service element (CASE) — This sets guidelines for the applications required quality of service and (*g*) Specific application service element (SASE) — deals with large amount of data, including database access.

11.5. TCP/IP

Transmission Control Protocol and Internet Protocol (TCP/IP) was developed not only to create LANs, but also for internetworking multiple LAN's. Data sharing and broadcasting are prominent features of LAN technology. An organization may create different LANs with different protocols. These networks are connected together by an internal gateway, in turn, connected to the external gateway of the internet. TCP/IP protocol is used to communicate among nodes. Today these protocols are the primary building blocks for the Internet. In 1974 TCP/IP was introduced by Advanced Research Project Agency (ARPA).

TCP/IP provides three sets of services. They are :

1. **Connectionless service.** This service is described as an unreliable (delivery is not guaranteed), packet delivery service. A packet may be lost, duplicated, delayed or delivered out of order, but the service will not detect such conditions. Here each packet is considered independently.

2. **Reliable transport services.** Work with any environment.

3. **Application services.** Interfaces to most services on other architectures.

11.5.1. TCP/IP Reference Model

TCP/IP architecture is a four layer stack that deals with the equivalent seven layer architecture of OSI, SNA and DNA. Fig. 11.12 shows the TCP/IP reference model.

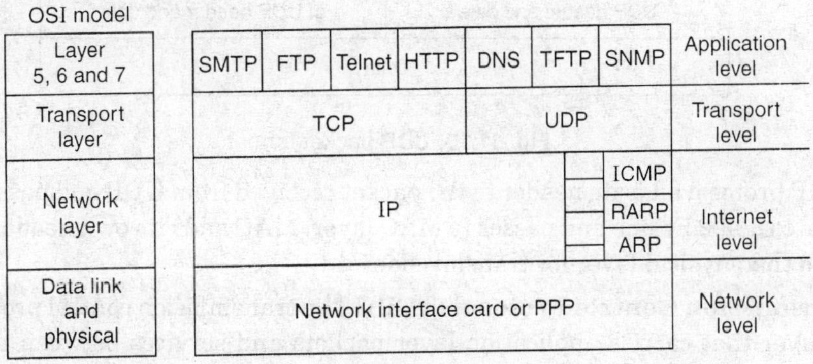

Fig. 11.12. TCP/IP reference model.

Application level : Some of the internet applications are SMTP, FTP, Telnet, HTTP, DNS, TFTP, SNMP.

SMTP. Simple Mail Transfer Protocol (SMTP) is used for E-mail. It is used for transferring messages between two hosts.

Telnet. It is one of the most important Internet applications. It enables one computer to establish a connection to another computer.

FTP. File transfer protocol (FTP) is an internet standard for file transfer. FTP establishes a connection to a specified remote computer using an FTP remote host address.

HTTP. Hypertext Transfer Protocol (HTTP) is an advanced file retrieval program that can access distributed and linked documents on the web. HTTP is a stateless protocol that treats each transaction independently.

DNS. Domain Name System (DNS) is used to identify and locate computers connected to the internet.

SNMP. Simple Network Management Protocol (SNMP) is used by the network administrator to detect problem in the network such as router and gateway.

Transport Level Protocols : This layer consists of User Datagram Protocol (UDP) and Transmission Control Protocol (TCP).

UDP. It provides unreliable service between hosts. UDP accepts infromation from the application layer and adds a source port, destination port, UDP length and UDP checksum. The resulting packet is called UDP datagram packet. The UDP allows applications to exchange individual packets over a network as datagrams. A UDP packet sends information to the IP protocol for delivery. Fig. 11.13 shows the UDP packet format.

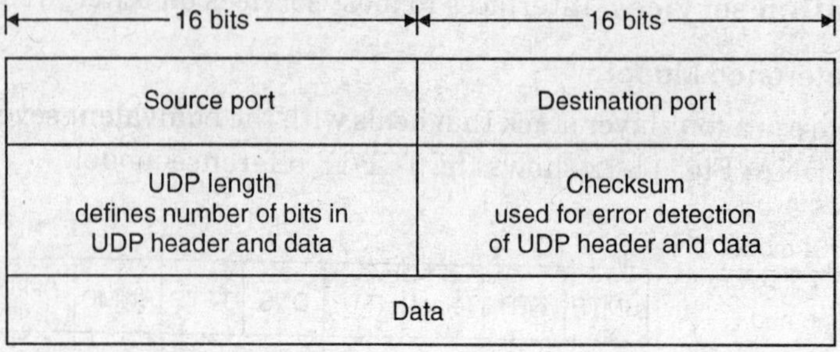

Fig. 11.13. UDP packet format.

The IP protocol adds its header to the packet received from UDP and passes to LLC. The LLC generates 802.2 Frame and passes to MAC layer. MAC adds its own header and transfers the frame to the physical layer for transmission.

Transmission Control Protocol (TCP) : The transmission control protocol (TCP) is a transport layer that carries application layer packets and services between two users. TCP offers reliable delivery of information through the internet. In TCP, connection between users is established before transmitting information. TCP assigns a sequence number to each packet. The receiving end checks the sequence number of all packets to ensure that they are received. When the receiving end gets a packet, it sends acknowledgment. If the sending node does not receive an acknowledgement within a given period of time, if retransmitts the previous packet. Fig. 11.14 shows the simple TCP operation.

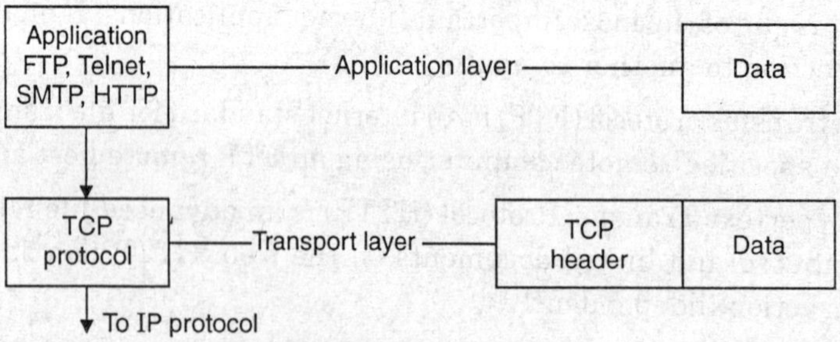

Fig. 11.14. TCP operation.

The TCP header includes valuable informations. Fig. 11.15 shows the TCP header.

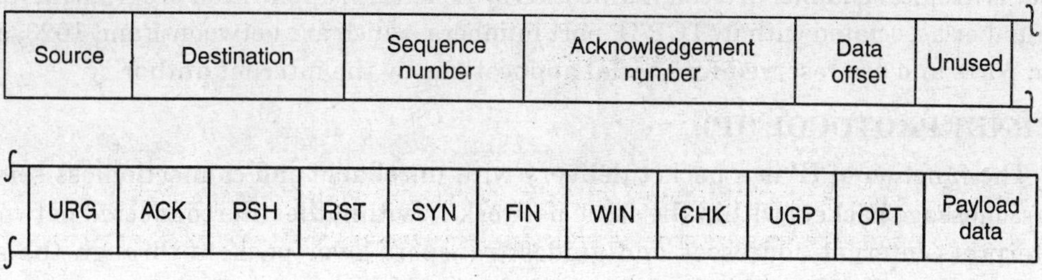

Fig. 11.15. TCP header.

Source and destination. These address provides routing information for the packet.

Sequence number. It is a number label for each packet sent by the source. It is 32 bit length. The order of the packets within the completed message frame work are maintained by the use of sequence numbers.

Acknowledgment number. It acknowledges the next packet expected to be received from the source. It is 32 bits length.

Data offset. The number of 32 bit words in the header is known as the data offset number, because it indicates how much the beginning of the data is offset by the header.

URG. Urgent pointer is set to '1' when the field contains urgent data. The urgent pointer indicates the offset value of the current segment and informs the receiver to find the urgent data indicated by URG.

ACK. Acknowledge flag bit is set to '1' to represent that the acknowledgement number is valid.

PSIF. This field set to '1' means that the receiver should push the data to an application as soon as possible.

RST. The reset bit causes the transport connection to be reset at the end of a session. It resets the connection.

SYN. A synchronization flag set to '1' when a node wants to establish a connection. A SYN indication is used to establish a connection in combination with the acknowledge (ACK) bit.

SYN = 1, ACK = 0 is a connection request and

SYN = 1, ACK = 1 is a connection acknowledgement.

FIN. The FIN flage set to '1' indicates that the incoming packet is the last packet. Otherwise, the final data segment is indicated by the FIN bit in the header.

WIN. The window specifies the number of data bytes the sender is willing to accept, inturn, from the host end. It is 16 bits length.

CHK. The TCP checksum is 16 bit long. It is used for error detection in TCP header and data field.

OPT. It is called option field

URG = 0 = end of option list

1 = no operation

2 = maximum segment size.

The TCP and UDP use port numbers to demultiplex messages to an application. A port number is a logical channel in a communication system. Each application program has a unique port number associated with it. TCP/IP port numbers, which are between 1 and 1023, are well known ports and are reserved for special applications by the internet authority.

INTERNET PROTOCOL (IP)

The function of IP is a packet delivery with unreliable and connectionless service. IP delivers message packets within the same network or within the interconnected networks the IP interprets internet addresses and guides transport layer packets through the system. Fig. 11.16 shows the IP header format.

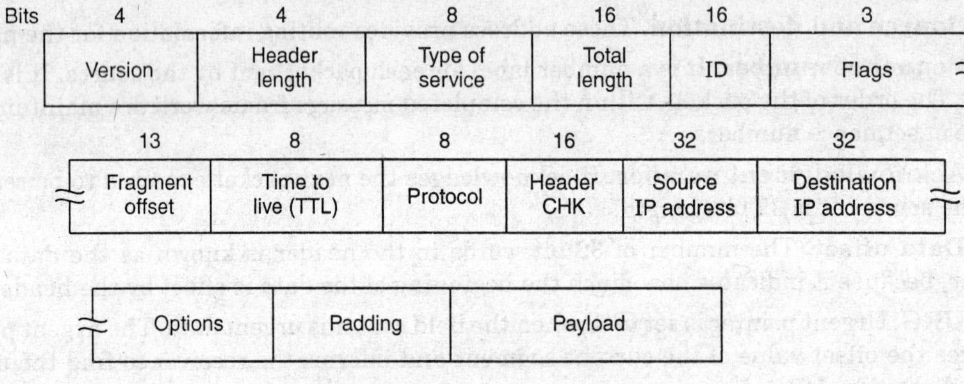

Fig. 11.16. IP packet format.

Version. Contains IP version number. The current version is IPV4. Due to the growth of the internet and the address limitations of IPVA, the internet Engineering task force (IETF) approved IPV6. The IPV4 address size is 32 bits and can connect upto 2^{32} = 4 billion users to the internet. The later version IPV6 has the address size of 128 bits. IPV6 has the features such as auto configuration (no human intervention), expanded addressing, simplified header format, support extension, flow labeling, authentication and privacy etc.

Header length. It represents the number of 32 bit words in the header. IP is an unreliable service. There is no acknowledgement from the destination to source. There is no physical connection between the source and destination. IP datagrams can arrive at the destination out of order.

Type of service (TOS). The format of TOS is shown in Fig. 11.17.

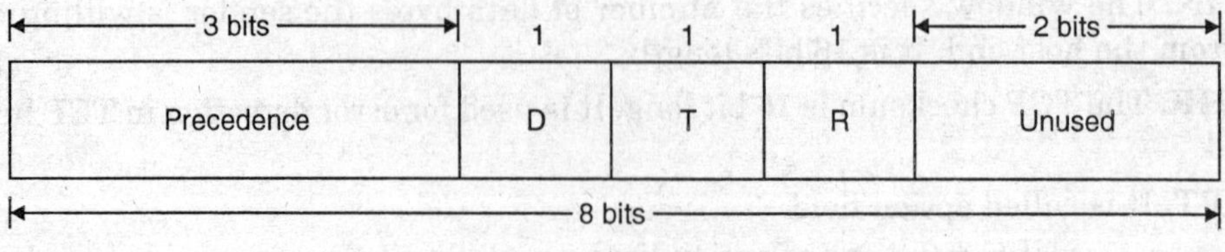

Fig. 11.17. Format of TOS.

The following table 11.3 shows the function of each field.

Table 11.3. Functions of TOS fields

Field	Specification/functions
Precedence	Indicates importance of the data gram 0—normal 1—next important
D	Delay, D = 0 Normal delay D = 1 Low delay
T	Throughput, T = 0 Normal T = 1 High
R	Reliability, R = 0 Normal R = 1 High

Flag. It has 3 bits, the first bit is unused. The other bits are DF = Do not fragment, and MF = More fragment. If DF = 1, datagram can not be fragmented. If MF = 1 fragmented.

Other fields. The fragmented offset field represents the offset of data in multiples of eight. The total length field identifies the total length of the datagram in bytes. ID is the identification number created by the sending node. This number is required when reassembling fragmented messages.

TTL field indicates the time in seconds that a datagram may remain on the network before it is discarded. This value (32 or 64) is set by the sender and is decremented by one everytime a router handles the datagram.

11.5.2. Protocols Related to TCP/IP

The following protocols are used to support TCP/IP :

1. **Internet Control Message Protocol (ICMP).** It controls transmission errors and control messages between host and gateways.

2. **Address Resolution Protocol (ARP).** The interior gateway (which connects the various LAN's of an organisation) must have the physical address of the host connected to one of its local networks. The gateway must record both the IP address and its physical address. If the gateway does not have the physical address of the host, it will send an ARP packet to get the host's physical address.

3. **Reverse ARP (RARP).**

4. **Simple Network Management Protocol (SNMP).** It is used for diagnostics purpose between hosts.

11.6. NETWORKING TECHNOLOGIES

A computer network is an interconnection of computers. Peripherals and data transfer devices with related software. Typical functions of the computer networks are sharing computer peripherals and equipments, communications among two or more computer or network devices and sharing information including software. Networks are generally classified by the size and distance between the network devices. They are local area network (LAN), metropolitan area network (MAN) and wide area network (WAN). In this section, the network topologies and types of networks are described in section 11.6.1. and 11.6.2. In subsequent sections, the industry standard LAN technologies such as Ethernet, Token ring and Token bus, are discussed.

11.6.1. Network Topology

The topology of a network describes the way computers are connected together. Topology is a major design consideration for cost and reliability. Common topologies found in computer networking are

1. Star
2. Ring
3. Bus
4. Fully connected network or mesh
5. Tree and
6. Hybrid.

The first five topologies are illustrated in Fig. 11.18. The hybrid topology is shown in Fig. 11.19. All the network topologies are explained below.

Star topology. Here, all computers are connected to central controller usually referred as hub as shown in Fig. 11.18 (*a*). All the communication between any computers are made through the hub only. Hence, if the hub breaks down, the entire network is disabled. But easy to set up, and simple procedure of expansion of network are the advantage.

Ring topology. It is also referred as IBM token ring as it was invented by IBM. In this topology (Fig. 11.18 (*b*)), the source station transfers information to the next station on the ring, which checks the address of the information. If the address matches, it copies the information and passes otherwise it passes the same to the next station without copying. The next stations repeats the process, till it reaches the source station. The source then removes the message from the ring. In figure shown, station A puts a message on ring to station D. While passing on the ring, the station D copies the message and then puts the same on ring. As the message is not entitled to remaining stations, the message is not copied and simply passed.

Although the ring topology has the advantage like easy installation, easy network expansion and use of fiber optic cable, it has worst disadvantages also. They are

1. If a link or a station breaks, the entire network is disabled.
2. Requirement of complex hardware. The network interface card is expensive.
3. Adding a new client distrupts the entire network.

Fully-connected topology. This topology connects every station to every other station (Fig. 11.18 (*c*)). Because of direct link, the topology offers high reliability and security. But this advantage is at the cost of large amount of wiring. The number of connection is determined by N(N – 1)/2, where N is the number of stations. If N = 100, no. of connection required is 4950. Hence this topology is not cost effective.

Bus topology. In this topology, all the work stations are connected to a common cable called bus (Fig. 11.18 (*d*)). Ethernet is one of the most popular LAN's that uses bus topology. This topology is simple, low cost and easily expandable. The disadvantage is that the breakdown of the bus brings the entire network down.

Tree technology. It uses an active hub or repeater to connect stations together. The functions of the hub is to accept information from one station and repeat the information to other stations and hub as shown in Fig. 11.18 (*e*). Different types of hubs are explained briefly below.

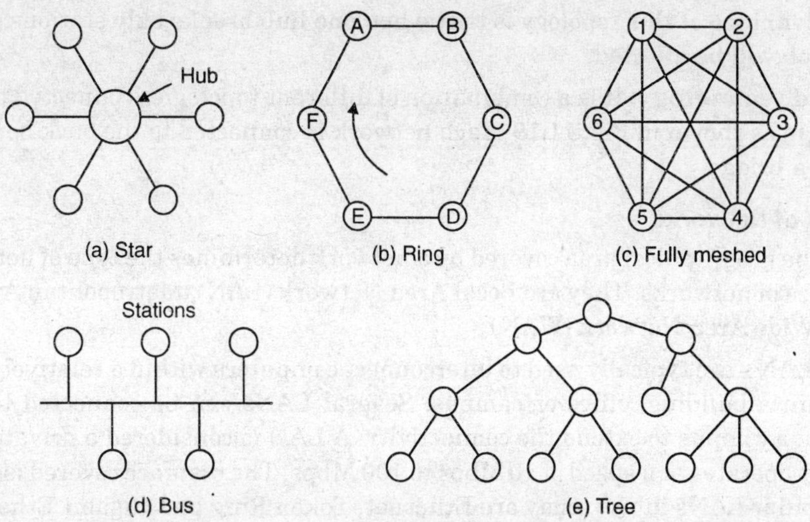

Fig. 11.18. Network topologies.

1. Manageable hub. The ports on the hub can be enabled or disabled by the network administrator through software.

2. Stand alone hub. It is used for work groups of computers that are separate from the rest of the network. They can not be linked together to represent a larger hub.

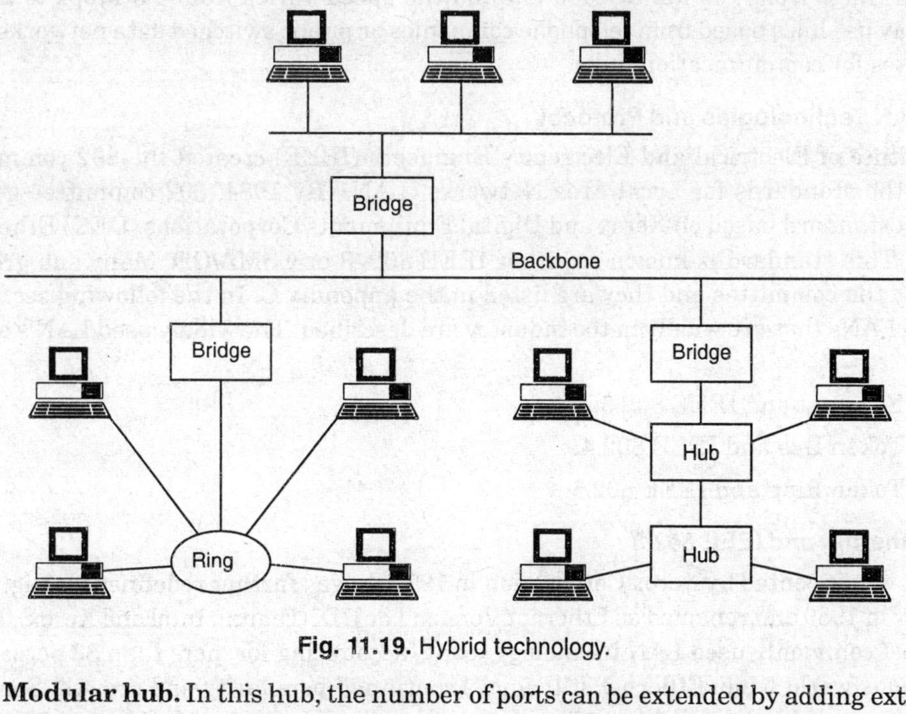

Fig. 11.19. Hybrid technology.

3. Modular hub. In this hub, the number of ports can be extended by adding extra cards.

4. Stackable hub. A stackable hub looks like a stand alone hub, but several of them can be stacked or connected together in order to increase the number of ports.

The advantage of this topology is that when one hub breaks only stations connected to the broken hub will be affected.

Hybrid technology. It is a combination of different topologies connected together by a backbone cable as shown in Fig. 11.19. Each network is connected to the back bone cable by a device called a bridge.

11.6.2. Types of Networks

Depends on the geographical area covered by a network determines the type of network. There are three types of networks. They are Local Area Network (LAN), Metropolitan Area Network (MAN), and Wide Area Network (WAN).

LAN. LAN's are typically used to interconnect computers within a relatively small area, such as within a building, office or campus. Several LANs can be connected together in a building or on a compus to extend the connectivity. A LAN is considered a private network. A LAN typically operates at a speed of 10 Mbps to 100 Mbps. The distance covered is up to 10 kM. The most popular LANs in use today are Ethernet, Token Ring and Gigabit Ethernet. It uses cable media.

MAN. As the name implies, this network covers a metropolitan city. It cover approximately 100 miles, connecting multiple networks which are located in different locations of a city or town. A typical MAN operates at a speed of 1.5 to 150 Mbps. It uses different hardware and transmission media.

WAN. WAN cover a large geographical area, such as an entire country or continent. A WAN can range from 100 km to 1000 km and the speed varies from 1.5 Mbps to 2.4 Gbps. WANs may use lines based from telephone companies or public switched data networks (PSDN) or satellites for communication links.

11.6.3. LAN Technologies and Protocol

The Institute of Electrical and Electronics Engineers (IEEE) created the 802 committee for drafting the standards for Local Area Networks (LAN). By 1984, 802 committee generated their first standard based on Xerox and Digital Equipments Corporations (DEC) Ethernet bus network. This standard is known as either IEEE 802–3 or CSMA/CD. Many sub groups are formed by the committee and they are listed in the appendix C. In the following sections, the standard LANs that are widely in the industry are described. The widely used LAN's are listed below.

1. Ethernet and IEEE 802.3.
2. Token Bus and IEEE 802.4.
3. Token Ring and IEEE 802.5.

11.6.4. Ethernet and IEEE 802.3

Ethernet was invented by Xerox Corporation in 1972. It was further redefined by Digital, intel and Xerox in 1980 and renamed as Ethernet Version I or DIX (Digital, Intel and Xerox). Ethernet is the most commonly used LAN technology today, accounting for more than 83 percent of the installations world wide. Ethernet still enjoys continued popularity and growth. Some of the reasons are :

1. Least expensive
2. It is fast enough for the vast majority of applications in use
3. It continue to keep pace with other LAN technologies
4. Its various standards supporting a wide variety of media
5. Sufficiently defined standards
6. Ethernet is considered to be the most user friendly.

Ethernet Reference Model. The IEEE developed the standards for Ethernet in 1984. It is called IEEE 802.3 and uses the bus topology. The function of Ethernet is as follows.

The Ethernet system consists of three basic elements. They are the physical media (the ether) used to carry signals, a set of rules, embedded in each Ethernet interface and the Ethernet frame that consists of bits used to carry data, control and address information. Each station in the network consists of Network Interface Card (NIC). This card is connected to the Ethernet cable via a transceiver cable. The NIC is also called as Ethernet controller. As there is no central controller, Ethernet LAN operates independently of all other stations.

When a station transmits a frame on the bus, all stations connected to the network will copy the frame. Ethernet signals are transmitted serially, one bit at a time over the bus. Each station checks the address of the frame. If it matches the station's NIC address, it accepts the frame, otherwise the station discards the frame.

Fig. 11.20 shows the Ethernet reference model. It shows, how Ethernet fits into the OSI model.

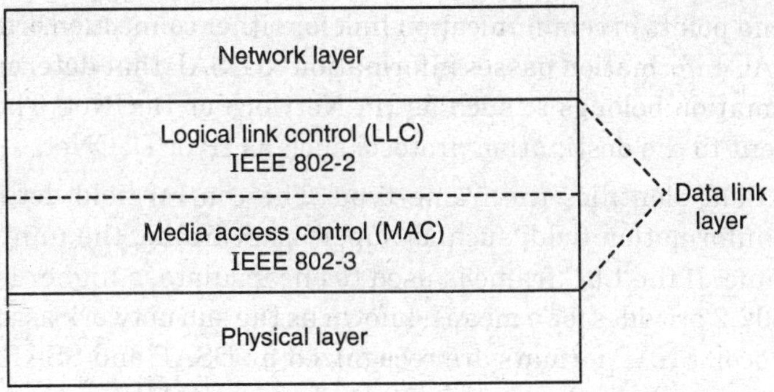

Fig. 11.20. Ethernet reference model.

The data link layer is divided into two sublayers. They are :

1. Logical link control (LLC) and
2. Media Access control (MAC).

LLC and MAC work together to formulate an ISO data-link layer protocol. The LLC and MAC field together with data and additional fields form the final message format.

LLC layer. The LLC layer is designed to establish a logical connection between source and destination. The IEEE standard for LLC is IEEE 802.2. The responsibilities of LLC sublayer are :

(*a*) establishing and terminating a communication link

(*b*) Supplying a frame format for the pay load

(*c*) detecting and correcting errors and

(*d*) Maintaining control over the traffic flow.

The establishment of communication link at LLC may be connection oriented (LLC 1) or connectionless (LLC 2). In LLC 1, data link connection is established before data is sent. LLC 2 establishes the connection path when the first frame being transmitted. The basic format of LLC is shown in Fig. 11.21.

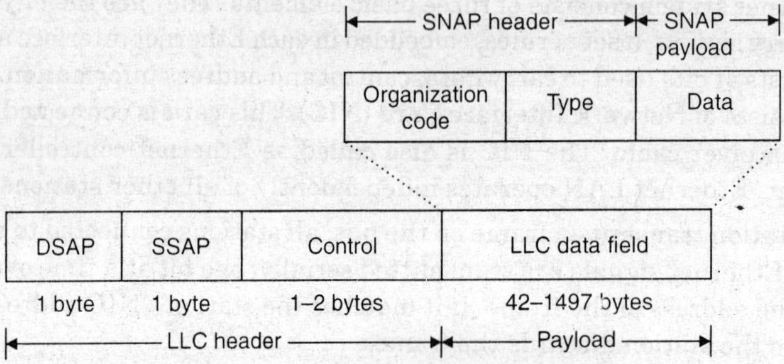

Fig. 11.21. LLC, IEEE 802.2 frame format.

Destination Service Access Point (DSAP) and Source Service Access Point (SSAP) defines the location of the end points in communication link for either connection oriented or connectionless service. The MAC information passes information to DSAP that determines which protocol the incoming information belongs to such as IP, Network or DeCNet. The SSAP determines which protocol is sent to the destination protocol such as IP or DeCNet.

The control field identifies the frame type. The control field determines the type of information in the information field, such as supervisory frame, the unnumbered frame and the information frame. If the LLC frame is used to encapsulate, a higher level protocol within the pay load field, 802.2 provides for a means, known as the sub network assess protocol (SNAP), to identify the protocol. SNAP formats are recognized by DSAP and SSAP. In response to the address, LLC protocol will interpret the payload as a SNAP format. Actual source and destination address are included in the protocol format of the data area in the payload.

MAC layer. The function of the MAC is to access the network. It defines, how different stations can access the transmission medium. The MAC uses CSMA/CD protocol. In an Ethernet network, each station uses CSMA/CD protocol to access the network in order to transmit information. The MAC and CSMA/CD are explained under the CSMA/CD protocol heading.

CSMA/CD Protocol. The MAC mechanism is based on the Carrier Sense Multiple Access with Collision Detection (CSMA/CD). The CSMA/CD works as follows:

1. If a station wants to transmit, the station senses the channel.

2. If the channel is busy, it continues the sensing of the channel. When the channel becomes idle, the station starts transmit data.

3. After sending data, the station senses for the collision, as there is a possibility that two station may send data at the same time.

4. If the collision detected, the station which detected first sends a jamming code (32 bits) on the bus, in order to indicate the other station that there is a collision on the bus.

5. If no collision, transmission completed.

6. The two stations, which enveloped in collision, wait according to back-off algorithm. Back off algorithm is a method used to generate waiting time for stations that were involved in collision.

When a collision occurs, the stations involved senses it and stops the transmission and resend the frames. This is called collision detection.

Fig. 11.22 shows the IEEE 802.3 MAC format. Each fields of the frame are explained below :

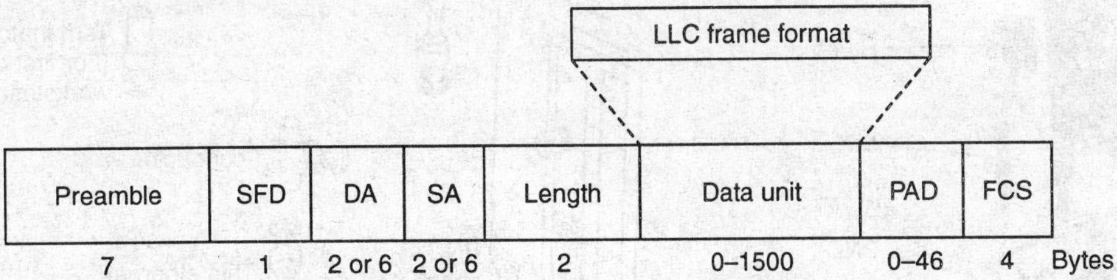

Fig. 11.22. IEEE 802.3 MAC format.

The preamble provides signal synchronization and consists of seven bytes of alternating 1 and 0 bits. The preamble and start Frame Delimiter (SFD) Synchronize the receiver by using a 64 bit sequence of alternating 1's and 0's ending with 11 (10101010 ... 11). The destination address (DA) is the hardware address (unique in the entire world). The hardware address of the NIC is also called a MAC address or physical address. IEEE assigns 22 bits of physical address to the manufacturers of NIC. The 46 bit address is burned into the ROM of each NIC and is called the universal administered number. Source address (SA) shows the address of the source from which the frame is originated.

The "length" field indicates frame length. This two byte field defines the number of bytes in the data field. The data field contains the actual information. The IEEE specifies that the minimum size of a data field must be 46 bytes and maximum sizes 1500 bytes. If information in the data field is less than 46 bytes, extra information is added in the PAD field to increase the size to 46 bytes. Frame check sequence (FCS) is used to detect errors and corrupted information during transmission. IEEE uses CRC-32 for error detection.

Ethernet Media. The Ethernet network uses four different media. They are 10 Base 5, 10 Base 2, 10 Base T and 10 Base-F.

10 Base 5 (Thick Net). It uses 10 Mbps Ethernet media with base band signalling with maximum segment lengths of 500 meters. The maximum length of the Attachment Unit Interface cable (AVI) is 50 meters. A maximum of five segments can be connected with a total of 2.5 km and not more than four repeaters in a path. The minimum distance between transceivers is 2.5 meters, allowing a maximum of 1000 stations at all. Both ends of the segments must be terminated by a 50 Ω resistor.

10 Base 5 is occasionally used for network back bones. Connector DIX is used for 10 Base 5 that come with an NIC.

10 Base 2 (ThinNet). It uses 10 Mbps Ethernet media with maximum segment lengths of 185 m. The maximum length of a network cable is 925 meters with four repeaters. The transceiver is built into the NIC. No more than 30 connections are allowed per segment. 10 Base 2 uses thin coaxial cable.

BNC connectors are used for 10 Base 2 connections. Hence, the connectors and cables in this case are NICs thin coaxial cable, and a T shaped device called BNC-T connector.

Fig. 11.23 shows the NIC and components of ThinNet.

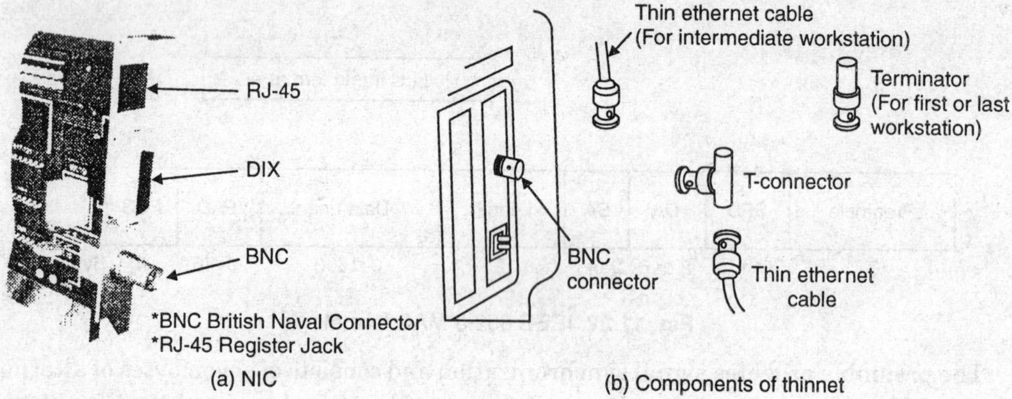

Fig. 11.23. *(a)* NIC *(b)* Components of thinnet.

10 Base T. It uses 10 Mbps 22 to 26 AWG twisted pair cable (unshielded) instead of coaxial cable. A maximum segment length of 100 meter is supported. The tranceiver is built into NIC. The devices are connected to 10 Base T hub in a star topology. 10 Base T allows a maximum of four repeaters connected together and the maximum diameter is 500 m.

RJ 45 connector is used for 10 Base T connections. Devices with standard AVI connectors may be attached to the hub by using a 10 Base T receiver. Many companies currently use 10 Base T media.

10 Base F. It uses 10 Mbps Ethernet media over fiber optic cable. Low cost, Interference free electrical characteristics and high speed are the advantage of this media.

High Speed Ethernets. Due to recent advances in microprocessor technology, end devices (computers) communicates large amount of data, high end graphics including 3-D images at a high speed. The transfer of this type of data between modern computers through Ethernet network is very slow. Many companies upgraded their networks to Fast Ethernet which offers a data rate of 100 Mbps. The goal of the fast Ethernet is to increase the bandwidth of Ethernet network while using the same CSMA/CD protocol. Rapid researches are under process to cope up with the advances in microprocessor technology. Following are the currently available fast Ethernet.

1. IEEE 802.3*u* is an extension of IEEE 802.3 and was approved as the standard for Fast Ethernet in 1995. Fast Ethernet offers a data rate of 100 Mbps. The media for Fast Ethernet are 100 Base T4, 100 Base TX and 100 Base FX.

2. Gigabit Ethernet is a new technology that is compatible with Ethernet and Fast Ethernet. It transfers data at one gigabit per second or 100 times faster than standard Ethernet. By adding Quality of Service (QoS), the Gigabit Ethernet is able to handle all types of data transmission, even voice and video information.

11.6.5. Token Bus and Token Ring Networking

The Ethernet system has two main disadvantages.

1. This system may not be suitable for a high traffic if excessive number of collisions are anticipated.

2. Due to number of retransmission, reduced throughput and increased delay results. The alternative for the Ethernet system are Token Bus (IEEE 802.4) and Token Ring (802.5). The token bus combines features of Ethernet and Token ring to provide a deterministic delay under heavy loads without causing collisions. Token ring is a powerful LAN technology that is designed to handle heavy loads. Both Token Bus and Token Ring Networking are described in this section.

TOKEN BUS NETWORKING (IEEE 802.4)

The Token bus system operates on a bus topology and is considered to be suitable for industry applications in which factory automation and process control are desired. It uses token passing for accessing the network. A token is a short message that specifies the station currently using the network and the next station that access after the current station finish its work.

The token is passed from station to station in descending order and it is not necessary for all the stations on the bus to be active at all times. The procedure for transmit a data by a computer is as follows.

1. The computer has to wait for the token. Once it possess the token, it can add its traffic to the data stream. It then includes the successors address and passes the token.

2. After passing the token, the sending computer monitors, whether the successor receives the token or not. If the successor had data, it sends in data stream and passes to the next successor. If it does not have any data to communicate, it simply modify the successor station address.

3. Similarly the token passes to each station and reaches the original station which sends message. The computer checks whether the data reached or not. Then if it has any data, it transfers to the destination, otherwise the token is passed to the successor address.

4. As the destinated system may not be active, the information sent to this computer may be kept in queue. After three continuous attempt, the source station removes the message meant for destination.

This technology has not been widely used because of its delayed property.

TOKEN RING (IEEE 802.5)

Token ring network was introduced in 1985, at a data rate of 4 Mbs. IBM introduced its second type of token ring in 1989 with a data rate of 16 Mbps. Currently, the term token ring is generally used to refer to both IBM's token ring network and IEEE 802.5 networks as there is a little difference between the two types. The main difference of token ring over token bus is that tokens and messages are passed around the ring to each station in the ring in a fixed sequence.

Token ring technology consists of a ring station and a transmission medium. A token ring network uses a wiring concentrator device called Multistation Access Unit (MAU). Physically, token ring is a star topology and electrically it is a ring topology. Each computer in the network connected to the MAU. Each MAU can accomodate upto eight stations. Shielded twisted pair cabling (STP) is used for a 16 Mbps token ring network and unshielded twisted pair cabling (UTP) is used for a 4 Mbps token ring network. Fig. 11.24 shows the token ring topology.

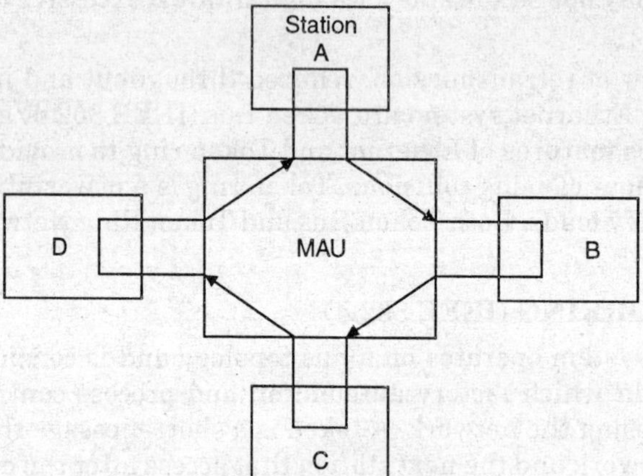

Fig. 11.24. Token ring topology.

Any station wants to send messages follow the procedures given below:

1. The station seize the token. The token is a three byte frame circulating around the network.

2. Once the station possesses the token, it inserts information into the token and transmits the frame on the ring.

3. The next station checks the destination address of the frame. If not matches, pass it to the next station. If the address matches, it performs the following function

(*a*) The destination station copies the message and sets the last two bits of the frame to inform the source that the frame was copied and the frame is retransmitted.

(*b*) The frame circulates on the ring until it reaches the source. Once it reaches the source, the source removes the frame from the ring.

(*c*) The source releases the token by changing the Token Bit (T bit) to one.

Token Frame format. The token is a three byte frame. Only one token is allowed on the network at anytime. Fig. 11.25 shows the token frame format.

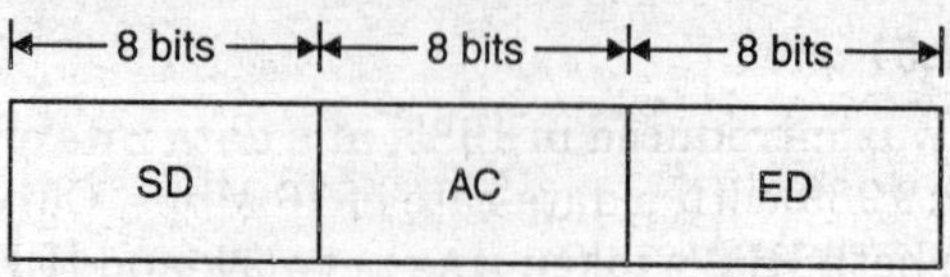

Fig. 11.25. Token frame format.

Start Delimiter (SD). It is set to JK0JK000. J and K are non-data bits. It follows violated Differential Manchester coding. The purpose of this pattern is to keep the SD byte (or ED type) from repeating in the information field of token ring frame.

End of Delimiter (ED). It is set to JK1JK10E. E is always set to zero. If any station detects an error, it will set E to 1.

Access control byte (AC). It contains three priority bits (PPP or more), a token bit (T), monitor bit (M) and remaining reserved bits. The priority bits varies 000 through 111. If token bit T = 0, data is in token or if T = 1, no data in token. The Monitor Bit (M) is used to prevent frames from circulating onto the ring.

IEEE 802.5 Frame format. Fig. 11.26 shows the Token ring format.

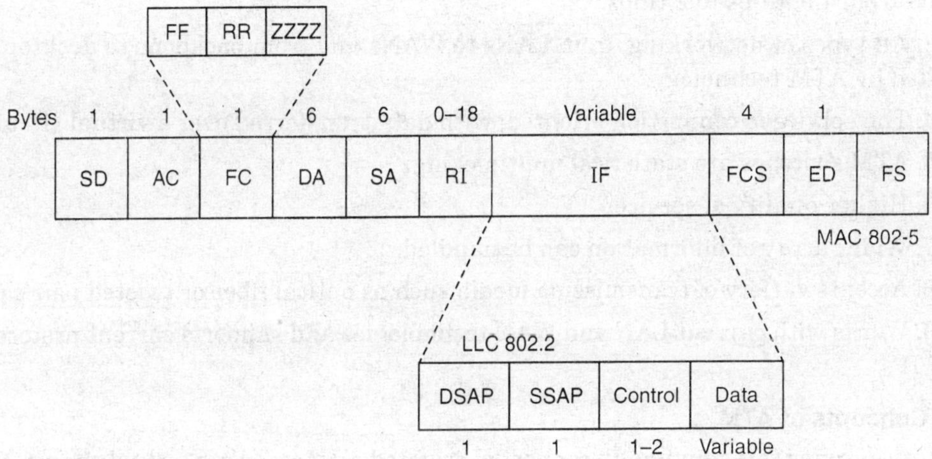

Fig. 11.26. Token ring format.

SD, AC and ED fields are already described. FC field is the Frame control (FC). In FC field, FF bits indicate the frame type. FF = 00 indicates MAC Frame, FF = 01 indicates LLC frame, and FF = 10 and 11 meant for reserved. RR set to 00 and they are reserved bits. ZZZZ = 0000 means a normal buffer and ZZZZ = 0001 indicates express buffer.

DA and SA are destination and source address fields respectively. DA uses the same address format of 802.3. Routing information (RI) is optional. If information field (IF) contains a MAC frame, the frame is called the MAC protocol data unit. If IF field contains an LLC frame, the field is called LLC protocol unit (LPDU). FCS is frame check sequence for error detection and Frame Status field (FS) have the field of 1 byte.

11.7. ASYNCHRONOUS TRANSFER MODE (ATM)

ATM is a high Performance, cell oriented switching and multiplexing technology that utilizes fixed length packets to carry different types of traffic. ATM is the next generation of networking technology to be used on the information superhighway. ATM is well suited for bursty traffic and allows communications between devices that operate at different speeds.

ATM is a technology that has its history in the development of broadband ISDN in the 1970s and 1980s. ATM is a telecommunications concept defined by ANSI and ITU committees for the transport of a broad range of user information. The ATM Forum has played a vital role in the ATM Market since its formulation in 1991. The forum writes specifications and definitions for ATM technology. ITU approves this specification. ITU-T and ATM forum are works closely.

11.7.1. Advantages of ATM

The most important advantages or benefits of the ATM are:

1. A much wider array of information can be transmitted using ATM technology, for example, voice, data, images, CATV scans, MRI images and video conferencing.

2. ATM delivers bandwidth on demand, is not dependent on applications and works at a data rate from 1.5 Mbps to 2 Gbps.

3. All types of networking, from LANs to WANs and from backbone to desktop can be integrated by ATM technology.

4. The service is connection oriented, with data transferred over a virtual circuit.

5. ATM switches are statistical multiplexing.

6. Higher quality of service.

7. Wider array of information can be handled.

8. Accepts variety of transmission media such as optical fiber or twisted pair cable.

9. Works with current LAN and WAN technologies and supports current protocols such as TCP/IP.

11.7.2. Concepts of ATM

Connection oriented service. In connected oriented service, over a virtual circuit, the data stream from origin to destination follows the same path. Virtual circuit (a type of packet switching) operate on the same concept as packet switching, but the routing of packet is specified before transmission. Data from different connections is distinguished by means of virtual path identifier (VPI) and virtual channel identifier (VCI). By virtual circuit, cells in the same connection reach the destination in the order they are sent. It eliminates the need for sequencing numbers and buffering packets. Also, each cell incurs an overhead corresponding to the length of VPI/VCI which is less than the length source/destination address nedded. VPI and VCI are called connection identifiers.

ATM Services. ATM forum specifies five types of services. They are:

1. **Constant bit rate (CBR).** This is used for emulating circuit switching. The cell rate is constant with time. Telephone traffic, videoconferencing and television are the examples that use CBR.

2. **Variable bit rate–non real time (VBR–NRT).** This service allows users to send traffic at a rate that varies with time depending on the availability of user information. Multimedia email is an example of VBR–NRT.

3. **Variable bit rate–real time (VBR–RT).** This service is similar to VBR–NRT, but it is designed for applications that are sensitive to cell delay variation. Examples for real time VBR are voice with speech activity detection (SAD) and interactive compressed video.

4. **Available bit rate (ABR).** This service provides rate based flow control and is aimed at data traffic such as file transfer and e-mail.

5. **Unspecified bit rate (UBR).** This class is widely used today for TCP/IP.

Traffic parameters:

1. **Peak Cell Rate (PCR).** PCR is the reciprocal of the minimum time between two cells.
2. **Sustained Cell Rate (SCR).** Long term average cell rate.
3. **Initial Cell Rate (ICR).** Rate at which a cell should he sent after an idle period.
4. **Cell Delay Variation Tolerance (CDUT).** Measures permissible departure from periodicity of the traffic.
5. **Burst tolerance.** Maximum number of cells in a burst of back-to-back cells.
6. **Minimum Cell Rate (MCR).** Reciprocal of the maximum time between two cells.

Quality of Service Parameters:

1. Cell loss ratio (CLR)
2. Cell delay variation (CDV)
3. Peak-to-peak cell delay variation (peak-to-peak CDV)
4. Maximum cell transfer delay (Max CTD)
5. Mean cell transfer delay (Mean CTD).

Two types of connection:

1. **Permanent virtual connection (PVC).** PVC is a connection that is setup and taken down manually by a network manager. A set of network switches between the ATM source and destination are programmed with predefined values for VCI/VPI.

2. **Switched virtual circuit (SVC).** SVC is a connection that is setup automatically by a signalling protocol. SVC is more widely used because it does not require manual setup, but it is not reliable.

ATM switch operation : ATM switch process cells at an extremly high rate or speed. Fig. 11.27 illustrates the basic operation of ATM switch.

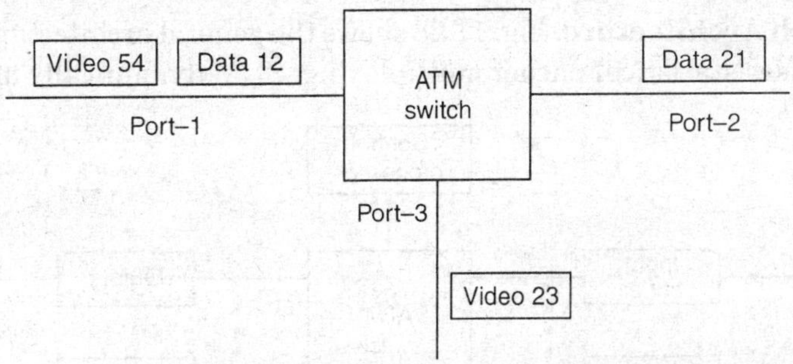

Fig. 11.27. ATM switch.

Fig. 11.27 shows the ATM switch with 3 ports. The cells enter the switch from port 1 and go out from the switch to ports 2 and 3 according to the routing table. The routing table is shown in table 11.4.

Table 11.4. Routing table for Fig. 11.27

Input		Output	
Port	VPI/VCI	Port	VPI/VCI
1	1/12	2	2/21
1	1/54	3	3/23

The ATM switch should have enough capacity to store incoming calls and the routing table. It should have 16 to 32 input and output ports and supports all AAL (ATM Adaptation layer). It should acts as support for congestion control. The switch should support cell switching at a rate of atleast 1 million cells per sec. The ATM switch operation is given below by step by step procedure.

1. The switch examines the VPI/VCI of the incoming cell to determine the output port to which the cell should be forwarded.

2. The ATM switch modifies the VPI/VCI fields to new value for the output port.

3. The Header error control (HEC) is used for error detection and correction in the header field of each cell. If the HEC can not correct the error, the ATM switch will discard the cell.

4. The ATM switch can modify the routable table using its control unit.

Traffic congestion : An ATM switch involved with two types of blocking of traffic congestion. They are

1. **Fabric blocking.** It occurs when the fabric capacity of a switch is less than the sum of its input data rate. In this case the switch must drop some of the cells. Some ATM switches are limited to 16 or 32 OC-3 input ports.

2. **Head of the line blocking.** It occurs when an output port is congested and a cell is waiting in the input port. The switch must drop some of the cells in the output port. Some switches randomly discard the cells, and all stations must retransmitt all the cells.

ATM Switch Architecture. Fig. 11.28 shows the general architecture of an ATM switch. The ATM switch uses statistical packet multiplexing. SDM dynamically allocates bandwidth

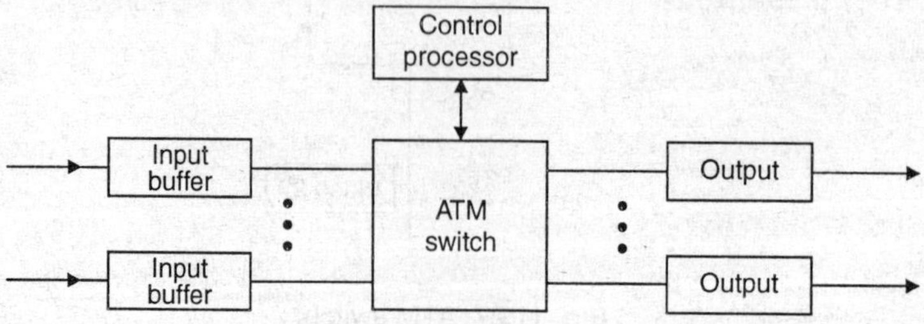

Fig. 11.28. ATM switch architecture.

to the active input channels, resulting in very efficient bandwidth utilization. In SDM, an idle channel does not receive any time allocation. SDM uses a store and forward mechanism in order to detect and correct error from incoming packets. SDM will not allocate a time slot to any idle input.

The control processor is used to control the input/output buffers and update the routing table of the switch. The switch must be capable of processing atleast one billion cells per second and supports atleast one million cells per second. The buffers are used to store the incoming data cell and the routing table.

ATM cell format. ATM uses VLSI technology to segment data to the cell at high speeds. Each cell consists of 53 bytes, in which there are a 5 byte header and a 48 byte payload. ATM cell format is shown in Fig. 11.29.

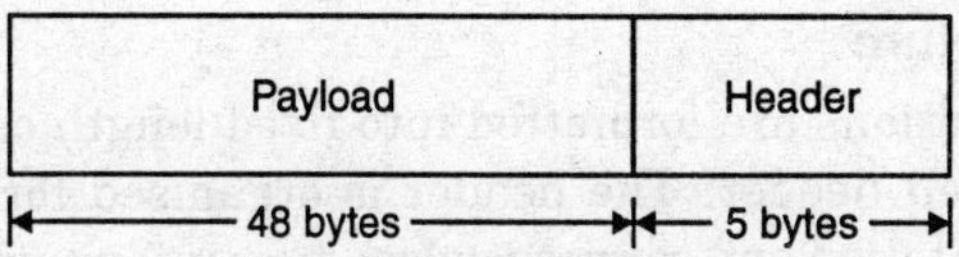

Fig. 11.29. ATM cell format.

The ATM protocol is not influenced by the actual configuration of the payload. The data is segmented into 48 bytes sizes and packaged into ATM cells. These cells can be multiplexed with other payload cells and transported through ATM routers and delivered to an end point network, where there are demultiplexed and reassembled into the original payload format and placed on to the end point network to be eventually routed to the user terminal or node. The fixed cell size ensures that time critical information such as voice or video is not adversely affected by long data frames or packets.

ATM Network Interface. Fig. 11.30 illustrates a typical ATM network. It consists of switches and end users with necessary interfaces.

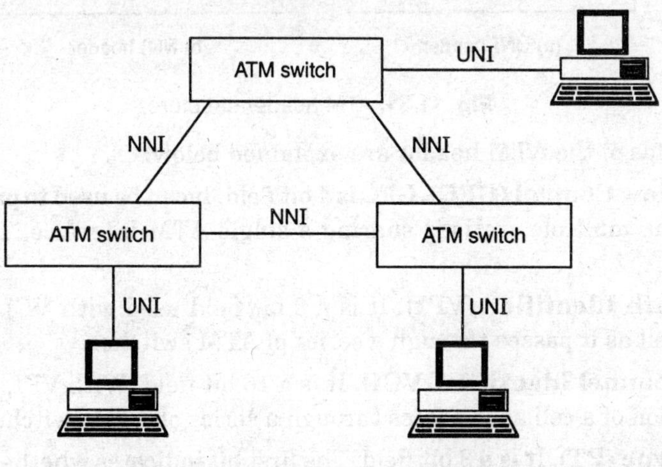

Fig. 11.30. ATM network interfaces.

ATM switches offer two types of interfaces. They are:

1. **Switch to Switch Interface or Network to Network Interface (NNI).** It is an interface between nodes within the network or between different sub network.

2. **Switch to User Interface or User to Network Interface (UNI).** UNI is the standard technical specification allowing ATM customer equipment (CEQ) from various manufacturers to communicate over a network provided by yet another manufacturer. It is the interface employed between ATM customer equipment and either ATM switch.

Another interface called ATM inter network interface (INI) used for intercommunication. It is also used for operational and administrative boundaries between interconnected networks. It is based upon NNI but include more features for ensuring security, control and proper administration of inter-carrier connections.

11.7.3. ATM Header Structure

As mentioned, all informations are formatted into fixed length cells consisting of 48 bytes of payload and 5 bytes of cell header. The header is organised for efficient switching in high speed hardware implementation and carries payload type information, virtual circuit identifiers and header error check. Fig. 11.31 shows the ATM header structure. It shows the header structure for both UNI and NNI interfaces.

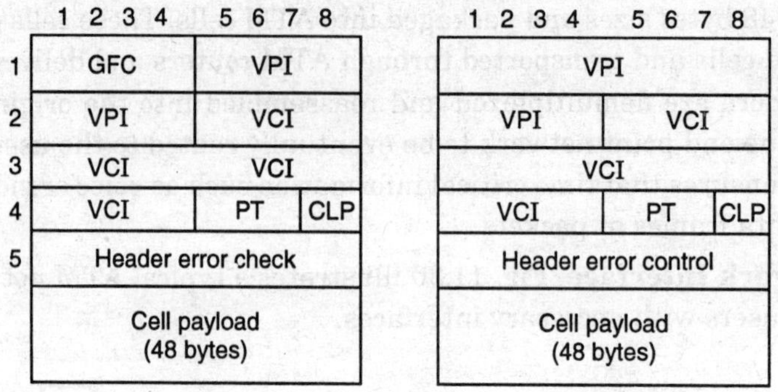

(a) UNI header (b) NNI header

Fig. 11.31. ATM header structure.

Various fields of the ATM header are explained below:

Generic Flow Control (GFC). GFC is 4 bit field. It can be used to provide local functions such as identifying multiple stations sharing a single ATM interface. This is currently not used.

Virtual Path Identifier (VPI). It is a 8 bit field used with VCI to identify the next destination of a cell as it passes through a series of ATM switches.

Virtual Channel Identifier (VCI). It is a 16 bit field. With VPI, it is used to identify the next destination of a cell as it passes through a series of ATM switches.

Payload type (PT). It is a 3 bit field. The first bit indicates whether the payload is data or control data. The second bit indicates congestion, and third bit indicates whether the cell is the last cell in the series.

Congestion loss priority (CLP). It is just a one bit field. It indicates whether a cell should be discarded if it encounters extreme congestion. This bit is used for QOS.

Header Error Control (HEC). It is an 8 bit field. This 8 bit CRC detect all single errors and certain multiple bit errors.

The only difference between UNI header and NNI header is that UNI has a GFC field of 4 bit and VPI of 8 bit and NNI has no GFC field but VPI is 12 bit field. The 4 bit GFC field in UNI as used to signal to the user the need for flow control. The NNI uses these bits to expand the VPI field.

11.7.4. ATM Layers

Fig. 11.32 shows an ATM end point operational model and an ATM switch operational model.

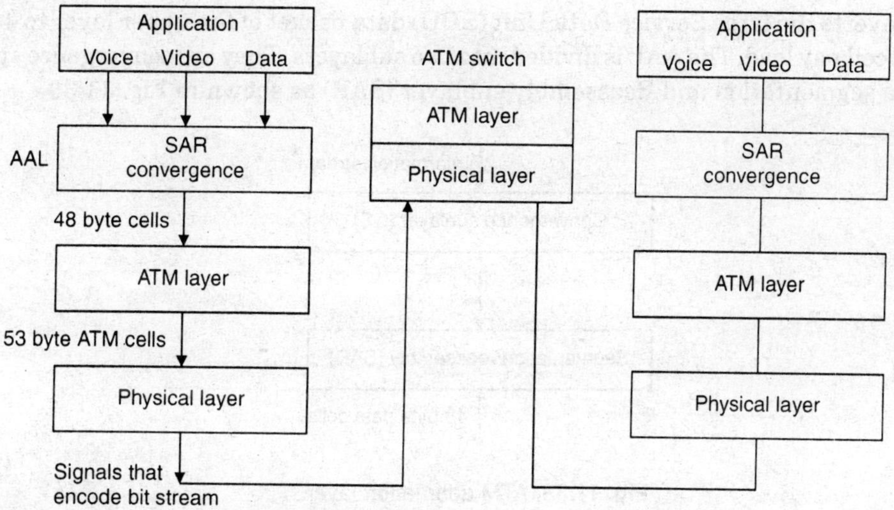

Fig. 11.32. Headers of ATM cell.

ATM end point operational model:

The ATM end point operational model consists of three layers

1. ATM adaptation layer (AAL)
2. ATM layer and
3. Physical layer

The ATM adaptation layer (AAL) is described in the section 11.7.5. Remaining two layers are explained below.

Physical layer. The physical layers are divided into two sublayers. They are Physical Medium Dependent (PMD) layer and the transmission convergence (TC) layer.

The PMD sublayer provides bit transmission, coding, electrical and optical conversion and bit timing. The functions of the TC sublayer are as follows:

1. Extracting the cell from physical layer
2. Scrambling the cell before transmission and descrambling the cell after transmission.

3. After receiving a cell from physical layer, TC calculate the HEC header and compares with the cell's HEC header. The results of the comparison are used for error correction in the cell header. If the error can not be corrected, the cell will be discarded.

ATM layer. There is no sublayers. It performs the following function.

1. It adds cell header and remove cell header.

2. The VPI/VCI values of incoming cell is translated to a new VPI/VCI value for the outgoing cell.

3. Generic flow control—to determine the destination of the receiving cell.

4. Multiplexes and demultiplexes cell.

11.7.5. ATM Adaptation Layer (AAL)

The AAL converts the large Service Data Unit (SDU) data packet of the upper layer to 48 bytes for the ATM cell pay load. The AAL is divided into two sublayers. They are convergence sublayer (CS) and the segmentation and Reassembly sublayer (SAR) as shown in Fig. 11.33.

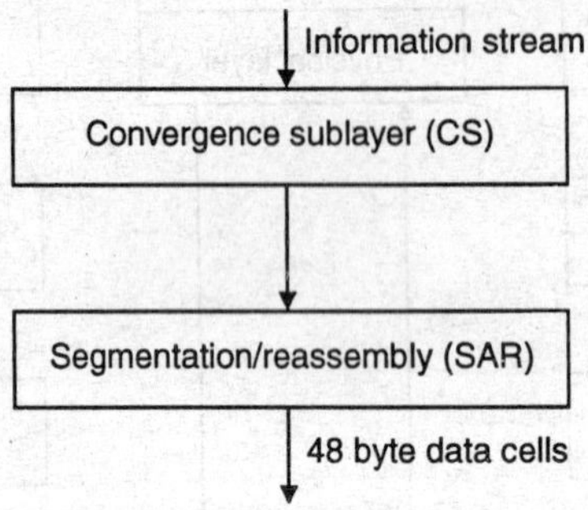

Fig. 11.33. ATM adaptation layer.

ATM can be used for various applications. Therefore different types of AALs are needed to provide service to upper layer applications. The AALs are divided into four classes of traffic (class A, class B, class C and class D). Also AAL is divided into four types of AAL protocol (Type 1, Type 2, Type 3/4 and Type 5). Table 11.5 shows the service classification of ATM AAL.

Table 11.5. Service classification of ATM

Application type	Class A	Class B	Class C	Class D
AAL type	AAL 1	AAL 2	AAL 3/4 AAL 5	AAL 3/4 AAL 5
Timing relation between source and destination	Required	Required	Not required	Not required
Bit rate	constant	variable	variable	variable
Connection type	connection oriented	connection oriented	connection oriented	connection-less
Applications	voice communications	compressed audio or video	data	data

The CS converts the information stream into four types of packet streams called AAL Type 1, Type 2, Type 3/4 and Type 5. The packet formats should match the five types of ATM services already described. CS tasks are likely to be application dependent. Hence CS is further subdivided into the upper, service specific CS (SSCS) and the lower common part SS (CPCS).

SAR tasks are quite standard, depending only on traffic type. The SAR segments the SDU received from CS and add necessary overhead and convert them into 48 byte cells or protocol data units (PDU).

Types of Adaptation Layers

The types of ATM adaptation layers are AAL type 1 or AAL 1, AAL type 2 or AAL 2, AAL type 3/4 or AAL 3/4 and AAL type 5 or AAL 5. Fig. 11.34 shows the cell format of all these types.

AAL 1. It is designed to carry class A traffic. The field SN (4 bits) is used as cell sequence number to detect any missing cell or disordered cell. The sequence number protection field (SNP) is 4 bit and used for error detection on an SN field using a CRC polynomial $X^3 + X^2 + 1$.

The SAR sublayer takes the (periodic) packet stream generated by the CS sublayer, segments it into 47 byte SDU, and prepends to each SDU, a 4 bit sequence number (SN) protected by a 4 bit SNP field.

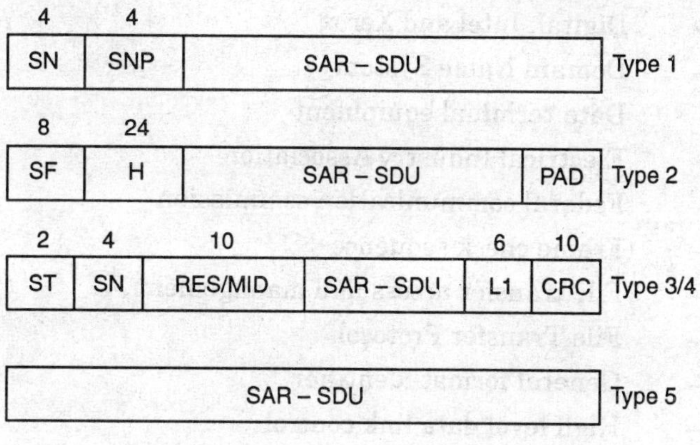

Fig. 11.34. Types of ATM adaptation layer.

AAL 2. It is designed to carry class B traffic. The 8 bit start field (SF) includes a 6 bit offset field (OSF), which indicates the number of bytes between the start field and the payload or between the start field. The 24 bit Header (H) includes the Length indicator (LI) for variable length payload.

AAL 3/4. It is started as two separate adaptation layers (AAL 3 and AAL 4). The specification of two AAL's merged and the newly formed type is called AAL 3/4. AAL 3/4 is designed to take variable length frames up to 64 kbytes and segment them into cells. The 2-bit ST or segment type is set as follows. ST = 10 for BOM, ST = 00 for COM, ST = 01 for EOM and ST = 11, if the packet fits in side a single cell. The 10 bit multiplexing identifier (MID) is used to distinguish among the cells originating from different data grams.

AAL 5. It is an efficient way to transfer information in ATM. The IP packet upto 64 k bytes is packaged into a CS packet that contains a length indicator and a 32 bit CRC calculated over the complete CS packet with the generator. AAL 5 provides services similar to those AAL 3/4, but uses fewer control fields.

ACRONYMS

AAL	—	ATM Adaptation layer
ARP	—	Address resolution protocol
ARPA	—	Advanced Research Project Agency
ASN	—	Abstract Syntax Notation
ATM	—	Asynchronous Transfer Mode
CASE	—	Common application service element
CRC	—	Cyclic redundancy check
CSMA/CD	—	Carrier sense multiple access with collision detection
DARPA	—	Defence Advanced Research Projects Agency
DCE	—	Data Communication Equipment
DIX	—	Digital, Intel and Xerox
DNS	—	Domain Name System
DTE	—	Data terminal equipment
ETA	—	Electrical Industry Association
FCC	—	Federal communication commission
FCS	—	Frame check sequence
FTAM	—	File transfer access and management
FTP	—	File Transfer Protocol
GFI	—	General format identifier
HDLC	—	High level data link control
HTTP	—	Hypertext Transfer Protocol
ICMP	—	Internet Control Message Protocol
IP	—	Internet protocol
ISO	—	Organization for standards
LAN	—	Local area network
LAPB	—	Link Access Procedure Balanced
LCN	—	Logical channel number
LCU	—	Line control unit
LGN	—	Logical channel group number
LLC	—	Logical link control
LLC	—	Logical link control

MAC	—	Media Access Control
MAN	—	Metropolitan area network
MID	—	Multiplexing identifier
NIC	—	Network interface card
NNI	—	Network to Network Interface
OSI	—	Open System interconnection
PAD	—	Packet assembler/disassembler
PTI	—	Packet type identifier
PVC	—	Permanent virtual connection
QoS	—	Quality of service
RARP	—	Reverse ARP
SASE	—	Specification and service element
SDLC	—	Synchronous data link control
SMTP	—	Simple Mail Transfer Protocol
SNA	—	Synchronous Network Architecture
SNAP	—	Sub network accesses protocol
SNMP	—	Simple Network Management Protocol
SUSE	—	Specific user service element
SVC	—	Switched virtual circuit
TCP	—	Transmission control protocol
UART	—	Universal asynchronous receiver transmitter
UDP	—	User datagram protocol
UNI	—	User to Network Interface
USART	—	Universal Synchronous/Asynchronous Rceiver Transmitter
VCI	—	Virtual circuit identifiers
VPI	—	Virtual path identifier
WAN	—	Wide Area Network

RELATED WEBSITES

http://www.iso.ch

http://www.ganges.cs.tcd.le/4ba2

http://www.salford.ac.uk/iti/books/osi/

http://www.cit.ac.nz/smac/dc/ooww/k_ooo.htm

http://www.wizard.com/users

http://www.linktionary.com/n/network_connection.html

http://www.IPPrimer.2nd level.net/

http://www.faqs.org/faqs/internet/tcp-Ip/resource_list

http://www.itprc.com/tcp-ip.htm

http://www.CISIO.com/warp/public/

http://www.techfest.com/networking/lan.htm

http://www.lantronix.com/tramy/turtorials

http://www.allied.avnet.com

http://www.atm forum.com

CHAPTER REVIEW QUESTIONS

1. Define Baud.
2. Relate Baud and bit rate.
3. Explain with neat diagram the concept of data communication link.
4. What is the packet switching principle ?
5. Explain with neat diagram the concept of (*a*) datagram and (*b*) virtual circuit in packet switching.
6. Write short notes on packet size.
7. Compare the circuit switching and packet switching.
8. Explain all the fields of the packet format.
9. What is X-25 ?
10. Explain the X-25 interface with neat diagram.
11. List the OSI layer specifications.
12. With neat sketch, explain the OSI network architecture.
13. Explain network support layer of OSI model.
14. Explain support layers of OSI model.
15. What are three sets of services provided by TCP/IP ?
16. Sketch the TCP/IP reference model and explain.
17. Explain the operation of TCP with TCP Header.
18. Explain the IP packet format.
19. What are the protocols related to TCP/IP ?
20. Compare the merits and demerits of various network topologies.
21. Draw the Ethernet reference model and explain.
22. What are the Ethernet media ? Explain each media.
23. What are the high speed Ethernets ? Write briefly.
24. Explain the Token passing ring with necessary diagrams.
25. What is ATM ?
26. List the advantages of ATM.
27. Explain with neat diagram, the formats of ATM Header structure.
28. Sketch ATM layers and explain.
29. Tabulate the service classification of ATM.
30. Give the concept of VPI and VCI.
31. What are the types of adaptation layer ? Explain each with its frame format.

12

ISDN

12

ISDN

12.1. INTRODUCTION

12.1.1. Overview

The integrated Services Digital Network (ISDN) is a set of digital transmission standards which are used for end-to-end digital connectivity. "Integrated Services" referring to its ability to sustain numerous applications. "Digital Network" relating to its end-to-end digital connection. In general, ISDN networks extend from the local telephone exchange to the remote user and include all the telecommunications and switching equipments in between. ISDN supports voice and data. In the past, video, audio, voice and data services required atleast four separate networks. ISDN integrates all four over the same network.

History. ISDN is based on technology developed during the 1970's designed to address the problem of how to transport digital services across a telephony infrastructure based on copper wiring originally intended to carry analog signals only. To meet the customer needs, initially, the telephone companies created Integrated Digital Networks (IDN). IDN is a combination of networks available for different purposes. Access to these networks is by digital pipes which are time multiplexed channels sharing very high speed paths. By this network, customers can use their local loops to transmit both voice and data to their telephone company's central office. The office then directs these calls to the appropriate digital networks via the digital pipes. ISDN integrates customer services with the IDN. With ISDN, all customers will become digital rather than analog.

Standards. ISDN technology is standardised according to recommondations of the CCITT, now ITU, which describe the protocols and architectures to implement a world wide digital communication network. The CCITT was comprised to study groups (SGs). Each SG has its own area of expertise. The following are the SGs related ISDN.

SG VII — Public data network (X.25) X-series standards
SG VIII — Terminal equipment for telematic services
SG XI — ISDN and telephone network switching and signalling
SG XII — Transmission performance of telephone networks and terminals
SG XV — Transmission systems
SG XVII — Data transmission over public telephone networks
SG XVIII — Digital networks including ISDN.

Types of ISDN. There are two types of ISDN:

Narrow band ISDN (N-ISDN) and Broadband ISDN (B-ISDN)

N-ISDN — Carry data rating upto 64 kbps, ranging up to T1 rates. Sometimes used to refer regular telephone and nonvideo capable systems.

B-ISDN — The communication standards are being dveloped by the ITU to handle the high bandwidth applications such as video. B-ISDN will use ATM technology over SONET based transmission units to provide data rates of 155 Mbps to 622 Mbps and beyond.

12.1.2. ISDN Services

ISDN services generally fall into three catageries. They are bearer services, teleservices and supplementary services.

Bearer services. ISDN works on the principle of transport services known as bearer services. The bearer service offers the capability to transport digital voice or nonvoice services using this standard. The basic operation of the bearer service is the 64 kbps channel capacity. Bearer services provide the means to transfer information (voice, data and video) between users. The network does not need to process the information. Bearer service belongs to the first three layers of OSI model. These services can be provided with circuit switched, packet switched, frame switched or cell switched networks.

Tele services. In this service, the network may change or process the contents. This service correspond to layers 4–7 of the OSI model. Tele services include telephony, telefax, videofax, telex and teleconferencing.

Supplementary services. It provides additional functionality to the bearer service and teleservices. Supplementary services include call waiting, Reverse charging, and message handling.

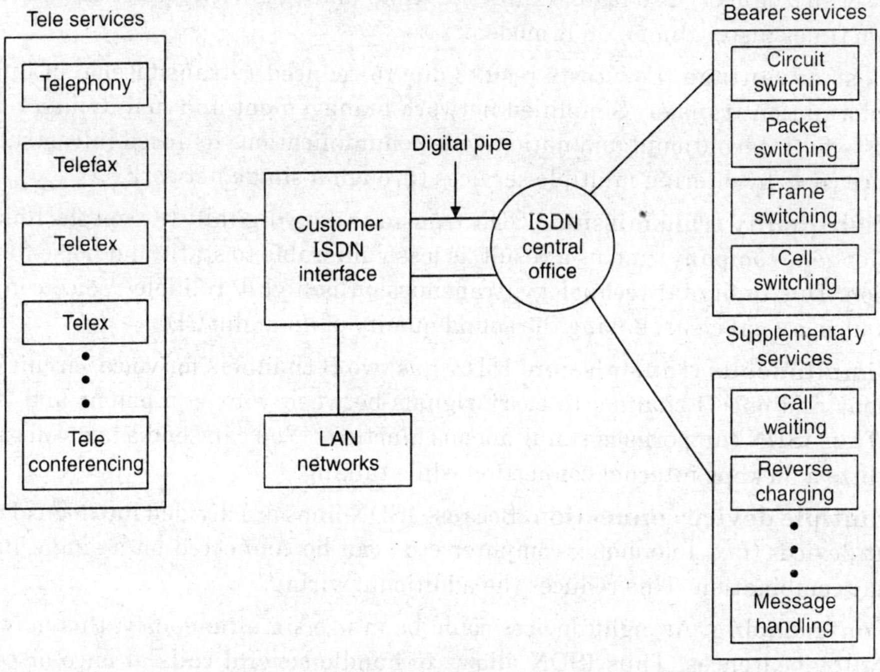

Fig. 12.1. Conceptual view of ISDN.

The conceptual view of ISDN is shown in Fig. 12.1.

The objective of ISDN was introduced in 1979. The concept is the digital end to end connectivity to support a wide range of services. The reason for this concept is the obsolescence and cost prohibitive nature of the analog technology. ISDN uses the existing networks to provide the above discussed services. To access to the various services, the user had to subscribe to a particular network.

The existence of integrated networks will allow for user management of the entire telecommunication functions through the single thread to the outside world. Hence ISDN provide the user with easy access to multiple service over a single connection to the network.

12.1.3. Advantages of ISDN

1. **High speed service.** ISDN is fast. As there is no need of conversion of analog to digital inside a digital network, the speed is high. Table 12.1 shows the speed comparison of various devices.

Table 12.1. Speed comparison

Type	Data rate
MODEM	28.8 kbps
ISDN 1B channel	64 kbps
ISDN 2B channel	128 kbps

Before ISDN, normal phone carries only 2.4 kbps. After ISDN, digital phone lines can carry 128 kbps over the same wire. Hence ISDN's speed allows very quick file transfers. The ISDN call setup (connect) also faster (2 to 4 sec) than analog devices (15 to 30 sec). ISDN can transfer two times faster than a 56 K modem.

2. **Cost advantage.** Low costs results due to reduced retransmission of information and fast information transfer. Simplified network management and maintenance results in reduced costs for international and nation-wide communication. Reduced infrastructure and maintenance costs by offering multiple services through a single network.

3. **High quality transmission.** ISDN transmits data digitally (except the link between you and telephone company) and as a result, is less vulnerable to static and noise than analog transmission. Due to digital technology, transmission is highly reliable. Voice conversation over ISDN also crystal clear, having the sound quality of an audio CD.

4. **Simultaneous transmission.** ISDN has two B channels for voice, circuit or packet conversations and one D channel to carry signals between your equipment and the phone company. Thus ISDN can perform simultaneous functions. You can send a fax, you can receive a fax or utilize a 56 kbps internet connection while talking.

5. **Multiple device connection.** Because ISDN lines are divided into logical channels, up to eight devices (fax, telephone, computer etc.) can be connected on a single Basic Rate ISDN in any combination. This reduces the additional wiring.

6. **Conferencing.** As eight devices could be in use simultaneously, this may result in multiple call appearences. Thus ISDN allows to handle several calls at once or conference them together with one number. This multiple call appearence is useful for small office with large number of outside sales people.

7. ISDN provides clear, quiter voice telephone service and easy to use call control features. Its caller identification features can screen incoming calls.

8. Call management features.

(*a*) **Call forwarding.** Forwards call to a preselected number.

(*b*) **Call pickup.** Call can be picked at another phone or station.

(*c*) **Directed call pickup.** Calls from specific extension be automatically forwarded to a second number.

(*d*) **Message Waiting indicator.**

(*e*) **Directed Dial.** Incoming calls can be automatically forwarded to a central office, car etc.

(*f*) **Ringing options.**

(*g*) **Additional call offering.** Allows the user to use a distinctive ring to a particular call.

12.1.4. ISDN Scenario in India

ISDN services were thrown open to the DOT subscribers in 1996. The DOT provides ISDN service on demand basis without waiting. This is the only value added service remaining with DOT, while all other value added services are given to private operators. The DOT has approved different customer's premises equipment (CPE) that any subscriber can use in his ISDN line. The only customer premises equipment provided by DOT is the network terminal NT.

NT performs network termination functions, communication physical layer connectivity, power feeding, multiplexing and interfacing functions with one U-interface port and one S interface port. All other equipment and interfaces are provided to the subscriber according to his requirement.

12.2. ISDN INTERFACES

ISDN is available with two maintypes of interfaces. They are Basic Rate Interface (BRI) and Primary Rate interface (PRI). These two interfaces are described below.

12.2.1. Basic Rate Interface (BRI)

BRI is made up of two B-channels (Bearer channels) and one D channel. Therefore the total rate is 2B + D. B channels are 64 kbps and can be used for voice and data communications. The D channel is 16 kbps and is used for call initialization and signalling connections. Fig. 12.2 shows ISDN BRI.

BRI is the most appropriate type of ISDN service for computer connections and individual use. Of three digital channels (2B + D), each of the channels can be used simultaneously. Thus a subscriber can perform several communication tasks at the same time. BRI is designed to carry the most data possible to the home through existing copper phone lines.

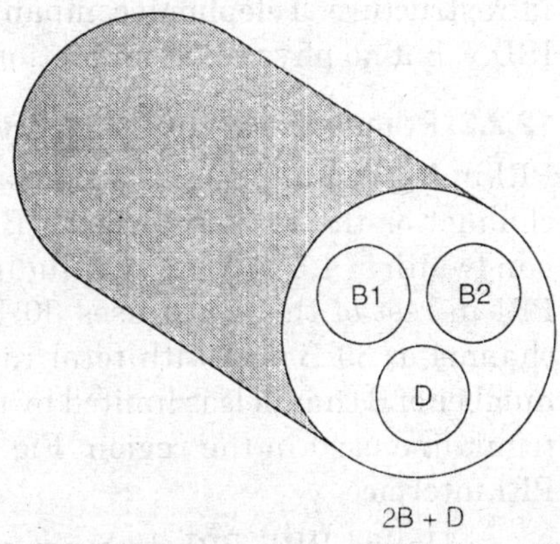

Fig. 12.2. ISDN BRI.

For every ISDN line, Network Terminator type 1 (NT 1) and a power supply are required. A special terminal adapter can combine the two B-channels to create a 128 kbps channel which can then be connected to a computer. The NT1 connected to ISDN line acts as a multiplexer and demultiplexer. Fig. 12.3 illustrates the BRI concept.

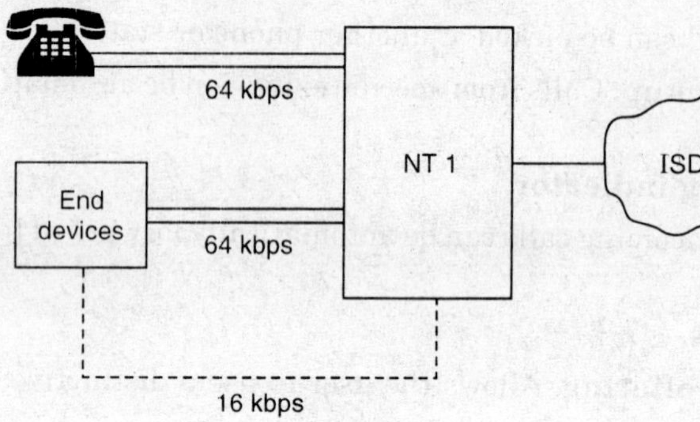

Fig. 12.3. BRI concept.

ISDN BRI frame format. The format for BRI is shown in Fig. 12.4. The size of a frame is 48 bits.

Bits	8	1	8	1	8	1	8	1
	B1	D	B2	D	B1	D	B2	D

B1 : 16 bits B2 : 16 bits D : 4 bits **overhead : 12 bits (not shown)**
BR1 = 4000 frames/sec = 4000 × 48 bits/frame = 192 kbps

Fig. 12.4. BRI frame format.

Each B channel is sampled two times during each frame (8 bits per sample). D channel is sampled four times during each frame (one bit per sample). The overhead is not shown. The reason for interleaved sampling is to create long frame. This is useful in many cases, for example the size of the BRI frame matches for a data portion of an ATM cell.

BRI functionality is attainable without any modification to the existing telephony infrastructure. Telephone companies must change the signalling on the local loop to support ISDN, but no physical modifications are required.

12.2.2. Primary Rate Interface (PRI)

PRI in North America has 23 B channels and one 64 K D channel or the total rate is 23 B + 1D, having a total bandwidth of 1.544 Mbps. (including 8 kbps of overhead) PRI in rest of the world uses 30 B channels and one D channel or 30 B + D with total rate of 2.048 Mbps. The number of B channels is limited by the size of the standard trunk line used in the region. Fig. 12.5 shows the ISDN PRI interface.

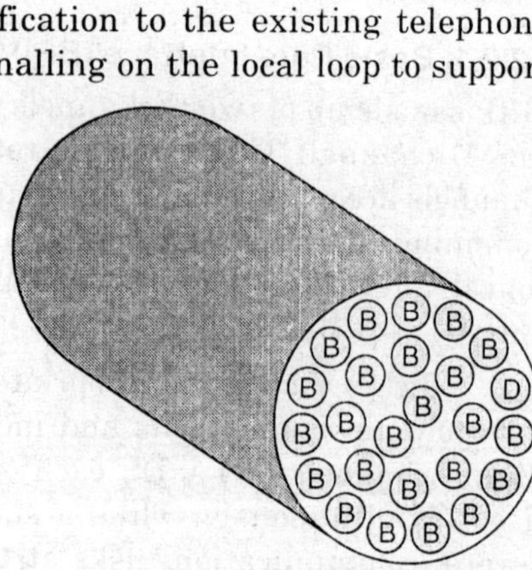

23 B + D or 30 B + D

Fig. 12.5. ISDN PRI.

Unlike BRI, PRI does not support a bus configuration and only one device can be connected to a PRI line. A PBX, however can reallocate ISDN PRI resources on to multiple BRI buses. Fig. 12.6 shows the ISDN PRI concept. A single PRI connection is usually much less expensive than obtaining the equivalent number of B channels through multiple BRI connections. The 1.544 Mbps of a PRI can be divided up in many ways to meet the requirements of many users. This indicates combination of 64 kbps B channels or all capacity of B channels (In LAN to LAN case) can be used.

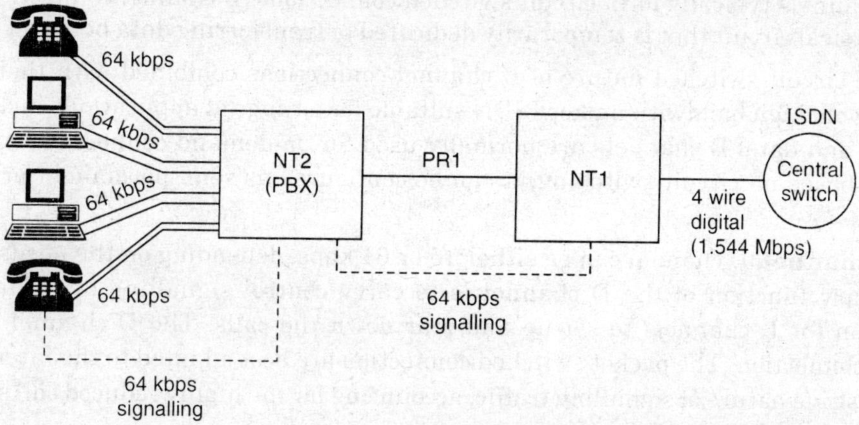

Fig. 12.6. ISDN PRI concept.

Network terminator 2 (NT 2). An application of the PRI is to connect two central switches together to use them as a T1 link (Fig. 12.6). The devices which handle switching and multiplexing (such as PBX) are called network terminator 2 (NT 2). ISDN PRI can connect the customer directly using an NT 2 device, while ISDN BRI requires an NT1 device. NT2 performs functions at the physical, data link and network layers of OSI model.

An NT2 provides intermediate signal processing between the data generating devices and an NT1. NT2 coordinates transmissions from a number of incoming links (user phone lines) and multiplexes them transmittable by an NT1.

PRI frame format. The B and D channels are multiplexed using synchronous TDM to create PRI frame as shown in Fig. 12.7.

PRI = 8000 frames/sec = 8000 × 193 bits/frame = 1.544 Mbps.

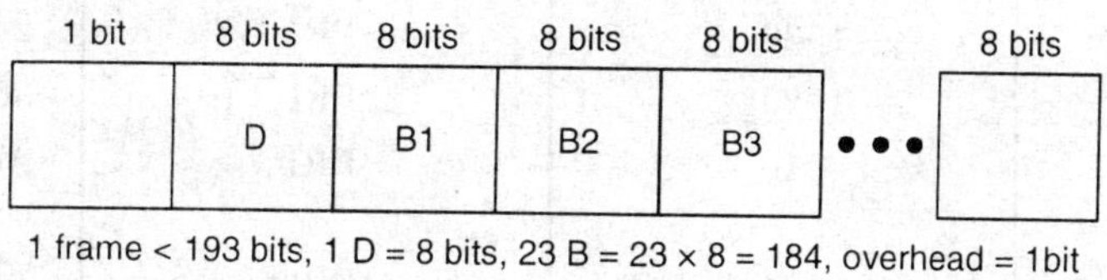

Fig. 12.7. PRI frame format.

12.3. ISDN CHANNELS

ISDN consists of three types of communications channels. They are:

1. Bearer channel (B channel)
2. Data channel (D channel), and

3. Hybrid channels (H channel).

These three ISDN channels are described below.

B channel. B channels are logical digital "pipes" which exist on a single ISDN line. B channel carry data and services at 64 kbps. It carries data in full duplex mode. Each B channel provides a 64 kbps clear channel, clear meaning that the entire bandwidth is available for data, B channels typically form circuit switched connections. B channel connection is an end-to-end physical circuit that is temporarily dedicated to transferring data between two devices.

The circuit switched nature of B channel connections combined with their reliability and relatively high bandwith makes ISDN suitable for a range of applications including voice, video, fax and data. B channels are normally used for on-demand connection. As B channel operation based on circuit switching, it can be configured as semi permanent or "nailed up" connections.

D channel. D channel can be either 16 or 64 kbps, depending on the needs of the user. The primary function of the D channel is to carry control signalling and administrative information for B channels to set up and tear down the calls. The D channel uses packet switched connection. The packet switched connection are best adapted to the intermittent but latecy sensitive nature of signalling traffic, accounting for the highly reduced call setup time of 1 to 2 seconds on ISDN calls.

Unlike the B-channel, which can function as a simple 'pipe', the D channel is associated with higher level protocols at layers 2 and 3 of OSI model which form the packet switched connections. The D channel provides the signalling information that is required for caller identification. It also includes low-rate data transfer and applications such as telemetry and alarm transmission.

H channels. H channels are suitable for high data rate applications such as video, teleconferencing and so on. Table 12.2 gives ISDN channel and its specifications.

Table 12.2. ISDN channels specifications

Channel	Bit rate (kbps)	Interface	Purpose
B	64	BRI	Bearer services
H0	384	PRI	6 B channels
H11	1536	PRI	24 B channels
H12	1920	PRI	30 B channels
D	16	BRI	Administrative and control signalling
D	64	PRI	,,

12.4. ISDN SWITCHING, FUNCTIONAL GROUPING AND REFERENCE POINTS

12.4.1. ISDN Switching

The process of moving data through a network is called switching. ISDN takes advantage of two types of switching. Circuit switching and packet switching. Circuit switching routes voice or data. Packet switching routes multiple data packets. Both switching are explained earlier in this book. The circuit switching and packet switching related to ISDN switching are described below.

Circuit switched voice (CSV) service is a digital voice service that offers many of the capabilities of a business centrex, such as call waiting, speed calling and call transfer over an ISDN digital subscriber line (DSL). For point to point data connections, circuit switched data (CSD) service is used. CSD service provides end-to-end digital service to pass data or video information over the public network,

D channel uses packet switching. Using D-channel, it is possible to implement various low bandwidth services for communicating with ISDN users. Packet switching can also be used on the B channels, although this is generally for X.25 or similar networks. X-31 extends X.25 to provide dial up packet switched services in an ISDN.

12.4.2. Functional Grouping

The devices that enable users to access the services of the BRI or PRI in ISDN switching are collectively called functional groupings. The functional groupings used at the subscriber premises include Network termination 1 and 2 (NT1 and NT2), Terminal equipment 1 and 2 (TE1 and TE2) and Terminal adaptors (TA). The NT1 and NT2 were described already in previous sections. The TE1, TE2 and TA alone are descibed below. Fig. 12.8 shows the functional grouping.

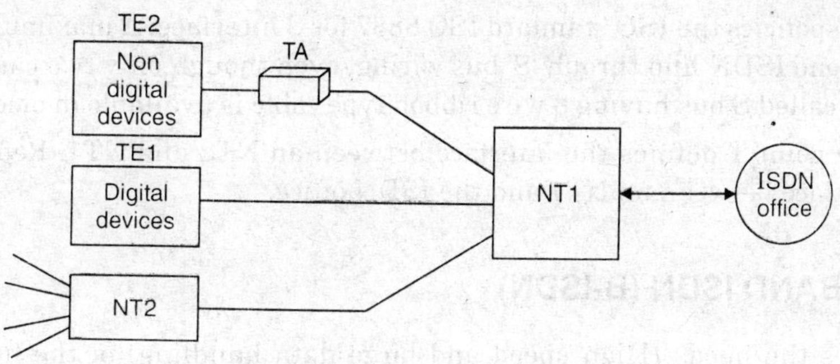

Fig. 12.8. Functional groupings.

Terminal Equipment 1 (TE1). TE1 are ISDN terminals like video conferencing equipment, Group-4, fax, feature telephone which are digital and can be directly connected to NT through 'S' bus interface.

Terminal Equipment 2 (TE 2). TE2 are non ISDN terminals such as analog phone, PC, G3, FAX which are non digital and can not be directly connected to NT1. They require another interface called Terminal adaptor (TA).

Terminal adapter (TA). TA enables analog to digital conversion and vice versa. Otherwise the purpose of TA is to convert the bipolar signalling from the public network to the unipolar signalling used by computers. The most prevalent use of TA is to connect a computer to an ISDN line.

12.4.3. Reference Points

Reference piont refers to the lable used to identify individual interfaces between two elements of an ISDN installation. It defines how two elements are connected. There are four reference points generally used. They are R, S, T and U. Fig. 12.9 illustrates the use of reference point.

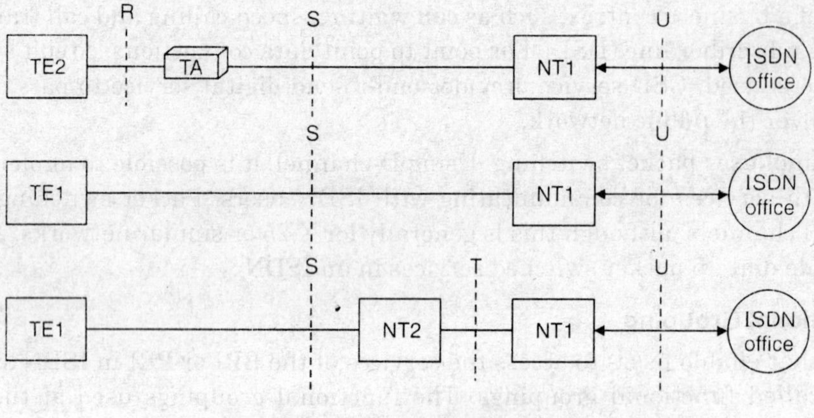

Fig. 12.9. Reference points.

The reference point R defines connection between a TE2 and a TA. R interface is not defined by the ISDN. A subscriber can use any of the EIA standards.

The reference point S defines the connection between a TE1 or TA and TA and an NT1 (or NT2). ITU-7 speficics the ISO standard ISO 8887 for S interface. A maximum of 8 TE's can be connected to one ISDN line throuh 'S' bus wiring, even though only two can be in use at a time. RJ_{45} cable called S bus having 8 wire ribbon type cable is available in electronic market.

Reference point T defines the interface between an NT2 and NT1. Reference point U defines the interface between an NT1 and the ISDN office.

12.5. BROADBAND ISDN (B-ISDN)

B-ISDN provides the needs (High speed and large data handling) of the next generation technology. B-ISDN is a digital service with speed above 1.544 Mbps. The original ISDN is called narrow band ISDN (N-ISDN). B-ISDN uses fiber at all levels of telecommunications. B-ISDN provides two types of services. They are interactive (conversational or messaging or retrieval). The interactive service is bidirectional. The distributive services are unidirectional (with user control or without user control).

Several forms of B-ISDN exists. Some are listed below:

Frame relay service. Frame relay is considered to be a B-ISDN service. Frame relay is a packet switching protocol service offered by telephone corporations to replace the X-25 protocol. It is a WAN network.

Switched Multimegabit Digital Service (SMDS). SMDS is a digital service that provides a high speed digital path. The transport speed of SMDS is usually 155 Mbps.

ATM. The transport speed of most ATM applications are 155 Mbps.

B-ISDN uses the same functional groupings of N-ISDN. But named as B-NT1, B-NT2, B-TE1, B-TE2 and B-TA. B-ISDN uses the same R, S, T and U reference points. However B-ISDN uses three different access methods to satisfy the user needs. They are symmetrical 155.52 Mbps, asymmetrical 155.520 Mbps and symmetrical 622.080 Mbps.

ACRONYMS

B-ISDN	—	Broad band ISDN
BRI	—	Basic Rate Interface
CPE	—	Customer Premises Equipment
CSV	—	Circuit Switched Voice
IDN	—	Integrated Digital Networks
ISDN	—	Integrated Services Digital Network
N-ISDN	—	Narrowband ISDN
NT	—	Network Termination
PRI	—	Primary Rate Interface
SMDS	—	Switched Multimegabit Digital Service
TA	—	Terminal Adapters
TE	—	Terminal Equipment

RELATED WEBSITES

http://www.linktionary.com/i/id.isdn.html
http://www.nivf.nist.guv/
http://www.natinalisdn council.com/
http://www.aluminaca/tech.edu/~dank/isdn/
http://www.isdnzone.com
http://www.cs.rull.nl/wals/html/na-dir/isdn-faq.

CHAPTER REVIEW QUESTIONS

1. List the types of ISDN.
2. Explain the ISDN services.
3. Explain the concept of ISDN with neat diagram.
4. List the advantages of ISDN.
5. Explain with neat diagram, the BRI concept. Explain the BRI frame format.
6. Explain with neat diagram, the PRI concept. Explain the PRI frame format.
7. Explain three types of ISDN channels. Tabulate the specifications of all the channels.
8. What is functional grouping ?
9. What is reference point ?
10. Explain the importance of B-ISDN.

APPENDIX – A

BASIC EQUATIONS FOR MICROPHONE AND EARPHONE.

The quantitative study of the microphone shows the microphone functions as an amplitude modulator. Similarly the study on earphone indicates that it can recover the speech signals from the varying current received.

Microphone as amplitude modulator: When the sound waves impinge on the diaphragm, the instantaneous resistance is given by,

$$r_i = r_q - r_m \sin \omega t \quad \text{...(A.1)}$$

when r_i = instantaneous resistance

r_q = quiescent resistance of the microphone when there is no speech signal.

r_m = maximum variation in resistance offered by the carbon granules $r_m < r_q$

$w = 2\pi f$, f = frequency measured in Hz.

The negative sign indicates the decrease in resistance when the carbon granules are compressed and vice versa. At ideal condition, the instantaneous current in the microphone is given by

$$i = \frac{V}{r_q - r_m \sin \omega t} \quad \text{...(A.2)}$$

$$i = \frac{V}{r_q\left(1 - \frac{r_m}{r_q}\sin \omega t\right)} = \frac{I_q}{(1 - m \sin \omega t)} \quad \text{...(A.3)}$$

$$i = I_q (1 - m \sin \omega t)^{-1} \quad \text{...(A.4)}$$

where I_q = quiescent current in the microphone

$$m = \frac{r_m}{r_q}\ ;\ m < 1 \quad \text{...(A.5)}$$

By binomial theorem, Eq. (A.5) may be expressed as

$$i = I_q (1 + m \sin \omega t + m^2 \sin^2 \omega t + ...) \quad \text{...(A.6)}$$

As the amplitude of the higher order terms are smaller those terms are neglected. Thus

$$i = I_q (1 + m \sin \omega t) \quad \text{...(A.7)}$$

The equation (A.7) resembles the amplitude modulation equation and hence the microphone acts as modulator.

Earphone as sound detector. The variations in current through the coils wound on the electromagnet results in change in flux. This instantaneous flux linking the poles of the electromagnet and the diaphragm is given by

$$\phi_i = \phi_q + \phi_m \sin \omega t \quad \text{...(A.8)}$$

where ϕ_i = instantaneous flux

ϕ_q = flux due to quiescent current

ϕ_m = maximum amplitude of flux variation, $\phi_m < \phi_q$

The instantaneous force exerted on the diaphragm is proportional to the square of the instantaneous flux linking the path. Thus

$$F = K(\phi_q + \phi_m \sin \omega t)^2 \qquad \text{...(A.9)}$$

where k = proportionality constant

$$F = K(\phi_q^2 + \phi_m^2 \sin^2 \omega t + 2\phi_q\phi_m \sin \omega t)$$

neglecting the second order term, we have

$$F = K(\phi_q^2 + 2\phi_q\phi_m \sin \omega t)$$

$$= K\phi_q^2 \left(1 + \frac{2\phi_m \sin \omega t}{\phi_q}\right)$$

$$= K\phi_q^2 (1 + K_1 I_0 \sin \omega t) \qquad \text{...(A.10)}$$

where K_1 = constant

$I_0 \sin \omega t$ = the current through the coil.

Thus the force experienced by the diaphragm is in accordance with the signals produced by the microphone.

APPENDIX – B

Erlang-B loss

erl	m = 1	m = 2	m = 3	m = 4	m = 5	m = 6	m = 7	m = 8	m = 9	m = 10
0.5	0.333	0.077	0.013	0.002	0.000	0.000	0.000	0.000	0.000	0.000
1.0	0.500	0.200	0.063	0.015	0.003	0.001	0.000	0.000	0.000	0.000
1.5	0.600	0.310	0.134	0.048	0.014	0.004	0.001	0.000	0.000	0.000
2.0	0.667	0.400	0.211	0.095	0.037	0.012	0.003	0.001	0.000	0.000
2.5	0.714	0.472	0.282	0.150	0.070	0.028	0.010	0.003	0.001	0.000
3.0	0.750	0.529	0.346	0.206	0.110	0.052	0.022	0.008	0.003	0.001
3.5	0.778	0.576	0.402	0.260	0.154	0.082	0.040	0.017	0.007	0.002
4.0	0.800	0.615	0.451	0.311	0.199	0.117	0.063	0.030	0.013	0.005
4.5	0.818	0.648	0.493	0.357	0.243	0.154	0.090	0.048	0.024	0.010
5.0	0.833	0.676	0.530	0.398	0.285	0.192	0.121	0.070	0.037	0.018
5.5	0.846	0.699	0.562	0.436	0.324	0.229	0.153	0.095	0.055	0.029
6.0	0.857	0.720	0.590	0.470	0.360	0.265	0.185	0.122	0.075	0.043
6.5	0.867	0.738	0.615	0.500	0.394	0.299	0.217	0.150	0.098	0.060
7.0	0.875	0.754	0.638	0.527	0.425	0.331	0.249	0.179	0.122	0.079
7.5	0.882	0.768	0.658	0.552	0.453	0.362	0.279	0.207	0.147	0.100
8.0	0.889	0.780	0.675	0.575	0.479	0.390	0.308	0.230	0.173	0.122
8.5	0.895	0.792	0.692	0.595	0.503	0.416	0.336	0.263	0.199	0.145
9.0	0.900	0.802	0.706	0.614	0.525	0.441	0.362	0.289	0.224	0.168
9.5	0.905	0.811	0.720	0.631	0.545	0.463	0.386	0.314	0.249	0.191
10.0	0.909	0.820	0.732	0.647	0.564	0.485	0.409	0.338	0.273	0.215

erl	m = 10	m = 20	m = 30	m = 40	m = 50	m = 60	m = 70	m = 80	m = 90	m = 100
5	0.018	0.000	0.000	0.000	0.000	0.000	0.000	0.000	0.000	0.000
10	0.215	0.002	0.000	0.000	0.000	0.000	0.000	0.000	0.000	0.000
15	0.410	0.046	0.000	0.000	0.000	0.000	0.000	0.000	0.000	0.000
20	0.538	0.159	0.008	0.000	0.000	0.000	0.000	0.000	0.000	0.000
25	0.622	0.280	0.053	0.001	0.000	0.000	0.000	0.000	0.000	0.000
30	0.681	0.380	0.132	0.014	0.000	0.000	0.000	0.000	0.000	0.000
35	0.725	0.459	0.220	0.054	0.003	0.000	0.000	0.000	0.000	0.000
40	0.758	0.521	0.299	0.116	0.019	0.001	0.000	0.000	0.000	0.000
45	0.784	0.571	0.367	0.185	0.054	0.005	0.000	0.000	0.000	0.000
50	0.805	0.612	0.425	0.250	0.105	0.022	0.001	0.000	0.000	0.000
55	0.822	0.646	0.473	0.308	0.161	0.053	0.007	0.000	0.000	0.000
60	0.837	0.674	0.515	0.360	0.216	0.096	0.024	0.002	0.000	0.000
65	0.849	0.699	0.550	0.406	0.267	0.144	0.052	0.009	0.001	0.000
70	0.859	0.720	0.581	0.445	0.314	0.192	0.090	0.025	0.003	0.000

75	0.869	0.738	0.608	0.480	0.356	0.237	0.131	0.051	0.011	0.001
80	0.877	0.754	0.632	0.511	0.393	0.279	0.173	0.084	0.026	0.004
85	0.884	0.768	0.653	0.539	0.427	0.317	0.213	0.121	0.050	0.012
90	0.890	0.781	0.672	0.564	0.457	0.352	0.251	0.158	0.080	0.027
95	0.896	0.792	0.689	0.586	0.484	0.384	0.287	0.195	0.112	0.049
100	0.901	0.802	0.704	0.606	0.509	0.413	0.320	0.229	0.146	0.076

Line Utilizations

erl	m = 1	m = 2	m = 3	m = 4	m = 5	m = 6	m = 7	m = 8	m = 9	m = 10
0.5	0.333	0.231	0.165	0.125	0.100	0.083	0.071	0.062	0.056	0.050
1.0	0.500	0.400	0.313	0.246	0.199	0.167	0.143	0.125	0.111	0.100
1.5	0.600	0.517	0.433	0.357	0.296	0.249	0.214	0.187	0.167	0.150
2.0	0.667	0.600	0.526	0.452	0.385	0.329	0.285	0.250	0.222	0.200
2.5	0.714	0.660	0.598	0.531	0.465	0.405	0.354	0.312	0.278	0.250
3.0	0.750	0.706	0.654	0.595	0.534	0.474	0.419	0.372	0.332	0.300
3.5	0.778	0.741	0.698	0.647	0.592	0.535	0.480	0.430	0.386	0.349
4.0	0.800	0.769	0.732	0.689	0.641	0.589	0.536	0.485	0.439	0.398
4.5	0.818	0.792	0.761	0.724	0.681	0.634	0.585	0.535	0.488	0.445
5.0	0.833	0.811	0.784	0.752	0.715	0.673	0.628	0.581	0.535	0.491
5.5	0.846	0.827	0.803	0.776	0.744	0.707	0.666	0.622	0.578	0.534
6.0	0.857	0.840	0.820	0.796	0.768	0.735	0.699	0.659	0.617	0.574
6.5	0.867	0.852	0.834	0.813	0.788	0.759	0.727	0.691	0.652	0.611
7.0	0.875	0.862	0.846	0.827	0.805	0.780	0.751	0.719	0.683	0.645
7.5	0.882	0.870	0.856	0.840	0.820	0.798	0.772	0.743	0.711	0.675
8.0	0.889	0.878	0.865	0.851	0.834	0.814	0.791	0.764	0.735	0.703
8.5	0.895	0.885	0.874	0.860	0.845	0.827	0.807	0.783	0.757	0.727
9.0	0.900	0.891	0.881	0.869	0.855	0.839	0.821	0.800	0.776	0.749
9.5	0.905	0.897	0.887	0.877	0.864	0.850	0.833	0.814	0.793	0.768
10.0	0.909	0.902	0.893	0.883	0.872	0.859	0.844	0.827	0.808	0.785

erl	m = 10	m = 20	m = 30	m = 40	m = 50	m = 60	m = 70	m = 80	m = 90	m = 100
5	0.491	0.250	0.167	0.125	0.100	0.083	0.071	0.063	0.056	0.050
10	0.785	0.499	0.333	0.250	0.200	0.167	0.143	0.125	0.111	0.100
15	0.884	0.716	0.500	0.375	0.300	0.250	0.214	0.188	0.167	0.150
20	0.924	0.841	0.661	0.500	0.400	0.333	0.286	0.250	0.222	0.200
25	0.944	0.900	0.789	0.624	0.500	0.417	0.357	0.313	0.278	0.250
30	0.956	0.930	0.868	0.739	0.600	0.500	0.429	0.375	0.333	0.300
35	0.964	0.947	0.910	0.828	0.698	0.583	0.500	0.437	0.389	0.350
40	0.969	0.957	0.934	0.884	0.785	0.666	0.571	0.500	0.444	0.400
45	0.973	0.965	0.949	0.917	0.851	0.746	0.643	0.562	0.500	0.450

50	0.976	0.970	0.959	0.938	0.895	0.815	0.713	0.625	0.556	0.500
55	0.979	0.974	0.965	0.951	0.923	0.868	0.780	0.687	0.611	0.550
60	0.981	0.977	0.970	0.960	0.941	0.904	0.837	0.748	0.667	0.600
65	0.983	0.979	0.974	0.966	0.952	0.927	0.880	0.805	0.722	0.650
70	0.984	0.981	0.977	0.971	0.961	0.943	0.910	0.853	0.775	0.700
75	0.985	0.983	0.979	0.974	0.967	0.954	0.931	0.890	0.824	0.749
80	0.986	0.984	0.981	0.977	0.971	0.962	0.945	0.916	0.866	0.797
85	0.987	0.985	0.983	0.979	0.975	0.967	0.955	0.934	0.897	0.840
90	0.988	0.986	0.984	0.981	0.977	0.971	0.962	0.947	0.920	0.876
95	0.989	0.987	0.985	0.983	0.980	0.975	0.968	0.956	0.937	0.904
100	0.989	0.988	0.986	0.984	0.981	0.978	0.972	0.963	0.949	0.924

APPENDIX C

The Institute of Electrical and Electronic Engineers (IEEE) created the 802 committee in 1980. The committee drafts standards for LAN. The subcommittees of 802 addresses a specific LAN architectures. They are referred as LAN types or subgrouping. These groups and their responsibilities are listed below.

- 802.1 : The high level interface standard addresses matters related to network architecture, management and interconnection.
- 802.1*a* : Network Management Architecture
- 802. 1*b* : Network management protocols
- 802.1D : Bridging standard
- 802.1*d* : Source routing standard
- 802.1*g* : Remote bridge standard for WANs
- 802.1P : Prioritization of MAC layer bridges
- 802.1Q : Virtual LANs (VLAN)
- 802.2 : Logical link control (LLC) and media access control (MAC) are two sublayers that are equivalent to OSI data link layer
- 802.2*i* and *j* : Acknowledged connectionless LLC service
- 802.3 : Carrier sense multiple access with collision detection (CSMA/CD) standards cover a variety of architectures that are generally based on Ethernet.
- 802.3*i* : CSMA/CD over twisted pair and fiber obtics
- 802.3*u* : Fast Ethernet
- 802.3*x* : Gigabit Ethernet flow control
- 802.3*z* : Gigabit Ethernet
- 802.3*ab* : Gigabit on category 5 UTP cabling
- 802.4 : The token bus network standard describes how the token bus network operates.
- 802.5 : Token ring network.
- 802.5*j* : Fiber optic token ring
- 802.6 : The MAN standard describes the operations of networks covering large distances.
- 802.7 : The broadband Technical Advisory Group provides guidelines to other groups that are involved in establishing broadband LAN standards.
- 802.8 : The Fiber optic Technical Advisory group provides guidance to other groups that are involved in establishing LAN standards using fiber optic cable.
- 802.9 : Integrated data and voice networks

802.9*a*	:	Isochrononous Ethernet (ISONET)
802.10	:	A LAN security standard
802.11	:	Wireless LAN
802.12	:	Fast Ethernet
802.14	:	Cable TV Based Broad band communication network working group
802.15	:	Wireless personal area network (WPAN) working group.
802.16	:	Broad band wireless Access (BBWA) working group
802.17	:	Resilient Packet Ring Working Group (RPRWG) for use in LAN, MAN and WAN for transfer of data packets at rates scalable to many Gbps.

BIBLIOGRAPHY

1. A. Shalnmugam and S. Rajeev, Computer Communications Networks, ISTE, New Delhi, 2001.
2. Atkinson, J., Telephony, Vol. 2, Automatic exchange systems, The Era publishing Co., London 1950.
3. Bates, P., Practical Digital and Data communications, Reston, Virginia, 1987.
4. Behrouz A. Forouzan, Data Communications and Networking, 2nd ed., Tata McGraw Hill, New Delhi, 2000.
5. Bellamy, J.C., Digital Telephony, 2nd ed, John Wiley & sons, New York 1991.
6. Biswas, N.N, Principles of Telephony, Radiant Books, Bangalore, 1970.
7. Briley, B.E., Introduction to Telephone switching, Addison wesley, Reading (Mass), 1983.
8. Das. J., Review of Digital communications, Wiley Eastern, New Delhi, 1988.
9. Department of Tele communications, Forty years of Tele Communications in Independent India, New Delhi, 1988.
10. E.A. Lee and D.G. Messershnut, Digital communication, Kluwer Academic publisher, Norwell, MA, 1993.
11. Elahi, Network Communications technology, Delmar Thomson Learning, 2001.
12. Flood, J.E., Telecommunication Networks, Peter peragrinus Ltd, Herts (UK), 1977.
13. Freeman, R.L., Telecommunication system Engineering, 2nd end., Wiley-Interscience, New york ; 1989.
14. Freeman, R-L, Telecommunication Transmission Handbook, 2nd ed., Wiley-interscience, New York, 1981.
15. George Kennedy, Electronic communication systems, third edition, McGraw Hill, 1984.
16. Hebuterne, G., Traffic Flow in switching systems, Artech House; London, 1987.
17. Jain, B.N and Agarwala, A.K., Open system Interconnection. Its Architecture and protocols, Eisevier, Amsterdam, 1990.
18. Jean Warland and Pravin Varaiya, High performance communication networks, 2nd ed., Harcourt Asia PTE Ltd. Singapore 2000.
19. Jolley, E.H., Introduction to Telephony and Telegraphy, A.H. Wheeler and Co., Allahabad, 1967.
20. Keiser, G., Optical Fiber communication, McGraw. Hill, Singapore, 1983.
21. McDonald, J.C., Fundamentals of Digital switching, 2nd ed., Plenum Press, New York, 1989.
22. Mchael A-Miller, Data and Network Communications, Vikas publishing House, 2000.
23. Mike Sapien, Greg Piedno, Mastering ISDN, BPB Publications, New Delhi, 1997.
24. Regis J. 'B.U.P.' Bates and Donald W.Gregory, Voice and Data Communications Handbook, Tata McGraw Hill, New York, 2001.
25. Schindler, Jr, G.E. (ed.), A History of Engineering and Science in the Bell System: Switching technology (1925.75). Bell telephone laboratories, 1982.
26. Stalling, W., Local Networks, 3rd ed., MacMillan, New york, 1990.

27. Stallings, William, ISDN, An Introduction, MacMillan, New York, 1989.
28. Talley, D., Basic Telephone switching systems, 2nd ed, Hayden Inc., New Jercy, 1979.
29. Tanenbaum, A.S., Computer Networks, 2nd ed., Prentice Hall of India, New Delhi, 1990.
30. Taub, H and Schilling; D.L., Principles of communication systems, McGraw Hill, New York, 1986.
31. Thiagarajan Visawanathan, Telecommunication switching systems and networks, Pretice Hall of India, New Delhi, 2002.
32. Welch, S., Signalling in Tele communication networks, peter peragrinus Ltd, Herts (UK), 1979.
33. Youlu Zhong and Shakil Akhtar, Networks for computer Scientists and Engineers, Oxford University Press, New York, 2002.